Optical Properties of Crystalline and Amorphous Semiconductors:

Materials and Fundamental Principles

Optical Properties of Crystalline and Amorphous Semiconductors:

Materials and Fundamental Principles

By

Sadao Adachi

Department of Electronic Engineering
Gunma University

Kluwer Academic Publishers
Boston/Dordrecht/London

Distributors for North, Central and South America:
Kluwer Academic Publishers
101 Philip Drive
Assinippi Park
Norwell, Massachusetts 02061 USA
Telephone (781) 871-6600
Fax (781) 871-6528
E-Mail kluwer@wkap.com

Distributors for all other countries:
Kluwer Academic Publishers Group
Distribution Centre
Post Office Box 322
3300 AH Dordrecht, THE NETHERLANDS
Telephone 31 78 6392 392
Fax 31 78 6546 474
E-Mail <orderdept@wkap.nl>
Electronic Services <http://www.wkap.nl>

Library of Congress Cataloging-in-Publication Data

Adachi, Sadao, 1950-
Optical properties of crystalline and amorphous semiconductors : materials and fundamental principles / by Sadao Adachi
p. cm.
Includes bibliographical references and index.
ISBN 0-7923-8563-2 (acid-free paper)
1. Semiconductors--Optical properties. I. Title
QC611.6.06 A26 1999
537.6'226--dc21 99-23738
CIP

Printed on acid-free paper.

Printed in the United States of America

To

Yūki, Mai, and Kōya

CONTENTS

PREFACE

Knowledge of the refractive indices and absorption coefficients of semiconductors is especially important in the design and analysis of optical and optoelectronic devices. The determination of the optical constants of semiconductors at energies beyond the fundamental absorption edge is also known to be a powerful way of studying the electronic energy-band structures of the semiconductors.

The purpose of this book is to present an introduction to the fundamental optical properties of semiconductors. The aim is to develop an understanding of the optical response of crystalline and amorphous semiconductors over the entire spectral range.

This book presents tutorial articles in the categories of materials and fundamental principles (Chapter 1), optical properties in the reststrahlen region (Chapter 2), those in the interband transition region (Chapters 3 and 4) and at or below the fundamental absorption edge (Chapter 5).

This material is, for the most part, in a form which could serve to teach the underlying concepts of semiconductor optical properties and its implementation.

The bird's-eye view of optical properties over the wide spectral range afforded by the tables and graphs in the companion book "*Optical Constants of Crystalline and Amorphous Semiconductors: Numerical Data and Graphical Information*", serves as a good introduction to understanding a wide variety of absorption process and dispersion relation and, consequently, optical properties.

The extensive bibliography is included for those who wish to find additional information if required. It is hoped that the book will attract attention of not only device engineers, but also solid-state physicists and material scientists, and particularly students specializing in the fields of semiconductor physics and device engineering.

ACKNOWLEDGMENTS

The author wishes to thank the editors and authors of the following journals and books for permission to reproduce previously published figures: *John Wiley & Sons* for Fig. 1.1; *Physical Review* for Figs. 1.8, 1.11, 1.13, 1.15, 2.1, 2.4, 2.12(a), 2.16, 3.11–3.21, 3.24–3.26, 3.28, 3.29, 4.5, 4.12–4.14, 5.7(a), 5.8, 5.11, 5.15, 5.26, and 5.38; *North-Holland Publishing Company* for Fig. 1.14; *Journal of Applied Physics* for Figs. 1.17, 2.2, 3.22, 3.42–3.44, 4.6, 4.15, 4.16, 5.3, 5.7(b), 5.9, 5.14, 5.17, 5.18, 5.21–5.25, 5.27, 5.29, 5.30, and 5.36; *World Scientific* for Figs. 2.5, 2.6, 2.8, 2.9, and 2.12–2.14; *Solid State Communications* for Fig. 2.10; *Applied Optics* for Fig. 2.15; *Physica Status Solidi* for Fig. 2.17; *Journal de Physique* for Fig. 3.23; *Applied Physics Letters* for Figs. 3.27, 5.31, and 5.37; *Japanese Journal of Applied Physics* for Figs. 4.17, 4.18, 4.20–4.29, 4.31–4.38, and 5.19; *Academic Press* for Fig. 4.4; *Soviet Physics Semiconductors* for Fig. 5.20; *Journal of Physics: Condensed Matters* for Fig. 5.28; *Journal of Crystal Growth* for Fig. 5.35; *Journal of Physics and Chemistry of Solids* for Fig. 5.40; and *IEEE Journal of Quantum Electronics* for Fig. 5.41.

Contents of the Companion Book

Optical Constants of Crystalline and Amorphous Semiconductors:
Numerical Data and Graphical Information

Sadao Adachi, Author

ABBREVIATIONS AND ACRONYMS

a- — amorphous-
AFM — Atomic Force Microscopy

BZ — Brillouin Zone

c- — crystalline-
c- — cubic-
CP — Critical Point
CPA — Coherent-Potential Approximation
CVD — Chemical-Vapor Deposition

DFT — Density-Functional Theory
DHO — Damped Harmonic Oscillator
DOS — Density of States
DP — Deformation Potential

EMA — Effective-Mass Approximation
EMA — Effective-Medium Approximation
ENPM — Empirical Nonlocal Pseudopotential Method
EPM — Empirical Pseudopotential Method
ER — Electroreflectance
ETBT — Empirical Tight-Binding Theory

fcc — face-centered cubic

h- — hexagonal-
HH — Heavy-Hole
HOA — Harmonic Oscillator Approximation
HREELS — High-Resolution Electron-Energy Loss Spectroscopy

IR — Infrared

KK — Kramers–Kronig
KSM — Koster-Slater Method

LA — Longitudinal Acoustic
LCAO — Linear Combination of Atomic Orbitals
LDA — Local-Density Approximation

LEED — Low-Energy Electron Diffraction
LH — Light-Hole
LMTO — Linear Muffin–Tin Orbitals
LO — Longitudinal Optical
LPE — Liquid-Phase Epitaxy
LRA — Linear Regression Analysis

MBE — Molecular Beam Epitaxy
MDF — Model Dielectric Function
MOCVD — Metalorganic Chemical Vapor Deposition
MOVPE — Metalorganic Vapor Phase Epitaxy

PL — Photoluminescence
PR — Photoreflectance

RCA — Radio Corporation of America
rf — radio frequency
rms — root mean square
RS — Raman Scattering

SCP — Standard Critical Point
SC1(2) — Standard Cleaning 1(2)
SE — Spectroscopic Ellipsometry
SO — Spin-Orbit
STM — Scanning Tunneling Microscopy

TA — Transverse Acoustic
TO — Transverse Optical

UHV — Ultrahigh Vacuum
UV — Ultraviolet

VCA — Virtual-Crystal Approximation
VLSI — Very-Large-Scale Integration

w- — wurtzite-

XPS — X-Ray Photoelectron Spectroscopy

μc- — microcrystalline-

1D — One-Dimensional
2D — Two-Dimensional
3D — Three-Dimensional

Chapter 1

MATERIALS, PROPERTIES, AND BASIC FORMULAS

1.1 MATERIALS AND THEIR STRUCTURES

1.1.1 Crystalline Materials

(a) Diamond-, Zinc-Blende-, and Wurtzite-Type Crystals—The modern age of solid-state electronics is based upon materials which are neither metals nor insulators. Such materials are called semiconductors, and their electronic properties are intermediate between those of metals and insulators. Almost all the semiconductors of practical interest have the diamond, zinc-blende, or wurtzite (hexagonal) structure. In Fig. 1.1 we schematically illustrate the unit cell of these crystal structures [1]. Table 1.1 summarizes the crystal structure, space group, and lattice constant of some semiconductors considered in this book.

The atoms of certain elements are held together in the solid by strongly covalent bonds at tetrahedral angles of 109.5°. Each atom has four nearest neighbors and twelve next nearest neighbors, which is a consequence of each atom sharing one of its outer electrons with each of four neighbors. The typical structure so formed is that of diamond, as shown in Fig. 1.1 (a). The space lattice is face-centered cubic with pairs of atoms at (0,0,0) and (¼,¼,¼) forming a pattern unit.

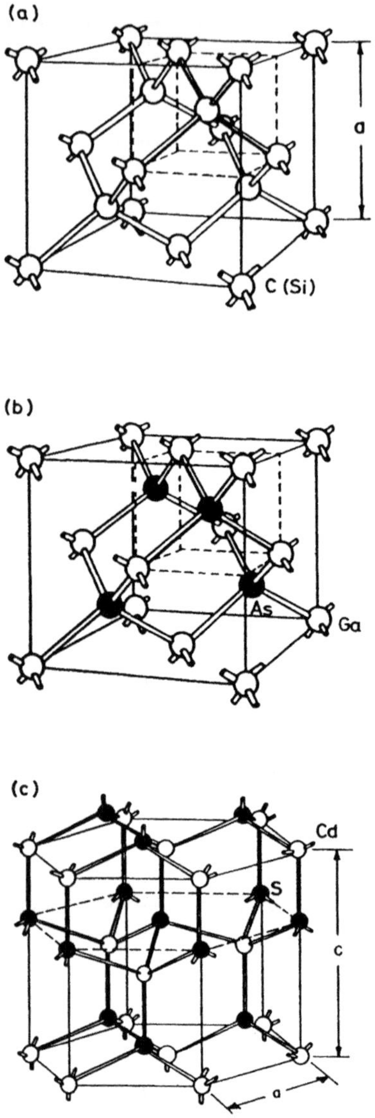

Figure 1.1 Some of the important crystal lattice structures. (a) diamond lattice [C (Si)], (b) zinc-blende lattice (GaAs), and (c) wurtzite lattice (*w*-CdS). (From Sze [1].)

In general, the atomic orbitals that are used to form hybridized bonding orbitals are not the same ones that are occupied in the ground state of the atom. For example, in Si the ground valence configuration of the atom is $3s^2 3p^2$ ($1s^2 2s^2 2p^6 3s^2 3p^2$), whereas the hybridized configuration appropriate for the diamond-type crystal structure is $3s^1 3p^3$ ($1s^2 2s^2 2p^6 3s^1 3p^3$, *tetrahedral coordination*).

TABLE 1.1 *Crystal structure, space group (point group), and lattice constant of various crystalline semiconductors considered in this book.*

Material	Crystal Structure	Space Group	Lattice Constant (Å)
C	Diamond	$Fd3m$ (O_h^7)	3.5668
Si	Diamond	$Fd3m$ (O_h^7)	5.4310
Ge	Diamond	$Fd3m$ (O_h^7)	5.6579
α-Sn	Diamond	$Fd3m$ (O_h^7)	6.4892
3C-SiC	Zincblende	$F\bar{4}3m$ (T_d^2)	4.3596
2H-SiC	Wurtzite	$P6_3mc$ (C_{6v}^4)	3.0763(*a*), 5.0480(*c*)
4H-SiC	Hexagonal	$P6_3mc$ (C_{6v}^4)	3.079(*a*), 10.254(*c*)
6H-SiC	Hexagonal	$P6_3mc$ (C_{6v}^4)	3.0817(*a*), 15.1183(*c*)
15R-SiC	Rhombohedral	$R3m$ (C_{3v}^5)	3.079(*a*), 37.78(*c*)
c-BN	Zincblende	$F\bar{4}3m$ (T_d^2)	3.6157
h-BN	Hexagonal	$P6_3/mmc$ (D_{6h}^4)	6.661(*a*), 2.504(*c*)
BP	Zincblende	$F\bar{4}3m$ (T_d^2)	4.538
BAs	Zincblende	$F\bar{4}3m$ (T_d^2)	4.777
AlN	Wurtzite	$P6_3mc$ (C_{6v}^4)	3.112(*a*), 4.979(*c*)
AlP	Zincblende	$F\bar{4}3m$ (T_d^2)	5.4625
AlAs	Zincblende	$F\bar{4}3m$ (T_d^2)	5.6611
AlSb	Zincblende	$F\bar{4}3m$ (T_d^2)	6.1355
α-GaN	Wurtzite	$P6_3mc$ (C_{6v}^4)	3.189(*a*), 5.185(*c*)
β-GaN	Zincblende	$F\bar{4}3m$ (T_d^2)	4.52
GaP	Zincblende	$F\bar{4}3m$ (T_d^2)	5.4512
GaAs	Zincblende	$F\bar{4}3m$ (T_d^2)	5.6533
GaSb	Zincblende	$F\bar{4}3m$ (T_d^2)	6.0959
InN	Wurtzite	$P6_3mc$ (C_{6v}^4)	3.537(*a*), 5.704(*c*)
InP	Zincblende	$F\bar{4}3m$ (T_d^2)	5.8688
InAs	Zincblende	$F\bar{4}3m$ (T_d^2)	6.0584
InSb	Zincblende	$F\bar{4}3m$ (T_d^2)	6.4794
MgO	Rocksalt	$Fm3m$ (O_h^5)	4.216
ZnO	Wurtzite	$P6_3mc$ (C_{6v}^4)	3.253(*a*), 5.213(*c*)
α-ZnS	Wurtzite	$P6_3mc$ (C_{6v}^4)	3.8226(*a*), 6.2605(*c*)
β-ZnS	Zincblende	$F\bar{4}3m$ (T_d^2)	5.4102
ZnSe	Zincblende	$F\bar{4}3m$ (T_d^2)	5.6676
ZnTe	Zincblende	$F\bar{4}3m$ (T_d^2)	6.1037
c-CdS	Zincblende	$F\bar{4}3m$ (T_d^2)	5.818
w-CdS	Wurtzite	$P6_3mc$ (C_{6v}^4)	4.1362(*a*), 6.714(*c*)
c-CdSe	Zincblende	$F43m$ (T_d^2)	6.077
w-CdSe	Wurtzite	$P6_3mc$ (C_{6v}^4)	4.2999(*a*), 7.0109(*c*)
CdTe	Zincblende	$F\bar{4}3m$ (T_d^2)	6.482
HgSe	Zincblende	$F\bar{4}3m$ (T_d^2)	6.084
HgTe	Zincblende	$F\bar{4}3m$ (T_d^2)	6.453

TABLE 1.1 *Continued.*

Material	Crystal Structure	Space Group	Lattice Constant (Å)
PbS	Rocksalt	$Fm3m$ (O_h^5)	5.9362
PbSe	Rocksalt	$Fm3m$ (O_h^5)	6.1243
PbTe	Rocksalt	$Fm3m$ (O_h^5)	6.4603
SnTe	Rocksalt	$Fm3m$ (O_h^5)	6.3268

The diamond cubic lattice is a consequence of the carbon valency of four. We can expect to find the same structure in compounds where one atom has more than four electrons and the other the same number less than four, so that a total of four valency electrons to each atom is maintained. If the compound is of the form of AB, this structure can be produced in two ways. The first is the cubic zinc-blende structure as shown in Fig. 1.1 (b), with four A (Ga) and four B (As) atoms per conventional unit cell.

The second method by which a structure is formed where each atom of one kind is surrounded by four of another is shown in Fig. 1.1 (c). This is the hexagonal CdS (w- or β-CdS) or wurtzite lattice, which differs only from the zinc-blende structure in the stacking sequence of the sulfur layers. Ideally, the wurtzite structure has the axial ratio $c/a=(8/3)^{1/2}=1.633$ (hexagonal close-packed structure). Most III–V compounds crystallize in the zinc-blende structure, however, many II–VI and some III–V compounds crystallize in the wurtzite structure (see Table 1.1).

In IIIb–Vb compounds, group-IIIb atoms have three electrons with an s^2p^1-configuration outside a core of closed shells, and group-Vb atoms five electrons in a s^2p^3-configuration. The IIIb and Vb atoms have, therefore, an average of four valence electrons per atom available for bonding. We might then expect that the covalent bonds are formed between tetrahedral s^1p^3-hybrid orbitals, *e.g.*, for GaAs:

$$\mathrm{Ga}\ (4s^24p^1) + \mathrm{As}\ (4s^24p^3) \rightarrow \mathrm{Ga}\ (4s^14p^3)^- + \mathrm{As}\ (4s^14p^3)^+ \qquad (1.1)$$

For such a covalent bonding each V atom donates an electron to a III atom, so that V^+ and III^- ions are formed, each with four valence electrons.

An ionic bond is due to Coulomb attraction between the excess positive and negative charges on ions formed by transfer of electrons from the metallic to the nonmetallic atom in the scheme:

$$\mathrm{Ga}\ (4s^24p^1) + \mathrm{As}\ (4s^24p^3) \rightarrow \mathrm{Ga}\ (4s^04p^0)^{+3} + \mathrm{As}\ (4s^24p^6)^{-3} \qquad (1.2)$$

The bonds in most III–V or II–VI compounds are not adequately described by any of these extreme types, but have characteristics intermediate to those usually associated with the terms covalent [Eq. (1.1)] and ionic [Eq. (1.2)].

(b) Silicon Carbide—It is well known that silicon carbide (SiC) is a semiconductor crystallizing in a large number of polytypes [2]. The various types of SiC differ one from another only by the order in which successive planes of Si (or C) atoms are stacked along the *c* axis; one polytype is cubic [3C, space group=$F\bar{4}3m$ (T_d^2)] while the remainder, including two of the more frequently occurring forms, 6H [$P6_3mc$ (C_{6v}^4)] and 15R [$R3m$ (C_{3v}^5)], possess uniaxial symmetry. Notice that in the polytype name, the integer refers to the number of Si (C) layers in the unit cell and C, H, and R indicate cubic, hexagonal, or rhombohedral (trigonal) symmetry. Of all the polytypes, type 6H is by far the most commonly occurring modification in commercial SiC. The next most common types are 15R and 4H, respectively. SiC can also crystallize in the wurtzite structure, 2H-SiC [$P6_3mc$ (C_{6v}^4)].

Figure 1.2 schematically shows the stacking sequences in 3C-, 2H-, and 6H-SiC. In the zinc-blende (3C) structure, the sequence involves three layers which are repeated periodically (ABC ABC ...). All the Si–C bond lengths are the same, and the angles are exactly tetrahedral (109.5°). In the wurtzite (2H) structure, only two layers are repeated (AB AB ...). The Si–C bond length along the stacking direction is not equal to that which is approximately perpendicular to it, and the angles are not exactly tetrahedral. In the 6H polytype, the basic sequence involves six layers (ABCACB ABCACB ...). Similarly, in the 15R polytype the basic sequence involves fifteen layers (ABCACBCABACABCB ...). The II–VI compound, α-HgS, also crystallizes in the red cinnabar structure (*rhombohedral*, D_3).

(c) Rocksalt-Type Crystals—The IV–VI compounds, such as PbS, PbSe, PbTe, and SnTe, crystallize in the rocksalt (NaCl) structure. The rocksalt structure shown in Fig. 1.3 is typical of ionic bonding. The Bravais lattice is face-centered cubic with

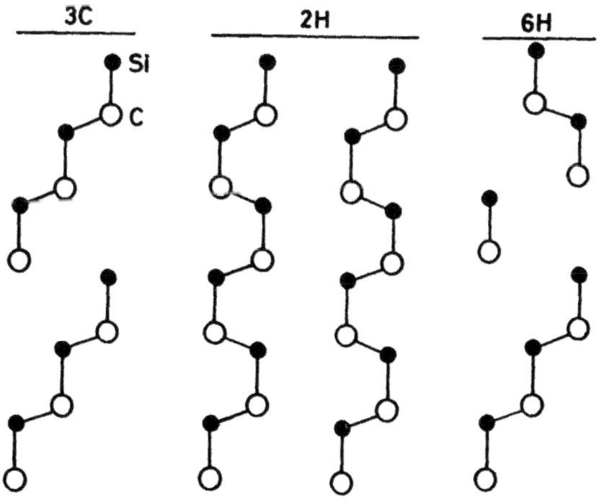

Figure 1.2 Stacking sequence in 3C-, 2H-, and 6H-SiC.

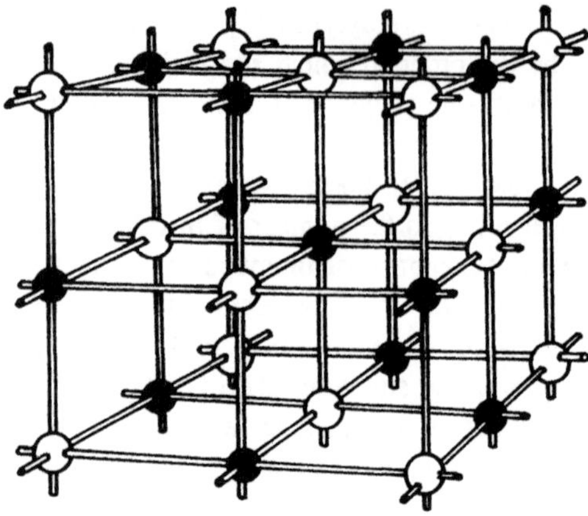

Figure 1.3 Rocksalt crystal structure (PbS).

the unit cell of atomic pattern consisting of one Pb and one S ion separated by one-half the body diagonal of the cube. Since each ion has six nearest neighbors of the opposite kind, the coordination number is six. Some II–VI (II*a*–VI*b*) compounds, such as MgO and CaO, also crystalline in the rocksalt structure.

1.1.2 Amorphous Materials

The simplest way to describe the structure of an amorphous material like *a*-Si consists of considering that short-range coordination is preserved whereas the long-range atomic network is random. This means that the local coordination is tetrahedral (sp^3 hybridization) as in crystalline Si (*c*-Si); the values of the first nearest neighbor distance, coordination number, or binding energy remaining more or less the same in the amorphous and crystalline phase.

Most models of amorphous solids usually fall into two categories: microcrystallite and random network [3]. A schematic representation of these structures is given in Fig. 1.4. In the microcrystallite model, atoms exist in small regions of crystalline order with disordered boundaries. The random network model, in principle, contains no regions of crystalline order. These two odels can also contain dangling bonds.

The microcrystallite model essentially involves a structure which can be viewed as an average over "*randomly*" oriented microcrystals which are connected by disorder boundaries. In this sense the properties of such a structure are determined, to a large extent, by the properties of the individual microcrystals. The random network model, on the other hand, describes a structure which is different topologically from the crystalline and microcrystalline structures.

The simplifications of Bloch theorem and the concept of wave vector ***k*** as a

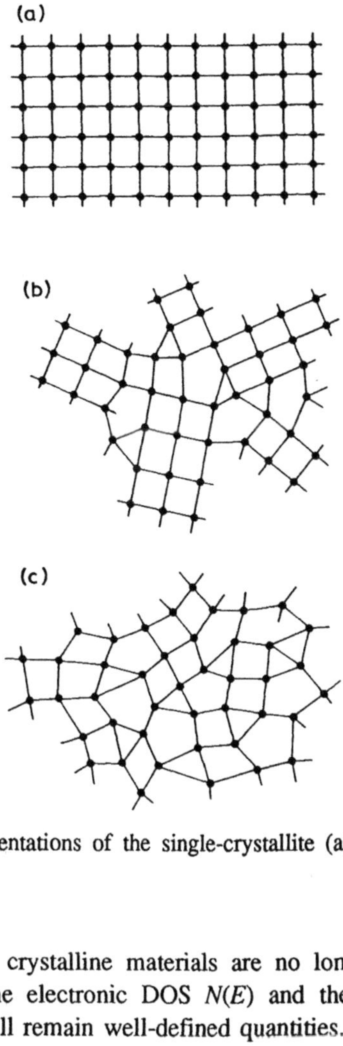

Figure 1.4 Schematic representations of the single-crystallite (a), microcrystallite (b), and random network (c) models.

good quantum number in crystalline materials are no longer valid in amorphous materials. Nevertheless, the electronic DOS $N(E)$ and the imaginary part of the dielectric function $\varepsilon_2(E)$ still remain well-defined quantities.

1.2 BRILLOUIN ZONES AND ELECTRONIC ENERGY-BAND STRUCTURE

1.2.1 Brillouin Zone

(a) Face-Centered Cubic Lattice—Both the electronic energy-band structure [energy (E)–momentum ($\boldsymbol{k}$) relation] and the wave functions of the crystal electrons are largely determined from the symmetry of the periodic crystal potential in which

the electrons move and thus by the symmetry of the crystal. The classification of electron states in a real crystal must, therefore, commence with a symmetry analysis of the crystal.

Let us consider the values of $k=|\boldsymbol{k}|$ at which the discontinuities in energy E occur. They occur whenever $\cos ka$ reaches its maximum value, that is, when $\cos ka=\pm1$ or $k=n\pi/a$ where a is the lattice period and n=1,2,3... . At these values of k a small increase in electron momentum, that is, in k, will make the energy of the electron jump discontinuously from the top of one allowed band to the bottom of the next. The region between the values of k where the first energy discontinuity takes place is called the *first BZ*. The region between the first and second values of k for which discontinuities occur is called the *second BZ*, and so on.

The symmetry of a crystal can be expressed by its space group consisting of all transformations which leave the crystal invariant. The unit cell of the reciprocal lattice is the BZ. The reciprocal lattice is invariant under the same point group as the corresponding real lattice. The groups of a wave vector $\boldsymbol{k}$ in the reciprocal lattice then consist of all transformations which leave $\boldsymbol{k}$ invariant or transform it into an equivalent wave vector $\boldsymbol{k}+\boldsymbol{K}_j$. The Hamiltonian for an electron in the periodic crystal potential is also invariant for all transformations of the point group. In this way the symmetry is transferred on to the Hamiltonian so that the irreducible representations of the point group characterize the eigenvalues of the operator and give their degeneracy and the symmetry of the associated wave functions [4]. The character tables of all thirty-two point groups have been given by Koster *et al.* [5].

Figure 1.5 shows the first two BZs for the fcc lattice. As shown in Fig. 1.5 (a), the first BZ for the cubic semiconductors is the truncated octahedron. It has fourteen plane faces; six square faces are along the <100> directions, and eight hexagonal faces are along the <111> directions. The coordinate axes of the BZ are the wave vectors of the plane waves corresponding to the Bloch states (electrons) or vibration modes (phonons). The points and directions of symmetry are usually referred to by the letters shown in Fig. 1.5 (a) [6]. The zone center is called the Γ point, the directions <100>, <110>, and <111> are called, respectively, Δ, Σ, and Λ directions and their intersections with the zone boundaries are called, respectively, X, K, and L points. Since the real lattice for the diamond, zinc-blende, and rocksalt crystal structures is the fcc lattice, the BZ for materials with either of these structures is the same as that shown in Fig. 1.5.

(b) Hexagonal Lattice—The real lattice for the wurtzite structure is the hexagonal lattice. The spaces of the first two BZs are shown in Fig. 1.6. As shown in the figure, the first BZ for the wurtzite lattice is the hexagonal prism. The wurtzite unit cell contains twice the number of the atoms of the zinc-blende cell, and hence the BZ has about half the volume of that of the zincblende. To make the picture consistent with that for the fcc lattice, it is usual to take a double BZ which is obtained by putting two wurtzite zones together in the <001> direction [4]. There

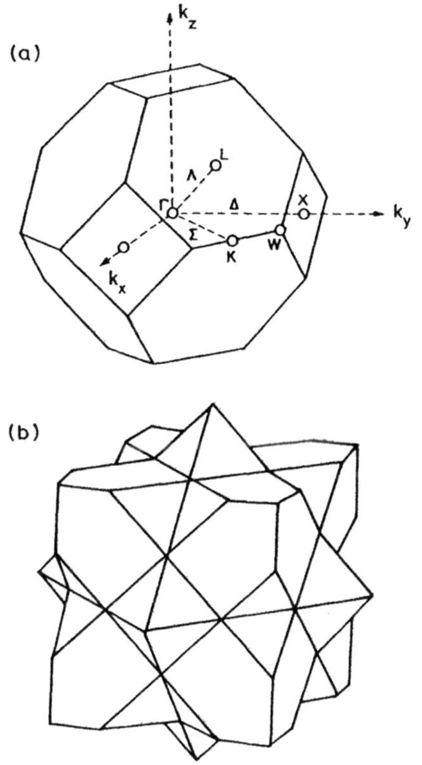

Figure 1.5 The first two BZs for the fcc lattice.

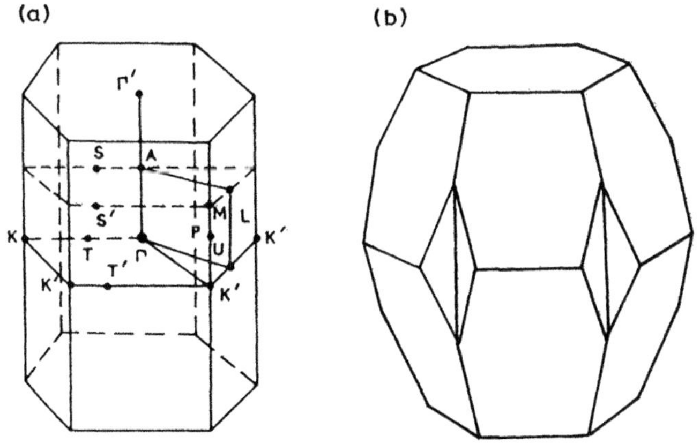

Figure 1.6 The first two BZs for the wurtzite lattice.

is then a correspondence between the vector ΓL in zincblende and the vector ΓAΓ' in wurtzite. In the reduced wurtzite zone Γ' coincides with Γ and hence the point L in zincblende is equivalent to Γ in wurtzite, i.e., the distance AΓ' is folded into AΓ. The number of Γ states is thus double that of zincblende. Half of the Γ states correspond directly to Γ states of zincblende; the other half of the states correspond to L states of zincblende.

(c) Rhombohedral Lattice—The structure of 15R-SiC is described in terms of a rhombohedral lattice with fifteen molecules in a unit lattice; the space-group symbol being $R3m\text{–}C_{3v}^{5}$. Figure 1.7 shows the first BZ for the rhombohedral lattice. It should be noted that the BZ for 15R-SiC is highly compressed parallel to the three-fold axis.

1.2.2 Electronic Energy-Band Structure

The characteristics of the electronic energy bands are usually indicated by plotting the energy eigenvalues of the electrons for different values of $\boldsymbol{k}$ in the BZ. The crystal potential is different in different directions because of the difference in the atomic spacing; as a result, the value of E depends both on the magnitude and on the direction of $\boldsymbol{k}$. In the following, we review the characteristic features of the $E\text{–}\boldsymbol{k}$ diagrams for semiconductors with three typical lattice structures; Si (diamond), GaAs (zincblende), and *w*-CdS (wurtzite).

(a) Diamond-Type Crystal: Si—In Fig. 1.8 we show the relativistic energy-band structure of Si as calculated by Schmid *et al.* [7] in the LDA but with the adjusting potentials included self-consistently. The electronic states are labeled using the

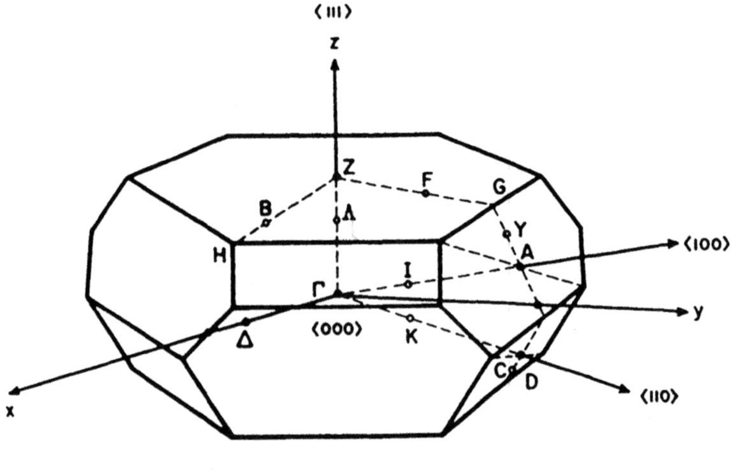

Figure 1.7 The first BZ for the rhombohedral lattice (15R-SiC).

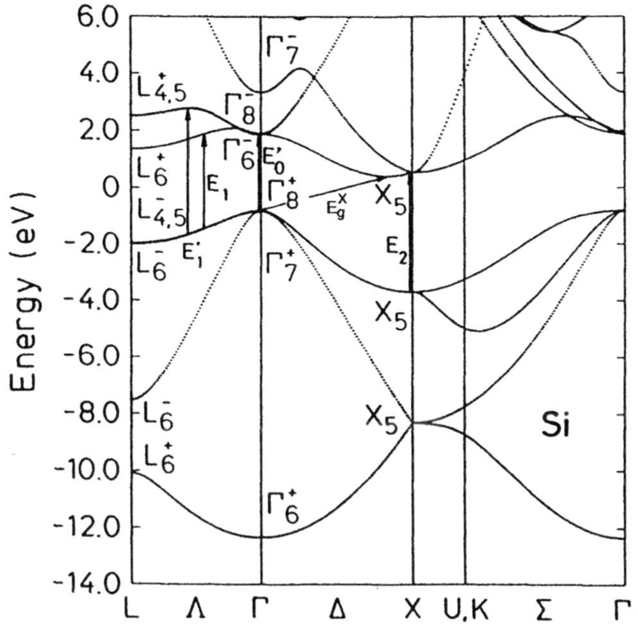

Figure 1.8 Electronic energy-band structure of Si along high-symmetry lines as calculated in the LDA but with the adjusting potentials included self-consistently. The electronic states are labeled using the notation for the representations of the double group of the diamond structure. (From Schmid *et al.* [7].) The main interband transitions (CPs) are indicated by the vertical arrows.

notation for the representations of the double group of the diamond structure. The locations of several interband transitions are indicated in the figure by the vertical arrows. These are transitions which may play an important role in the analysis of optical spectra [8]. We also list in Table 1.2 the indirect-band-gap (E_g^X) and CP energies observed in the optical spectra of Si at 300 K [8].

The fundamental absorption edge of Si corresponds to indirect transitions from the highest valence band at the Γ point to the lowest conduction band near X

TABLE 1.2 *Energies of the indirect band gap (E_g^X) and CPs in Si at 300 K (in eV).*

E_g^X	E_0'	E_1	E_0	$E_0+\Delta_0$	E_2	E_1'
1.12	3.281–3.5	3.38	4.06	4.13	4.27	5.32

E_g^X: $\Gamma_7 \rightarrow X_5$ transitions.

[i.e., $\Gamma_{25'}(\Gamma_8^+)\rightarrow X_1(X_5)$]. There is a clear difference in the ordering of the first two conduction bands between Si and Ge (α-Sn). In Si the lowest-lying conduction band at Γ, Γ_{15}, is *p*-like; however, in Ge and α-Sn the *s*-like $\Gamma_{2'}$ band is the lowest conduction band. A related effect is the decrease in the optical gap of Si from ~3.3 eV to identically zero is the semimetallic α-Sn.

As schematically shown in Fig. 1.9, the SO interaction splits the $\Gamma_{25'}$ valence band into Γ_8^+ and Γ_7^+ (double-group notation, splitting energy Δ_0), and the Γ_{15} conduction band into Γ_6^- and Γ_8^- (splitting energy Δ_0'). The corresponding transitions at or near k=0 are, respectively, labeled E_0 ($\Gamma_8^+\rightarrow\Gamma_7^-$), $E_0+\Delta_0$ ($\Gamma_7^+\rightarrow\Gamma_7^-$), E_0' ($\Gamma_8^+\rightarrow\Gamma_6^-$), $E_0'+\Delta_0'$ ($\Gamma_8^+\rightarrow\Gamma_8^-$), $E_0'+\Delta_0$ ($\Gamma_7^+\rightarrow\Gamma_6^-$, dipole forbidden), and $E_0'+\Delta_0'+\Delta_0$ ($\Gamma_7^+\rightarrow\Gamma_8^-$).

As seen in Fig. 1.10, the SO interaction splits the $L_{3'}$ valence band into $L_{4,5}^-$ and L_6^- and the L_3 conduction band into L_6^+ and $L_{4,5}^+$. The corresponding transitions at or near L [along the <111> direction (Λ) of the BZ] are, respectively, labeled E_1 ($L_{4,5}^-\rightarrow L_6^+$), $E_1+\Delta_1$ ($L_6^-\rightarrow L_6^+$), E_1' ($L_{4,5}^-\rightarrow L_6^+$), and $E_1'+\Delta_1'$ ($L_{4,5}^-\rightarrow L_{4,5}^+$). The E_2 transitions are also expected to occur along the <110> (Σ) or near X [X_5 (X_4)$\rightarrow X_5$ (X_1)].

The SO splitting energies Δ_0 and Δ_0' of Si are very small and have not been taken into consideration in the analysis of optical spectrum [8]. The SO splitting energies Δ_1 and Δ_1', like Δ_0 and Δ_0', of Si are extremely small, and thus no clear

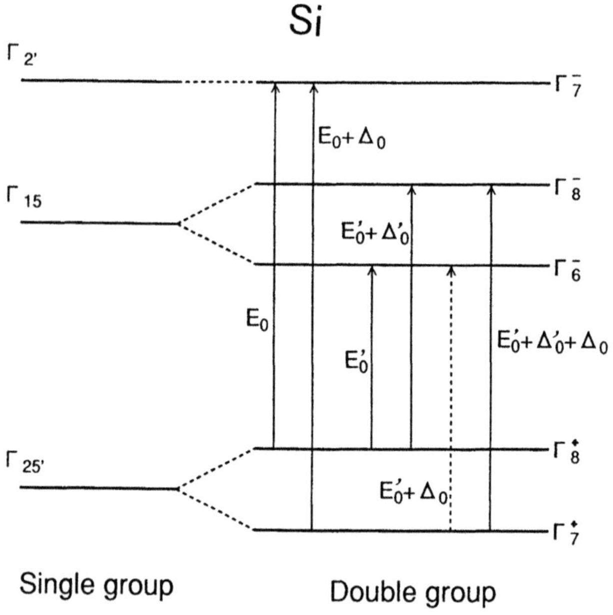

Figure 1.9 Energy gaps at or near the Γ point of Si without (single group) and with considering the SO interaction (double group). Note that the $E_0'+\Delta_0$ transitions are dipole forbidden.

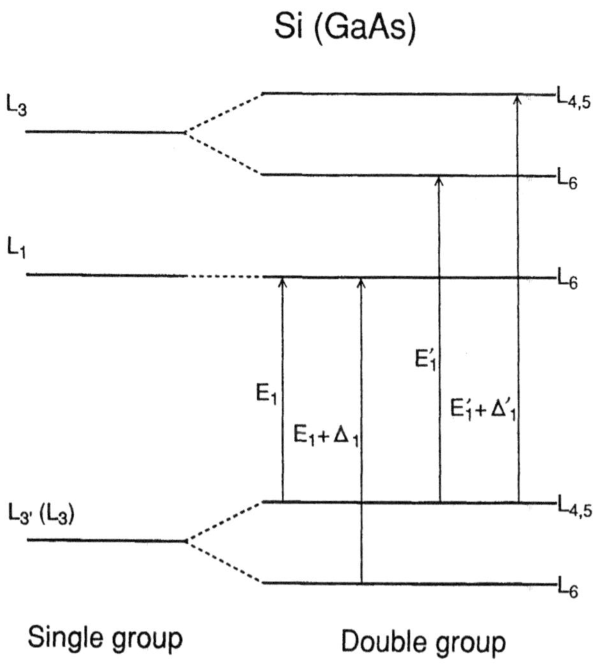

Figure 1.10 Energy gaps at or near the L point of Si (GaAs) without (single group) and with considering the SO interaction (double group).

observations relating to them have been reported so far. It is also hard to observe the $E_0/(E_0+\Delta_0)$ edges of Si because of their exceedingly weak structure in this material [9].

The lowest-direct-band gap of Si is the E_0' edge. The E_1 transitions in Si are nearly degenerate with the E_0' transitions, and therefore the most attention had been given to these complicated E_1-edge structures (3.1–3.4-eV region). The edge labeled E_1' is well separated in energy from other CPs (see Table 1.2).

(b) Zinc-Blende-Type Crystal: GaAs—GaAs crystallizes in the zinc-blende structure [space group=$F\bar{4}3m$ (T_d^2)]. The electronic energy-band structure of GaAs as calculated by an ENPM by Chelikowsky and Cohen [10] is reproduced in Fig. 1.11. The electronic states are labeled using the notation for the representations of the double group of the zinc-blende structure. Several interband transitions are included in the figure. In Table 1.3 also lists the CP energies and indirect-band-gap energies for GaAs at 300 K [11].

The fundamental absorption edge of GaAs corresponds to direct transitions from the highest valence band at the Γ point to the lowest conduction band at the Γ

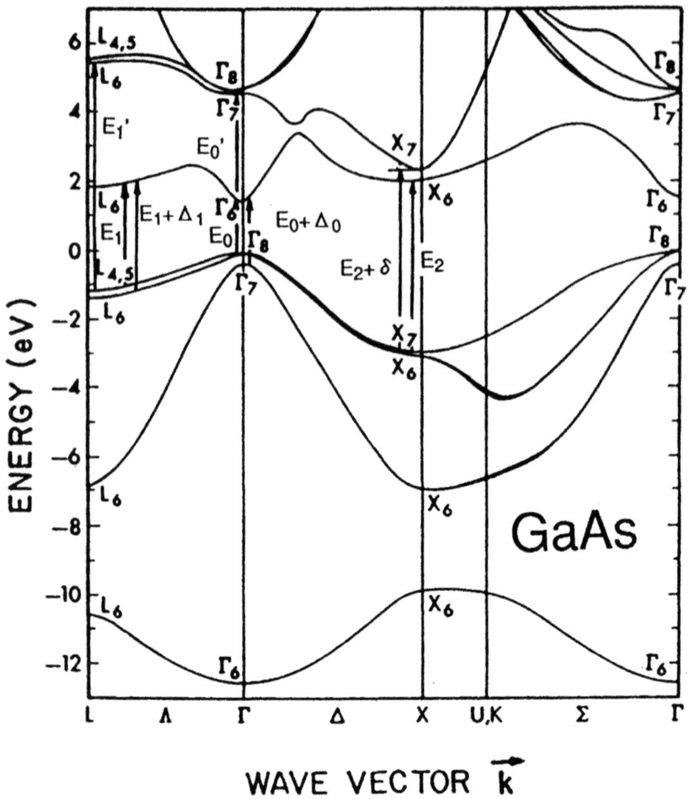

Figure 1.11 Electronic energy-band structure of GaAs as calculated by an ENPM. The electronic states are labeled using the notation for the representations of the double group of the zinc-blende structure. (From Chelikowsky and Cohen [10].) The main interband transitions (CPs) are indicated by the vertical arrows.

point (E_0, $\Gamma_8 \rightarrow \Gamma_6$). Transitions at this edge are dominated by the free excitons, a fact which is particularly evident at low temperatures [12]. The exciton transition energy is 3.6 meV less than the E_0 edge (see Table 5.1). The second lowest interband CP, $E_0+\Delta_0$, corresponds to transitions from the larger component of the SO-splitoff valence band to the lowest conduction band at the Γ point ($\Gamma_7 \rightarrow \Gamma_6$). In Fig. 1.12 we show the energy diagram indicating the SO splitting at or near the Γ point [E_0 doublet (E_0 and $E_0+\Delta_0$) and E_0' triplet (E_0', $E_0'+\Delta_0'$, and $E_0'+\Delta_0'+\Delta_0$)] of the zinc-blende-type semiconductors.

The optical-transition matrix element can be exactly calculated if the wave functions of initial ($|i>$) and final ($|f>$) states are exactly known. This will rarely be the case, but from the symmetry of the wave functions one can easily determine whether the transitions are allowed or forbidden, respectively.

TABLE 1.3 *Energies of the CPs and indirect band gaps (E_g^{ID}) in GaAs at 300 K.*

CPs, E_g^{ID}	eV
E_0	1.42
E_g^L	1.73
$E_0+\Delta_0$	1.76
E_g^X	1.91
E_1	2.91
$E_1+\Delta_1$	3.14
E_0'	4.45
E_2 Region	4.77
	5.00
E_1'	6.6

E_g^L: $\Gamma_8 \rightarrow L_6$ transitions.
E_g^X: $\Gamma_8 \rightarrow X_6$ transitions.

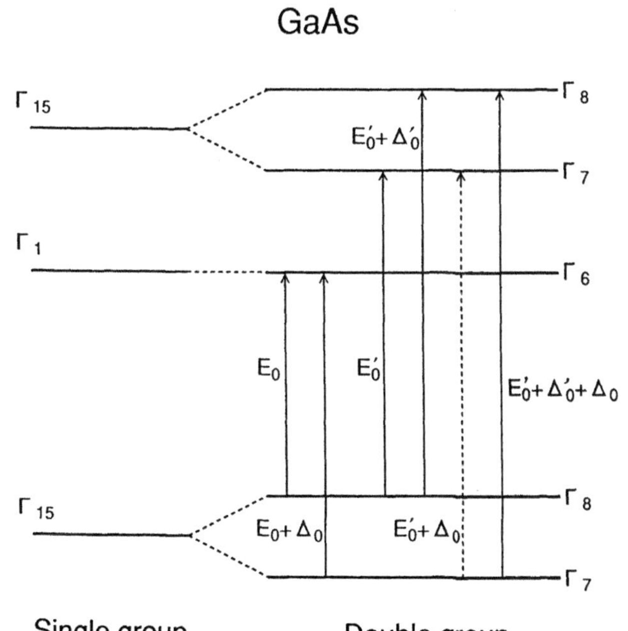

Figure 1.12 Energy gaps at or near the Γ point of GaAs without (single group) and with considering the SO interaction (double group). Note that the $E_0'+\Delta_0$ transitions are dipole forbidden.

The zinc-blende crystals have a T_d point-group structure. As seen from Fig. 1.11 (Fig. 1.12), at k=0 (Γ point) the conduction band has Γ_6 (s-like) or Γ_7 and Γ_8 (p-like) symmetries and the HH, LH, and SO-splitoff valence bands have, respectively, Γ_8, Γ_8, Γ_7 (p-like) symmetries. The polarization vector $\boldsymbol{E}\perp(x,y,z)$ of the point group T_d belongs to Γ_5 symmetry. The optical transitions between the conduction and valence bands can then be given by the direct product [13]:

$$\Gamma_8 \rightarrow \Gamma_6\ (E_0)\text{: } \Gamma_8 \times \Gamma_6 = \Gamma_3+\Gamma_4+\Gamma_5 \quad (1.3a)$$
$$\Gamma_7 \rightarrow \Gamma_6\ (E_0+\Delta_0)\text{: } \Gamma_7 \times \Gamma_6 = \Gamma_2+\Gamma_5 \quad (1.3b)$$
$$\Gamma_8 \rightarrow \Gamma_7\ (E_0')\text{: } \Gamma_8 \times \Gamma_7 = \Gamma_3+\Gamma_4+\Gamma_5 \quad (1.3c)$$
$$\Gamma_8 \rightarrow \Gamma_8\ (E_0'+\Delta_0')\text{: } \Gamma_8 \times \Gamma_8 = \Gamma_1+\Gamma_2+\Gamma_3+2\Gamma_4+2\Gamma_5 \quad (1.3d)$$
$$\Gamma_7 \rightarrow \Gamma_7\ (E_0'+\Delta_0)\text{: } \Gamma_7 \times \Gamma_7 = \Gamma_1+\Gamma_4 \quad (1.3e)$$
$$\Gamma_7 \rightarrow \Gamma_8\ (E_0'+\Delta_0'+\Delta_0)\text{: } \Gamma_7 \times \Gamma_8 = \Gamma_3+\Gamma_4+\Gamma_5 \quad (1.3f)$$

These products, except $\Gamma_7 \times \Gamma_7$ ($E_0'+\Delta_0$), contain the representation of Γ_5 symmetry. This means that the transitions at these gaps are dipole allowed. Only the transitions at the $E_0'+\Delta_0$ gap ($\Gamma_7 \rightarrow \Gamma_7$) are dipole forbidden.

The E_1 and $E_1+\Delta_1$ transitions occur along the Λ lines (<111> directions) or at the L point of the BZ (see Fig. 1.10). In the energy region from 4 to 5.5 eV in GaAs several CPs have been resolved [11]: The most prominent ones have been assigned to transitions near the Γ point (E_0' triplet). The E_2 and $E_2+\delta$ transitions are also expected in the 4–5.5-eV region to occur along the <110> (Σ) or near X [$X_7 \rightarrow X_6$ (E_2), $X_7 \rightarrow X_7$ ($E_2+\delta$)]. Note that the splitting energy δ at the X point is zero in the covalent, diamond-type semiconductors (Si, Ge, etc.).

(c) Wurtzite-Type Crystal: CdS—CdS normally crystallizes in the wurtzite structure [w-CdS, space group=$P6_3mc$ (C_{6v}^4)]. In Fig. 1.13 shows the electronic energy-band structure of CdS as calculated in terms of a nearest-neighbor ETBT by Kobayashi *et al.* [14]. The dashed lines show the results derived from the semiempirical pseudopotential method [15]. It should be noted that in the calculations no account had been taken of SO interaction effects. Table 1.4 lists the CP energies of w-CdS at 300 K [16].

Birman [17] discussed the relationship between the k=0 conduction and valence band states of the wurtzite structure and the corresponding states of the zinc-blende structure. As noted by him, the essential difference between the potential that an electron experiences in an ideal wurtzite lattice and that in a zinc-blende lattice is the relatively small difference in "*crystal field*" due to sites beyond the next nearest neighbors. The detailed differences in the energy bands then arise from the difference in the crystal field and from the difference in the BZs (see Figs. 1.5 and 1.6 below). The conduction- and valence-band structures proposed for wurtzite are illustrated in Fig. 1.14 [18]. As seen in the figure, the triplet states of zincblende (Γ_{15}) correspond to a doublet (Γ_5) and a singlet (Γ_1) of wurtzite. The Γ_5 and Γ_1

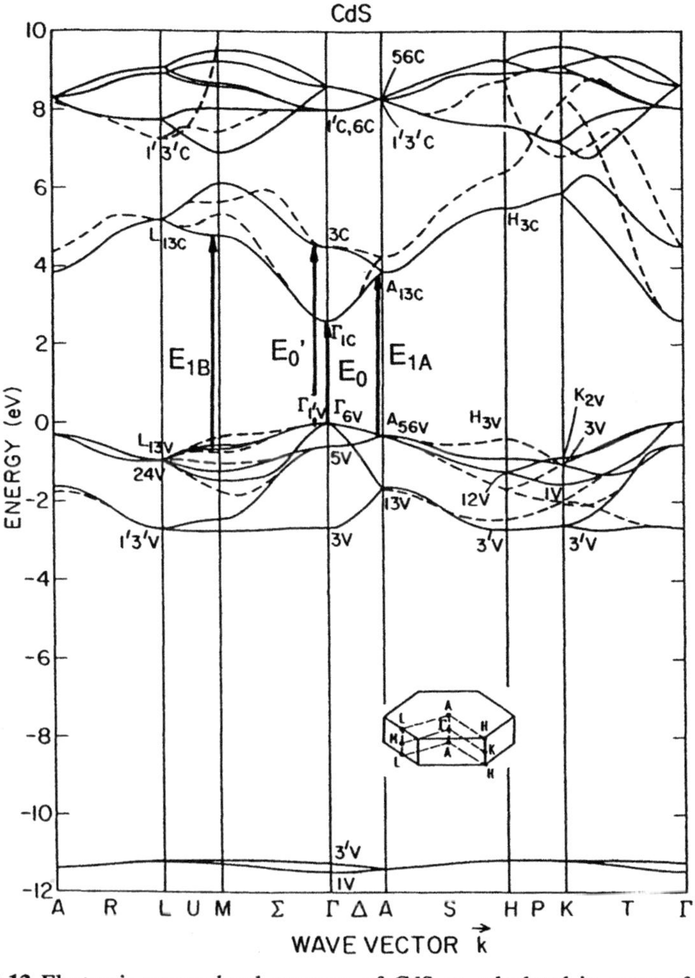

Figure 1.13 Electronic energy-band structure of CdS as calculated in terms of a nearest-neighbor ETBT. The dashed lines also show the results derived from the semiempirical pseudopotential method by Bergstresser and Cohen [15]. In the calculations, no account had been taken of SO interaction effects. (From Kobayashi *et al.* [14].) The locations of several interband transitions are also indicated by the vertical arrows. These are transitions which may play an important part in the analysis of optical spectra.

states are separated by Δ_{cr}, the crystal-field splitting. In the actual valence-band structure of a wurtzite crystal, the three valence bands illustrated in Fig. 1.14 (d) can be thought of as arising from Fig. 1.14 (a) by the combined effect of both "*spin-orbit*" and "*crystal-field*" perturbations, while the wave function for each band may be written as a linear combination of p_x, p_y, and p_z and spin functions, provided the interaction of the Γ_7 levels with the conduction band is neglected.

TABLE 1.4 *Energies of the CPs in w-CdS at 300 K (in eV).*

CPs	Polarization $\mathbf{E}\perp c$	$\mathbf{E}\parallel c$
E_{0A}	2.482	—
E_{0B}	2.496	2.496
E_{0C}	2.555	2.555
E_{1A}	4.84	—
E_{1B}	5.45	5.40
E_1'	6.2	6.3

This result is known as the *quasi-cubic* model [19].

Under the SO interaction, represented by the matrix element Δ_{so} and that of the crystal field represented by Δ_{cr}, the differences in energy of the splitoff valence bands are given by [19]

$$E_{BA}=E_{0B}-E_{0A}=\frac{\Delta_{so}+\Delta_{cr}}{2}-\left[\left(\frac{\Delta_{so}+\Delta_{cr}}{2}\right)^2-\frac{2}{3}\Delta_{so}\Delta_{cr}\right]^{1/2} \tag{1.4a}$$

$$E_{CA}=E_{0C}-E_{0A}=\frac{\Delta_{so}+\Delta_{cr}}{2}+\left[\left(\frac{\Delta_{so}+\Delta_{cr}}{2}\right)^2-\frac{2}{3}\Delta_{so}\Delta_{cr}\right]^{1/2} \tag{1.4b}$$

The corresponding wave functions of the three valence bands are

$$|A>:\ S_+\uparrow \tag{1.5a}$$
$$|B>:\ \alpha_B S_-\uparrow + \alpha_C S_0\downarrow \tag{1.5b}$$
$$|C>:\ \alpha_C S_-\uparrow + \alpha_B S_0\downarrow \tag{1.5c}$$

where $\uparrow$ and $\downarrow$ represent spin-up and spin-down, respectively, and S_+, S_-, and S_0 are functions defined by the *p*-like basis functions. The admixture coefficients α_B and α_C are now given by

$$\alpha_B=\left[1+\frac{1}{2}\left(2-\frac{3}{\Delta_{so}}E_{BA}\right)^2\right]^{-1/2} \tag{1.6a}$$

$$\alpha_C=\left[1+\frac{1}{2}\left(2-\frac{3}{\Delta_{so}}E_{CA}\right)^2\right]^{-1/2} \tag{1.6b}$$

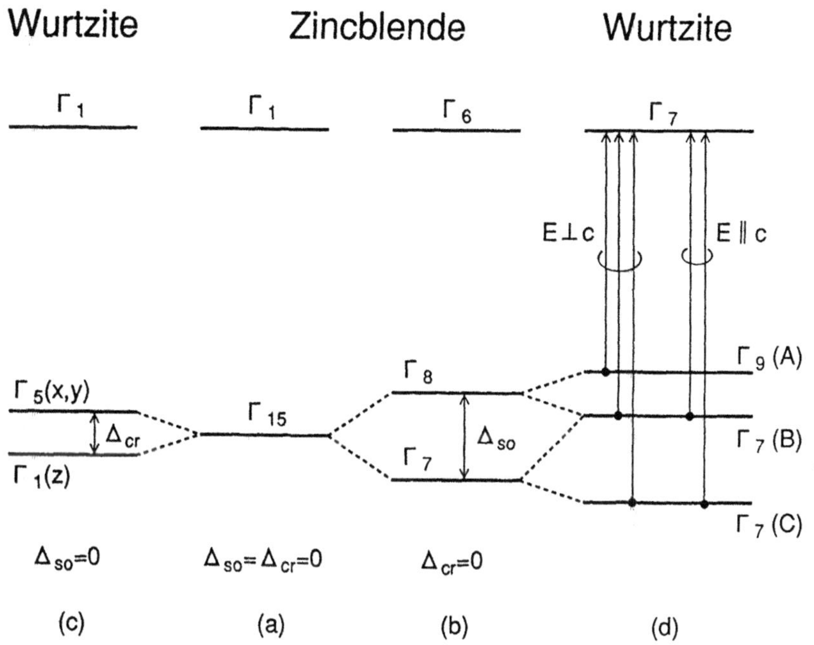

Figure 1.14 Relationships between the valence band splittings in the zinc-blende and wurtzite lattices and between the irreducible representations of the bands at the Γ point (k=0). (b) shows the actual splitting found in the zinc-blende crystals, while (d) gives that found in the wurtzite crystals. (a) and (c) show the relationship of these splittings to simpler cases where SO and/or crystal-field interaction vanish. (From Segall and Marple [18].)

with

$$\alpha_B^2 + \alpha_C^2 = 1 \qquad (1.7)$$

For CdS, we obtain Δ_{so}=62.45 meV, Δ_{cr}=24.5 meV, α_B=0.75, and α_C=0.67.

The corresponding momentum-matrix (p-matrix) elements $|<c|p|v>|^2$ between the p-like valence and s-like conduction bands are given by

$$P_{0A}(\perp)^2=P^2/2, \quad P_{0B}(\perp)^2=\alpha_B^2P^2/2, \quad P_{0C}(\perp)^2=\alpha_C^2P^2/2 \qquad (1.8a)$$

for the light polarization $\boldsymbol{E}\perp c$ (ordinary ray), and

$$P_{0A}(\parallel)^2=0, \quad P_{0B}(\parallel)^2=\alpha_C^2P^2, \quad P_{0C}(\parallel)^2=\alpha_B^2P^2 \qquad (1.8b)$$

for the light polarization $\boldsymbol{E}\parallel c$ (extraordinary ray).

As mentioned before, the wurtzite-type crystals have a C_{6v} point-group symmetry. At the center of the BZ, the conduction band has Γ_7 (*s*-like) symmetry and the A, B, and C valence bands have, respectively, Γ_9, Γ_7, and Γ_7 (*p*-like) symmetries. The polarization vectors $\boldsymbol{E}\perp c$ and $\boldsymbol{E}\|c$ of the point group C_{6v} belong to Γ_5 and Γ_1 symmetries, respectively. The direct products can now be given by

$$\Gamma_9 \rightarrow \Gamma_7 \text{ (A)}: \Gamma_9 \times \Gamma_7 = \Gamma_5+\Gamma_6 \tag{1.9a}$$

$$\Gamma_7 \rightarrow \Gamma_7 \text{ (B, C)}: \Gamma_7 \times \Gamma_7 = \Gamma_1+\Gamma_2+\Gamma_5 \tag{1.9b}$$

The direct product $\Gamma_7 \times \Gamma_7$ then contains the representations for both $\boldsymbol{E}\perp c$ and $\boldsymbol{E}\|c$, but $\Gamma_9 \times \Gamma_7$ contains only the representation for $\boldsymbol{E}\perp c$. This means that for $\boldsymbol{E}\|$c the optical transition between the A valence and conduction bands is forbidden while for $\boldsymbol{E}\perp c$ all the optical transitions are possible, in agreement with those obtained in Eq. (1.8).

The structures seen in CdS in the higher energy region than $E_{0\alpha}$ (α=A,B, and C, ~2.5 eV) are labeled $E_{1\alpha}$ (α=A and B, 4.8–5.5 eV). The original assignment of these structures was made by Cardona and Harbeke [20]; they correspond to the E_1 and $E_1+\Delta_1$ transitions of the zinc-blende-type materials, except for the existence of a strong polarization effect related to the optical anisotropy of wurtzite. For $\boldsymbol{E}\|c$ only the E_{1B} peak is seen, while the E_{1A} and E_{1B} peaks are seen for $\boldsymbol{E}\perp c$. The E_{1A} peak may be related to transitions along the $\Gamma\rightarrow$A direction of the BZ. The E_{1B} structure may be originated from transitions along the U direction in the BZ. It is known that in the case of CdSe the E_{1A} peak splits into two peaks by the SO interaction [21,22].

The E_0' structure (~6.3 eV) may originate from transitions at or near the Γ point of the BZ. Reflectivity [20,23] and SE measurements [24] also reveal additional weak peaks, F_1, E_2, E_1', F_3, and d, at energies higher than 7 eV. These transitions correspond to higher CPs and core excitons.

1.3 OPTICAL CONSTANTS AND SOME DISPERSION RELATIONS

1.3.1 Dielectric Constant: Tensor Representation

Let us consider the polarization that is the electric moment per unit volume (or the polarization charge per unit area taken perpendicular to the direction of polarization). The relationship of the *i*th spatial component of the polarization is expressed in terms of the dielectric field components by a power series of the form

$$P_i=\sum_j \chi_{ij}E_j+\sum_{j,k} \gamma_{ijk}E_jE_k+\ldots\ldots \tag{1.10}$$

With the advent of lasers, it is now quite common to observe nonlinear optical effects. However, the concern here is only with linear optics, and only linear terms will be retained in expressions such as Eq. (1.10).

The vectors are now connected by the relation

$$\boldsymbol{D} = \varepsilon_0 \boldsymbol{E} + \boldsymbol{P} \tag{1.11}$$

where $\boldsymbol{D}$ and $\boldsymbol{E}$ are, respectively, the electric displacement and field strength, and ε_0 is a scalar constant, the permittivity of a vacuum, with the numerical value 8.854×10^{-12} F/m. In many substances the polarization is directly proportional to the field strength, and thus we write

$$\boldsymbol{P} = \varepsilon_0 \varepsilon \boldsymbol{E} \tag{1.12}$$

Hence

$$\boldsymbol{D} = \varepsilon_0 (1+\varepsilon) \boldsymbol{E} \tag{1.13}$$

where ε is known as the *relative permittibity* or the *dielectric constant.*

The dielectric constant is a symmetric second-rank tensor. We have then, in stead of Eq. (1.13)

$$D_i = \varepsilon_0 (\delta_{ij} + \varepsilon_{ij}) E_j \tag{1.14}$$

where δ is the Kronecker delta. The dielectric or optical properties of a crystal may thus be characterized by the magnitudes and directions of the three principal dielectric constants, dielectric permittivities, or dielectric susceptibilities. These magnitudes and directions will, in principle, depend on the frequency of the electric field, but they must always, of course, conform to any restrictions imposed by crystal symmetry [25]. Table 1.5 summarizes the effect of crystal symmetry on dielectric properties represented by symmetric second-rank tensor.

Amorphous materials are, of course, optically isotropic. They then have only one dielectric-tensor component ε_{11}. However, it is known that it is not easy to prepare them in a macroscopically homogeneous form.

1.3.2 Optical Dispersion Relations and Sum Rules

A number of useful relations can be derived which link the real and imaginary parts of the dielectric function and optical constants. These so-called dispersion relations and sum rules have been extremely valuable in the analyzing and testing optical-constant data.

The complex dielectric function

$$\varepsilon(\omega) = \varepsilon_1(\omega) + i\varepsilon_2(\omega) \tag{1.15}$$

can describe the optical properties of the medium at all photon energies $E=\hbar\omega$ ($=h\nu$). From the causality and principle of superposition as applied to a linear

TABLE 1.5 *Effect of crystal symmetry on dielectric properties represented by symmetric second-rank tensors.*

Optical Classification	System	Tensor Form
Isotropic	Cubic	$\begin{bmatrix} \varepsilon_{11} & 0 & 0 \\ 0 & \varepsilon_{11} & 0 \\ 0 & 0 & \varepsilon_{11} \end{bmatrix}$
Uniaxial	Tetragonal Hexagonal Rhombohedral (Trigonal)	$\begin{bmatrix} \varepsilon_{11} & 0 & 0 \\ 0 & \varepsilon_{11} & 0 \\ 0 & 0 & \varepsilon_{33} \end{bmatrix}$
Biaxial	Orthorhombic	$\begin{bmatrix} \varepsilon_{11} & 0 & 0 \\ 0 & \varepsilon_{22} & 0 \\ 0 & 0 & \varepsilon_{33} \end{bmatrix}$
Biaxial	Monoclinic	$\begin{bmatrix} \varepsilon_{11} & 0 & \varepsilon_{13} \\ 0 & \varepsilon_{22} & 0 \\ \varepsilon_{13} & 0 & \varepsilon_{33} \end{bmatrix}$
Biaxial	Triclinic	$\begin{bmatrix} \varepsilon_{11} & \varepsilon_{12} & \varepsilon_{13} \\ \varepsilon_{12} & \varepsilon_{22} & \varepsilon_{23} \\ \varepsilon_{13} & \varepsilon_{23} & \varepsilon_{33} \end{bmatrix}$

medium, the KK relations linking ε_1 and ε_2 can be derived. These are [26]

$$\varepsilon_1(\omega) = 1 + \frac{2}{\pi}\int_0^\infty \frac{\omega'\varepsilon_2(\omega')}{(\omega')^2-\omega^2}d\omega' \qquad (1.16a)$$

$$\varepsilon_2(\omega) = -\frac{2\omega}{\pi}\int_0^\infty \frac{\varepsilon_1(\omega')}{(\omega')^2-\omega^2}d\omega' \qquad (1.16b)$$

The above relations are of fundamental importance.

The complex refractive index $n^*(E)$ is now given by

$$n^*(E) = n(E)+ik(E) = [\varepsilon_1(E)+i\varepsilon_2(E)]^{1/2} \qquad (1.17)$$

where $n(E)$ is the ordinary (real) refractive index and $k(E)$ is the extinction coefficient, also called the attenuation index. The optical constants, $n(E)$ and $k(E)$, are real and positive numbers and can be determined from optical measurements. From Eq. (1.17), it follows that

$$\varepsilon_1 = n^2-k^2 \ (1.18a), \qquad \varepsilon_2 = 2nk \ (1.18b)$$

$$n(E) = \left[\frac{[\varepsilon_1(E)^2+\varepsilon_2(E)^2]^{1/2}+\varepsilon_1(E)}{2}\right]^{1/2} \qquad (1.19a)$$

$$k(E) = \left[\frac{[\varepsilon_1(E)^2+\varepsilon_2(E)^2]^{1/2}-\varepsilon_1(E)}{2}\right]^{1/2} \qquad (1.19b)$$

The KK relations also link $n(E)$ and $k(E)$ in the manner

$$n(E) = 1 + \frac{2}{\pi}\int_0^\infty \frac{E'k(E')}{(E')^2-E^2}dE' \qquad (1.20a)$$

$$k(E) = -\frac{2}{\pi}\int_0^\infty \frac{n(E')-1}{(E')^2-E^2}dE' \qquad (1.20b)$$

The absorption coefficient $\alpha(E)$ depends not only on $\varepsilon_2(E)$ but also on both parts of the dielectric function through $k(E)$ as (see also Sec. 3.3.1)

$$\alpha(E) = \frac{4\pi}{\lambda}k(E) \qquad (1.21)$$

where λ is the wavelength of light in vacuum.

The normal-incidence reflectivity $R(E)$ can also be written as (see also Sec.

3.3.1)

$$R(E) = \frac{\left[n(E)-1\right]^2 + k(E)^2}{\left[n(E)+1\right]^2 + k(E)^2} \quad (1.22)$$

From the asymptotic behavior of $\varepsilon(\omega)$ at high frequencies

$$\varepsilon(\omega \rightarrow \infty) = 1 - \frac{\omega_p^2}{\omega^2} \quad (1.23)$$

together with the analyticity of $\varepsilon(\omega)$ and $\varepsilon(\omega)^{-1}$, one can obtain the following relations [27]

$$\int_0^{\infty} \omega \varepsilon_2(\omega) d\omega = \frac{1}{2}\pi \omega_p^2 \quad (1.24)$$

and

$$\int_0^{\infty} \omega \, \mathrm{Im}\, \varepsilon(\omega)^{-1} d\omega = -\frac{1}{2}\pi \omega_p^2 \quad (1.25)$$

where ω_p is the free-electron plasma frequency corresponding to the *total* electron density of the system.

Equation (1.24) is closely related to the Thomas–Reiche–Kuhn sum rule for an atom in an initial state i

$$\sum_j f_{ij} = Z \quad (1.26)$$

where the number of electrons in the atomic system is Z and the sum over j includes continuum states. Here, a dimensionless quantity f_{ij} is known as the oscillator strength and is defined by

$$f_{ij} = \frac{2m}{\hbar} \omega_{ij} (M_{ij})^2 \quad (1.27)$$

For a condensed rather than isolated atomic system, a sum rule analogous to Eq. (1.26) takes the form

$$\int_0^{\infty} f(\omega) d\omega = N \quad (1.28)$$

where N is the electron density in the condensed matter.

Philipp and Ehrenreich [28] have shown that how these rules can be applied to the valence electrons of a semiconductor such as Si in the absence of d bands. Since the results given by Eq. (1.25) for an infinite range of integration would involve a plasma frequency characteristic simply of four electrons per atom each having the free-electron mass, it is simplest to express the results of the integration over a finite range to E_m in terms of n_{eff}, an effective number of free electrons contributing to the optical properties in this range [28]:

$$n_{eff} = \frac{8\pi m\varepsilon_0}{Ne^2h^2}\int_0^{E_m} E\varepsilon_2(E)dE \qquad (1.29)$$

Similarly, the effective dielectric constant, $\varepsilon(0)_{eff}$, produced by interband transitions in this range may be written as

$$\varepsilon(0)_{eff} = 1 + \frac{2}{\pi}\int_0^{E_m} \frac{\varepsilon_2(E)}{E}dE \qquad (1.30)$$

where $\varepsilon(0)_{eff}$ is an expression for the static or optical dielectric constant. The static dielectric constant results if the IR lattice absorption is taken into account in the integration, and the optical dielectric constant is obtained otherwise.

We plot in Fig. 1.15 the n_{eff} and $\varepsilon(0)_{eff}$ values vs. E_m for Si and GaAs [28]. It should be noted that the n_{eff} curve for Si appears to saturate very nearly at a value of four electrons per atom. By contrast, the curve for GaAs extends appreciably above four. The increase above four is due to the *d*-band excitations (i.e., transitions between the filled *d* bands, lying below the valence band, and empty conduction-band states). The *d* bands are absent in Si, but not in GaAs.

From the tendency toward saturation for $\varepsilon(0)_{eff}$ at photon energies below ~5 eV [Fig. 1.15 (b)], it is clear that strong interband transitions at CPs below this energy are mainly responsible for the value of the long-wavelength dielectric constant. The curve for Si is seen to saturate at a value corresponding to the independently measured long-wavelength dielectric constant. By contrast, the curve for GaAs appcar to saturate at a value below the independently measured long-wavelength dielectric contrast showing in Fig. 1.15 (b) by dashed-line segment. This difference is due to the contribution of the *d* bands.

Regarding the considerable change of the optical spectra when going to the disordered state of semiconductors (see Fig. 1.18 below), it is interest to know whether the sum rule is fulfilled in the amorphous materials. Figure 1.16 shows the n_{eff} values plotted as a function of E_m for *a*-Si and *a*-GaAs. The experimental data are taken for *a*-Si from Philipp [29] and for *a*-GaAs from Stuke and Zimmerer [30]. At high energies we can find that n_{eff} reaches values which are close to those of the crystalline materials [Fig. 1.15 (a)]. Thus, for these amorphous materials the sum rule is fulfilled by the $\varepsilon_2(E)$ curves [Eqs. (1.29) and (1.30)], just as in the crystalline state.

1.3.3 Optical Spectrum and Its Classification into Several Regions

(a) Crystalline Materials—It is convenient to classify the optical spectrum of

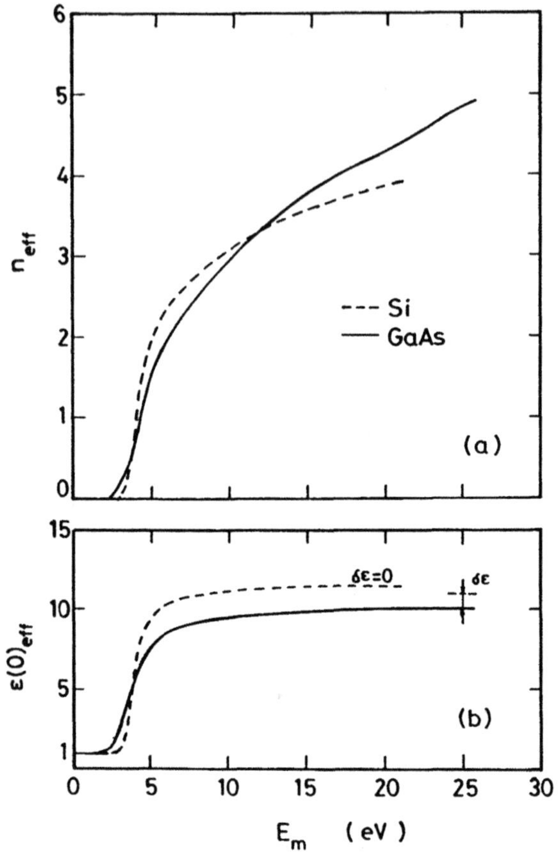

Figure 1.15 (a) n_{eff} vs. E_m for *c*-Si and *c*-GaAs. (b) $\varepsilon_1(0)_{eff}$ vs. E_m for *c*-Si and *c*-GaAs. The long-wavelength dielectric constant (ε_∞) is indicated by dashed-line segment (*c*-GaAs). (From Philipp and Ehrenreich [28].)

crystalline semiconductor into several photon-energy regions based on their own optical transition mechanisms. Figure 1.17 represents a rough sketch of how the components ε_1 and ε_2 of $\varepsilon(E)$ for GaAs vary through the spectral range 10 meV to a few tens of eV [31]. Three spectral regions are distinguished.

Region 1 is the so-called *reststrahlen* region. The reststrahlen region is the region where the radiation field interacts with the fundamental lattice vibrations. Two effects resulted: (1) absorption or emission of the electromagnetic wave due to the creation or annihilation of lattice vibrations and (2) scattering of the electromagnetic wave by the lattice vibrational modes. Below the reststrahlen range in optical spectra, the real part of the dielectric constant asymptotically approaches the *static* (or *low-frequency*) dielectric constant ε_s. The optical constant connecting the reststrahlen–near-IR range is called the *high-frequency* (or *optical*) dielectric constant ε_∞. The strong dispersion properties of GaAs in this region are indicated crudely in Fig. 1.17, where the curves of ε_1 and ε_2 have been drawn ten times broader than in reality, and with the amplitudes of excursion correspondingly reduced, in order that the S shape of dispersion of ε_1 and the inverted V shape of absorption of ε_2 should be visible. It is noted that in covalent-diamond-type semiconductors, such as Si and Ge, the fundamental vibration has no dipole moment and is then IR inactive.

The second region is the region where the material is considered to be primarily transmitting (i.e., near or below the fundamental absorption edge). Appropriately doped samples can show free-carrier, interconduction-, and intervalence-band absorption in this spectral region. Knowledge of the refractive indices and absorption coefficients in this spectral region is especially important in the design and analysis of various optical and optoelectronic devices.

The third region is the strongly absorbing or opaque region. Various opportunities for direct electron-hole pair creation ensure that dispersion and absorption are strong in this region. According to Philipp and Ehrenreich [28], this region can be classified into three subregions. The first subregion (A) is characterized by sharp structures associated with valence to conduction band transitions (CPs). The second subregion (B) is marked by a rapid decrease of the reflectance which is reminiscent

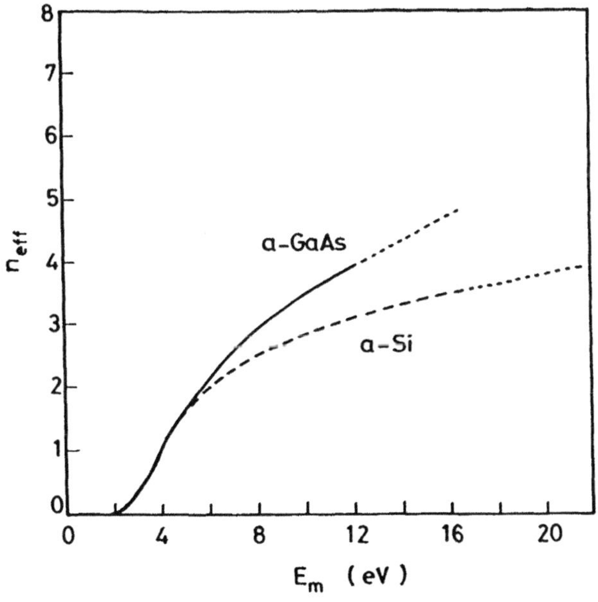

Figure 1.16 $n_{\rm eff}$ vs. $E_{\rm m}$ for *a*-Si and *a*-GaAs. The experimental data are taken for *a*-Si from Philipp [29] and for *a*-GaAs from Stuke and Zimmerer [30].

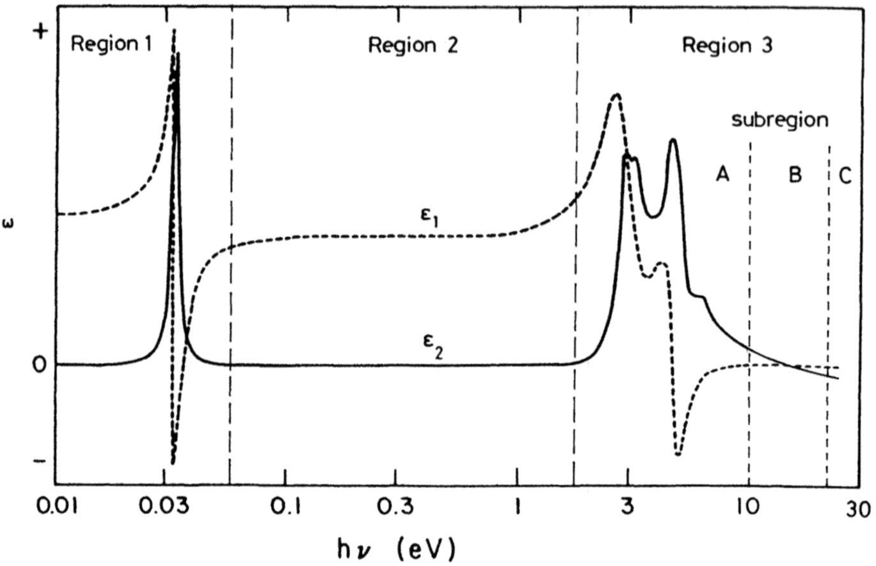

Figure 1.17 Symbolic representation of the real (ε_1) and imaginary (ε_2) parts of the complex dielectric constant for GaAs, from the "*low-frequency*" regime well below the reststrahlen region of the IR, through to the highly absorbing visible and UV regions of interband transitions. Note that the artistic license has been used in drawing the dispersion (ε_1) and absorption (ε_2) curves in the reststrahlen range ten times wider (and with peaks and valleys ten times reduced) than reality. (From Blakemore [31].)

of the behavior of certain metals in the UV region [32]. The function -Im ε^{-1}, describing the energy loss of fast electrons traversing the material, shows sharp peak in this subregion. Sharp maxima in this function (-Im ε^{-1}) have been frequently associated with the existence of plasma oscillations [33]. In the third subregion (C), the reflectance again rises indicating the onset of additional optical absorption. As mentioned in Sec. 1.3.2, this absorption is associated with transitions between the filled *d* bands and empty conduction states. Because of the absence of the *d* bands in Si, we cannot recognize Subregion C in this material [28].

(b) Amorphous Materials—The sharp structures observed in the fundamental optical spectra of crystals, both electronic and vibrational, can be classified and interpreted by symmetry arguments based explicitly on the existence of long-range order. If the long-range order is destroyed as in amorphous materials, the sharp structural detail may disappear. However, the broad features of the spectra are similar if the short-range order is similar.

We show in Fig. 1.18 (a) the $\varepsilon(E)$ spectrum for *a*-GaAs measured at room temperature. The result for *c*-GaAs (subregion A) is also shown in Fig. 1.18 (b).

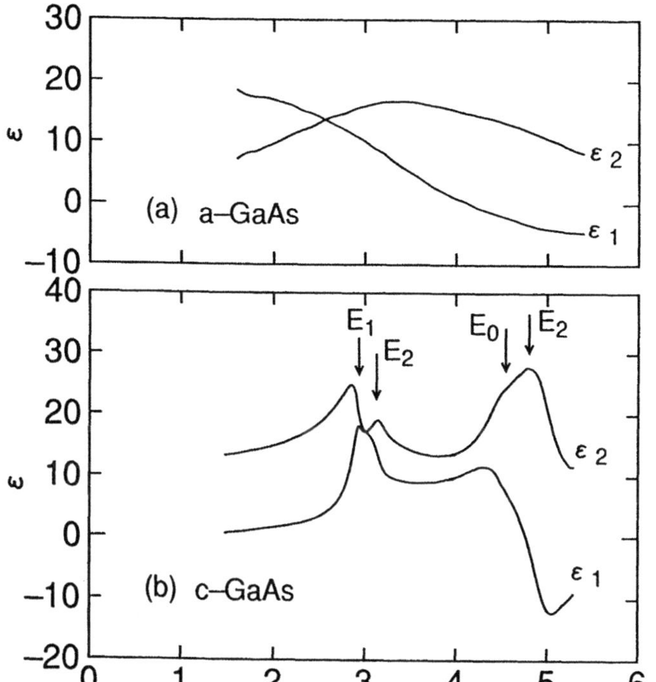

Figure 1.18 $\varepsilon(E)$ spectra for (a) *a*-GaAs and (b) *c*-GaAs measured at room temperature. The experimental data are from Erman *et al.* [34] for *a*-GaAs and from Jellison [35] for *c*-GaAs.

The experimental data are taken from Erman *et al.* [34] for *a*-GaAs and from Jellison [35] for *c*-GaAs. A single broad band found in the experimental $\varepsilon_2(E)$ spectrum [Fig. 1.18 (a)] is typical behavior of tetrahedrally-coordinated amorphous semiconductors. For *c*-GaAs, the first-order allowed optical transitions are ***k***-conserved direct transitions, and therefore the interband spectrum reflects the joint DOS in specific parts of the BZ. The prominent E_1, $E_1+\Delta_1$, E_0', and E_2 structures appear in *c*-GaAs in this spectral region. The disappearance of such CP features in *a*-GaAs is due to the breakdown of crystal periodicity in the amorphous material. The experimental *a*-GaAs data give a maximum value of $\varepsilon_2\sim17$ near 3.4 eV. The dielectric function exhibiting the largest ε_2 at its peak is most representative of a bulk sample of completely coordinated amorphous material. The value of $\varepsilon_2\sim17$ has been shown to be an apparently limiting peak value for *a*-GaAs, i.e., the corresponding spectrum is though to be the "*intrinsic*" dielectric properties of *dense a*-GaAs.

In tetrahedrally-coordinated covalent materials, linear combination of atomic orbitals leads to bonding and antibonding molecular states which then broaden into the valence and conduction bands when the solid is formed [36]. As a result, we can see that the fundamental absorption bands are often single smooth bands (Fig. 1.18 (a) and Ref. [37]). On the other hand, chalcogenide semiconductors (such as As_2S_3 and As_2Se_3) have three molecular orbitals: bonding, nonbonding, and antibonding states which produce three local bands in the solid. The resultant optical spectra show two strong absorption bands in the interband transition regions [38].

If the reststrahlen band can be ascribed to a vibration of two neighboring ions or all ions of one kind relative to the other, then one can also expect to observe it in the corresponding amorphous form. Also, the relaxation of the $\boldsymbol{k}$-selection rule in the amorphous material actives RS and IR absorption of phonon modes which are usually forbidden in the corresponding crystalline material. Then, going from the crystalline to the amorphous state has the effect of (1) broadening the absorption of the optical-like modes which is allowed in the crystal and (2) introducing an additional absorption band by the other phonon modes which is forbidden in the crystal (see Sec. 2.2.2).

REFERENCES

1. S. M. Sze, *Physics of Semiconductor Devices* (Wiley, New York, 1981).
2. A. R. Verma and P. Krishna, *Polymorphism and Polytypism in Crystals* (Wiley, New York, 1966).
3. J. D. Joannopoulos and M. L. Cohen, in *Solid State Physics*, edited by H. Ehrenreich, F. Seitz, and D. Turnbull (Academic, New York, 1976), Vol. 31, p. 71.
4. D. L. Greenaway and G. Harbeke, *Optical Properties and Band Structure of Semiconductors* (Pergamon, Oxford, 1968).
5. F. Koster, J. O. Dimmock, R. G. Wheeler, and H. Statz, *Properties of the Thirty-Two Point Groups* (MIT Press, Cambridge, 1963).
6. H. Jones, *The Theory of Brillouin Zones and Electronic States in Crystals* (North-Holland, Amsterdam, 1975).
7. U. Schmid, N. E. Christensen, and M. Cardona, *Phys. Rev. B* **41**, 5919 (1990).
8. S. Adachi, *Phys. Rev. B* **38**, 12966(1988); *J. Appl. Phys.* **66**, 3224 (1989).
9. D. E. Aspnes and A. A. Studna, *Solid State Commun.* **11**, 1375 (1972).
10. J. R. Chelikowsky and M. L. Cohen, *Phys. Rev. B* **14**, 556 (1976).
11. S. Adachi, Phys. Rev. B **35**, 7454(1987); S. Ozaki and S. Adachi, *J. Appl. Phys.* **78**, 3380 (1995).
12. M. D. Sturge, *Phys. Rev.* **127**, 768 (1962).
13. The group-theoretical symbols used here are taken from Ref. 5.

14. A. Kobayashi, O. F. Sankey, S. M. Volz, and J. D. Dow, *Phys. Rev. B* **28**, 935 (1983).
15. T. K. Bergstresser and M. L. Cohen, *Phys. Rev.* **164**, 1069 (1967).
16. S. Ninomiya and S. Adachi, *J. Appl. Phys.* **78**, 1183 (1995).
17. J. L. Birman, *Phys. Rev.* **115**, 1493 (1959).
18. B. Segall and D. T. F. Marple, in *Physics and Chemistry of II–VI Compounds*, edited by M. Aven and J. S. Prener (North-Holland, Amsterdam, 1967), p. 317.
19. D. G. Thomas and J. J. Hopfield, *Phys. Rev.* **116**, 573(1959); J. J. Hopfield, *J. Phys. Chem. Solids* **15**, 97 (1960).
20. M. Cardona and G. Harbeke, *Phys. Rev.* **137**, A1467 (1965).
21. R. L. Hengehold and C. R. Fraime, *Phys. Rev.* **174**, 808 (1968).
22. S. Logothetidis, M. Cardona, P. Lautenschlager, and M. Garriga, *Phys. Rev. B* **34**, 2458 (1986).
23. M. Cardona, *Phys. Rev.* **129**, 1068 (1963).
24. Ph. Hofmann, K. Horn, A. M. Bradshaw, R. L. Johnson, D. Fuchs, and M. Cardona, *Phys. Rev. B* **47**, 1639 (1993).
25. J. F. Nye, *Physical Properties of Crystals* (Clarendon, Oxford, 1972).
26. See, for instance, G. Harbeke, in *Optical Properties of Solids*, edited by F. Abelès (North-Holland, Amsterdam, 1972), p. 21.
27. P. Nozieres and D. Pines, *Phys. Rev.* **113**, 1254 (1959).
28. H. R. Philipp and H. Ehrenreich, *Phys. Rev.* **129**, 1550 (1963).
29. H. R. Philipp, *J. Phys. Chem. Solids* **32**, 1935 (1971).
30. J. Stuke and G. Zimmerer, *Phys. Status Solidi B* **49**, 513 (1972).
31. J. S. Blakemore, *J. Appl. Phys.* **53**, R123 (1982).
32. See, for instance, F. Wooten, *Optical Properties of Solids* (Academic, New York, 1972).
33. L. Marton, *Rev. Mod. Phys.* **28**, 172 (1956).
34. M. Erman, J. B. Theeten, P. Chambon, S. M. Kelso, and D. E. Aspnes, *J. Appl. Phys.* **56**, 2664 (1984).
35. G. E. Jellison, Jr., *Opt. Mat.* **1**, 151 (1992).
36. R. Zallen, *The Physics of Amorphous Solids* (Wiley, New York, 1983).
37. S. Adachi, *Phys. Rev. B* **43**, 12316 (1991).
38. R. E. Drews, R. L. Emerald, M. L. Slade, and R. Zallen, *Solid State Commun.* **10**, 293 (1972).

Chapter 2

THE RESTSTRAHLEN REGION

2.1 STATIC AND HIGH-FREQUENCY DIELECTRIC CONSTANTS

2.1.1 General Consideration

The strong IR absorption band found in heteropolar semiconductors is essentially bonded by the TO and LO frequencies for long-wavelength vibrations. The difference between these frequencies can be related to the amount of polar character in the chemical bond in these compounds. Below the reststrahlen range in optical spectra, the real part of the dielectric constant asymptotically approaches the static (or low-frequency) dielectric constant ε_s. The optical constant connecting the reststrahlen–near-IR range is called the high-frequency (or optical) dielectric constant ε_∞. The high-frequency dielectric constant ε_∞, thus, measured for frequencies well above the long-wavelength LO phonon frequency but below the fundamental absorption edge.

The dielectric constant ε_s can, in principle, be written as

$$\varepsilon_s = \varepsilon_\infty + \frac{4\pi N e_T{*}^2}{\omega_{TO}^2 M} \qquad (2.1)$$

where M is the reduced mass of the crystal, N is the number of unit cells per unit volume, $e_T{*}$ is the transverse effective charge, and ω_{TO} is the long-wavelength TO phonon frequency. The second term on the right-hand side of Eq. (2.1) is the lattice contribution which arises because the LO mode in heteropolar semiconductors produces a macroscopic electric moment separating it in energy from the TO mode [1]. The dielectric constants ε_s and ε_∞ are also related to the long-wavelength TO and LO phonon frequencies ω_{TO} and ω_{LO} by the Lyddane–Sachs–Teller relation:

$$\frac{\varepsilon_s}{\varepsilon_\infty} = \left(\frac{\omega_{LO}}{\omega_{TO}}\right)^2 \qquad (2.2)$$

In Fig. 2.1, we show the phonon dispersion curves, together with the corresponding DOS, for (a) Si and (b) GaAs [2]. When comparing these phonon dispersion curves it becomes apparent that, while the general shapes in these figures are similar, there is one important difference. At q=0 (Γ) the TO and LO branches are degenerate in Si, while $\omega_{TO}<\omega_{LO}$ in GaAs. The origin of this splitting in GaAs is the

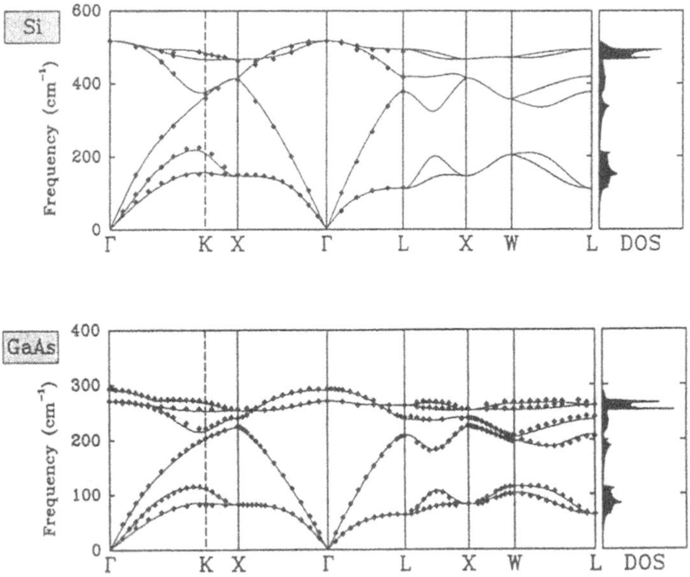

Figure 2.1 Phonon dispersion curves along several symmetry lines, together with the corresponding DOS, for (a) Si and (b) GaAs. (From Giannozzi *et al.* [2].)

electrostatic field created by the long-wavelength (q=0) modes of lattice vibrations. Thus, from Eq. (2.2) we note that $\varepsilon_s \sim \varepsilon_\infty$ for homopolar crystals (such as Si and Ge) and $\varepsilon_s > \varepsilon_\infty$ for heteropolar crystals (such as GaAs and ZnSe).

Figure 2.2 plots the values of ε_s and ε_∞ as a function of the lowest-direct-gap energy E_0 for some of the III–V, II–VI compounds, and alkali halides [3]. Table 2.1 also lists the values of ε_s and ε_∞ for some of the group-IV, III–V, II–VI, and IV–VI semiconductors. We can easily find from this figure that both ε_s and ε_∞ decrease with increasing E_0 values. Note that $\varepsilon_\infty^{1/2}$ corresponds to the long-wavelength refractive index of the material. We can then conclude that the smaller E_0-gap material has a larger value of the refractive index. If the refractive index in the

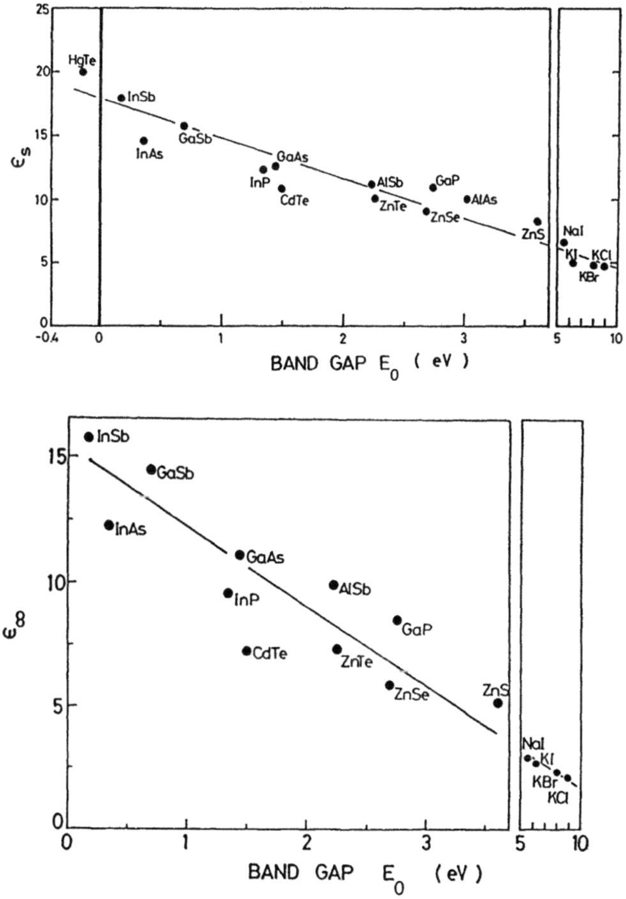

Figure 2.2 Static and high-frequency dielectric constants, ε_s and ε_∞, as a function of the lowest-direct-gap energy E_0 for some of the III–V, II–VI compounds, and alkali halides. (From Adachi [3].)

TABLE 2.1 *Values of ε_s and ε_∞ for the group-IV semiconductors, III–V, II–VI, and IV–VI semiconductors at T=300 K.*

Material	Crystal Structure	ε_s	ε_∞	Remark
C	Diamond	5.70 [a]	—	
Si	Diamond	11.6 [b]	—	
Ge	Diamond	16.04 [c]	—	
α-Sn	Diamond	24 [d]	—	
3C-SiC	Zincblende	9.52 [e]	6.38 [e]	
2H-SiC	Wurtzite		6.51 [f]	$E \perp c$
			6.84 [f]	$E \parallel c$
4H-SiC	Hexagonal		6.56 [g]	$E \perp c$
			6.78 [g]	$E \parallel c$
6H-SiC	Hexagonal	9.66 [h]	6.520 [h]	$E \perp c$
		10.09 [h]	6.742 [h]	$E \parallel c$
15R-SiC	Rhombohedral	9.66 [h]	6.520 [h]	$E \perp c$
		10.09 [h]	6.742 [h]	$E \parallel c$
c-BN	Zincblende	6.80 [i]	4.46 [i]	
h-BN	Hexagonal	6.85 [j]	4.95 [j]	$E \perp c$
		5.06 [j]	4.10 [j]	$E \parallel c$
BP	Zincblende	11 [k]		
AlN	Wurtzite	9.14 [l]	4.84 [l]	$E \perp c$
		9.14 [l]	4.84 [l]	$E \parallel c$
AlP	Zincblende	9.8 [m]	7.6 [m]	
AlAs	Zincblende	10.06 [n]	8.16 [n]	
AlSb	Zincblende	11.21 [o]	9.88 [o]	
α-GaN	Wurtzite	8.9 [p]	5.2 [p]	$E \perp c$
		9.8 [p]	5.2 [p]	$E \parallel c$
β-GaN	Zincblende	9.40 [q]	5.35 [q]	
GaP	Zincblende	11.15 [r]	9.09 [r]	
GaAs	Zincblende	12.90 [s]	10.86 [s]	
GaSb	Zincblende	15.7 [t]	14.4 [t]	
InN	Wurtzite		6.6 [u]	
InP	Zincblende	12.50 [v]	9.66 [v]	
InAs	Zincblende	14.60 [v]	12.25 [v]	
InSb	Zincblende	17.72 [w]	15.68 [w]	
MgO	Rocksalt	9.9 [x]	3.3 [y]	
ZnO	Wurtzite	8.1 [z]	3.95 [z]	$E \perp c$
		9.0 [z]	4.05 [z]	$E \parallel c$
α-ZnS	Wurtzite	9.6 [aa]	5.7 [aa]	$E \perp c$
		9.6 [aa]	5.7 [aa]	$E \parallel c$
β-ZnS	Zincblende	8.9 [aa]	5.7 [aa]	
ZnSe	Zincblende	9.3 [ab]	6.2 [ab]	
ZnTe	Zincblende	9.1 [aa]	6.7 [aa]	
c-CdS	Zincblende		5.3 [ac]	
w-CdS	Wurtzite	8.4 [ad]	5.3 [ad]	$E \perp c$
		8.9 [ad]	5.4 [ad]	$E \parallel c$
w-CdSe	Wurtzite	9.29 [ae]	6.20 [ae]	$E \perp c$
		10.16 [ae]	6.30 [ae]	$E \parallel c$
CdTe	Zincblende	10.5 [af]	7.3 [af]	
HgSe	Zincblende		8.0 [ag]	

TABLE 2.1 *Continued.*

Material	Crystal Structure	ε_s	ε_∞	Remark
HgTe	Zincblende	21.0 [ah]	15.2 [ah]	
PbS	Rocksalt	150 [ai]	16.8 [ai]	
PbSe	Rocksalt	203 [aj]	23 [aj]	
PbTe	Rocksalt	400 [ak]	32.8 [ak]	
SnTe	Rocksalt	380 [al]	45 [am]	

[a]J. Fontanella *et al.*, *Appl. Opt.* **16**, 2949 (1977).
[b]K. Seeger, *J. Appl. Phys.* **63**, 5439 (1988).
[c]M. N. Afsar *et al.*, *IEEE Trans. Micro. Theor. Tech.* **MTT-25**, 505(1977).
[d]R. E. Lindquist and A. W. Ewald, *Phys. Rev.* **135**, A191(1964).
[e]W. J. Moore *et al.*, *J. Appl. Phys.* **78**, 7255(1995).
[f]Estimated from J. A. Powell [*J. Opt. Soc. Am.* **62**, 341 (1972)].
[g]Estimated from P. T. B. Shaffer [*Appl. Opt.* **10**, 1034 (1971)].
[h]A. N. Pikhtin *et al.*, *Opt. Spectrosc.* **43**, 420 (1977).
[i]M. I. Eremets *et al.*, *Phys. Rev. B* **52**, 8854 (1995).
[j]R. Geick *et al.*, *Phys. Rev.* **146**, 543 (1966).
[k]T. Takenaka *et al.*, *Jpn. J. Appl. Phys.* **15**, 2021 (1976).
[l]A. T. Collins *et al.*, *Phys. Rev.* **158**, 833 (1967).
[m]B. Monemar, *Solid State Commun.* **8**, 1295 (1970).
[n]R. E. Fern and A. Onton, *J. Appl. Phys.* **42**, 3499 (1971).
[o]W. J. Turner and W. E. Reese, *Phys. Rev.* **127**, 126 (1962).
[p]H. Sobotta *et al.*, *Phys. Status Solidi B* **174**, K57 (1992).
[q]G. Mirjalili *et al.*, *Infrared Phys. Technol.* **37**, 389 (1996).
[r]D. F. Parsons and P. D. Coleman, *Appl. Opt.* **10**, 1683 (1971).
[s]W. J. Moore and R. T. Holm, *J. Appl. Phys.* **80**, 6939 (1996).
[t]M. Hass and B. W. Henvis, *J. Phys. Chem. Solids* **23**, 1099 (1962).
[u]K. L. Westra and M. J. Brett, *Thin Solid Films* **192**, 227 (1990).
[v]S. Adachi, *Physical Properties of III–V Semiconductor Compounds: InP, InAs, GaAs, InGaAs, and InGaAsP* (Wiley-Interscience, New York, 1992).
[w]R. B. Sanderson, *J. Phys. Chem. Solids* **26**, 803 (1965).
[x]*Numerical Data and Functional Relationships in Science and Technology*, edited by K.- H. Hellwege and O. Madelung, Landolt–Börnstein, New Series, Group III, Vol. 17, Pt. b (Springer, Berlin, 1982).
[y]S. Cunsolo *et al.*, *Infrared Phys.* **33**, 539 (1992).
[z]E. F. Venger *et al.*, *Phys. Status Solidi* **188**, 823 (1995).
[aa]A. Manabe *et al.*, *Jpn. J. Appl. Phys.* **6**, 593 (1967).
[ab]A. Deneuville *et al.*, *Phys. Rev. B* **43**, 6544 (1991).
[ac]Estimated from M. Cardona *et al.* [*Phys. Rev.* **140**, A633 (1965)].
[ad]M. Balkanski, in *Optical Properties of Solids*, edited by F. Abelés (North-Holland, Amsterdam, 1972), p. 529.
[ae]R. Geick *et al.*, *J. Appl. Phys.* **37**, 1994 (1966).
[af]L. K. Vodop'yanov *et al.*, *Sov. Phys. Solid State* **16**, 912 (1974).
[ag]S. Einfeldt *et al.*, *Phys. Rev. B* **51**, 4915 (1995).
[ah]T=77-K value [J. Baars and F. Sorger, *Solid State Commun.* **10**, 875 (1972)].
[ai]R. Geick, *Phys. Lett.* **10**, 51 (1964).
[aj]H. Burkhard *et al. Phys. Status Solidi B* **63**, 89 (1974).
[ak]G. Bauer and H. Krenn, in *Handbook of Optical Constants of Solids*, edited by E. D. Palik (Academic, Orlando, 1985), p. 535.
[al]See, H. Lehmann *et al.*, *Appl. Phys.* **25**, 291 (1981).
[am]R. F. Bis and J. R. Dixon, *Phys. Rev. B* **2**, 1004 (1970).

active region of an injection laser is larger than the index of the cladding layer on both sides, the effect is like that of a waveguiding configuration which confines radiation to the neighborhood of the active region. This simple rule is known to be very useful for the design of various optical waveguiding devices.

2.1.2 Alloys

To our knowledge, there have been reported no detailed experimental data on the dielectric constants of alloy semiconductors. For computational purpose, it is convenient to express the dielectric constant as a function of alloy composition. Let us consider a ternary alloy of the form $A_xB_{1-x}C$. We now take a linear representation for the composition dependence of the dielectric constant. The most simplest linear interpolation method is the use of the equation:

$$\varepsilon(x) = x\varepsilon(\mathrm{AC}) + (1\text{-}x)\varepsilon(\mathrm{BC}) \qquad (2.3)$$

The dielectric constant ε of a medium containing a mixture of dipoles, with the ith type having polarizability α_i and occurring with a density N_i, can be given by the Clausius–Mosotti relation:

$$\frac{\varepsilon-1}{\varepsilon-2} = \frac{1}{3}\sum_i N_i\alpha_i \qquad (2.4)$$

Based on the Clausius–Mossotti relation, we obtain another interpolation expression:

$$\frac{\varepsilon(x)-1}{\varepsilon(x)-2} = x\frac{\varepsilon(\mathrm{AC})-1}{\varepsilon(\mathrm{AC})-2} + (1-x)\frac{\varepsilon(\mathrm{BC})-1}{\varepsilon(\mathrm{BC})-2} \qquad (2.5)$$

In Fig. 2.3, we show the interpolated ε_s and ε_∞ as a function of x for $In_{1-x}Ga_xAs$ ternary alloy. The solid and dashed lines are obtained from Eqs. (2.4) and (2.5), respectively. It is evident from the figure that these two interpolation schemes give nearly the same numeric values. It is easy to extend these expressions to more complex alloy systems, such as quaternary and pentanary systems.

2.1.3 Temperature and Pressure Effects

We show in Fig. 2.4 (a), as an example, the temperature dependence of ε_s for GaAs measured by Samara [4] using a radio-frequency capacitance technique at different frequencies (1~100 kHz) at atmospheric pressure. Below ~150 K there is no clear frequency dependence, but relatively strong frequency dependence is observed at higher temperatures. Samara considered that this is associated with dipolar contributions becoming activated at higher temperatures. The $\varepsilon_s(T)$ response becomes frequency independent at 100 kHz up to ~300 K and this represents the true $\varepsilon_s(T)$ of the material.

Some data [5] also showed that at $T \geq 100$ K the dielectric constant ε_s increases with increasing temperature according to the relation:

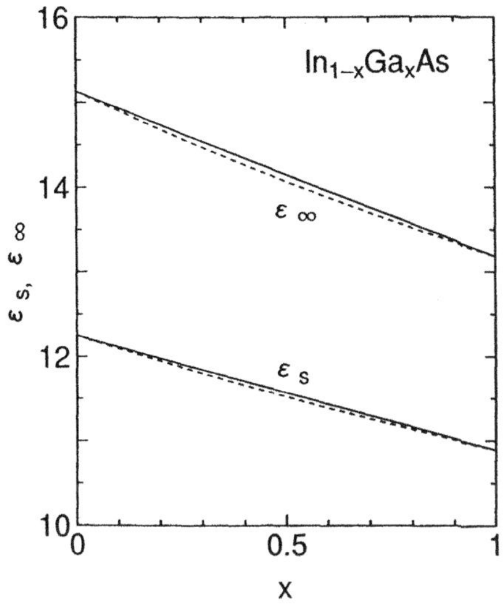

Figure 2.3 Static and high-frequency dielectric constants, ε_s and ε_∞, as a function of x for $In_{1-x}Ga_xAs$ ternary alloy. The solid and dashed lines are obtained from Eqs. (2.4) and (2.5), respectively.

$$\varepsilon_s(T) = \varepsilon_s(T{=}0\ \mathrm{K})\,(1 + \alpha T) \qquad (2.6)$$

The linear temperature coefficients α determined were $(2.01\pm0.02)\times10^{-4}$ K^{-1} [$\varepsilon_s(T{=}0$ K)=12.35±0.09] for GaAs, $(2.27\pm0.02)\times10^{-4}$ K^{-1} [$\varepsilon_s(T{=}0$ K)=10.31±0.08] for CdTe, and $(1.71\pm0.02)\times10^{-4}$ K^{-1} [$\varepsilon_s(T{=}0$ K)=8.80±0.07] for ZnSe [5]. As suggested by Samara [4], it is not likely that the quantities ε_s and ε_∞ are truly linear dependence on temperature over any very wide range.

Figure 2.4 (b) shows the pressure effect on ε_s and ε_∞ for GaAs as measured by Samara [4] using a maraging steel pressure cell up to ~1.2 GPa. The dielectric constants decreased almost linearly with increasing pressure. He reported logarithmic pressure derivatives for GaAs to be $d(\ln\varepsilon_s)/dp$=-16.7±0.3 (at 75.6 K) and -17.3±0.3 (at 300 K), and $d(\ln\varepsilon_\infty)/dp$=-14.0 (at 300 K) (all are in units of 10^{-3} GPa^{-1}).

We can expect that the static and high-frequency dielectric constants vary in a very similar way as in the case of the band-gap energy. This comes from the fact that the dielectric constant can be simply expressed as by a sum of two different contributions [4]: *intrinsic* term (no temperature dependence) and *lattice vibronic* term (temperature-dependent term). Although the sign of the lattice vibronic contribution is opposite, it is quite similar to an expression obtained by Varshni [6]

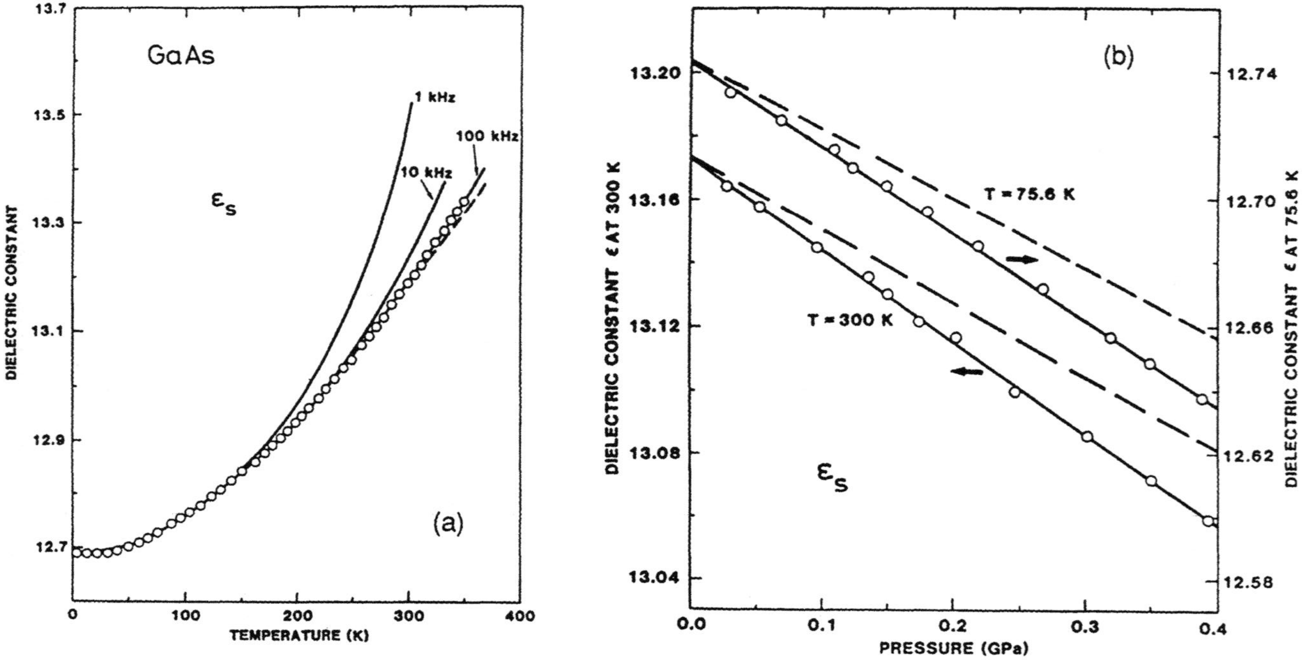

Figure 2.4 (a) Temperature dependence of the static dielectric constant ε_s for GaAs at different frequencies (1, 10, and 100 kHz) at atmospheric pressure. (b) Hydrostatic-pressure dependence of the static dielectric constant ε_s for GaAs at two different temperatures (T=75.6 and 300 K). The dashed lines represent the response corrected for the changes in the dimensions of the sample due to the finite compressibility of GaAs. (From Samara [4].)

for the calculation of the temperature variation of the band-gap energy. The temperature and pressure effects on the long-wavelength dielectric constant (refractive index) will also be discussed in Sec. 5.2.2.

2.2 RESTSTRAHLEN SPECTRA

2.2.1 Crystalline Materials

(a) Zinc-Blende-Type Crystals—As we have already shown in Fig. 1.5, the first BZ for the cubic semiconductors is the truncated octahedron. Phonons representing lattice vibrations occurring at a frequency ω_q have an energy $\hbar\omega_q$, where $\hbar$ is the reduced Planck constant, and have a momentum $\hbar \boldsymbol{q}$. In the primitive cell, if there are N different types of atoms either of differing mass or ordering in space, $3N$ vibration modes will result. In general, three of these branches, namely the acoustic branches, will disappear at the zone center (Γ). The remaining $3(N\text{-}1)$ branches will be optical branches. For the cubic crystals, GaAs or Si, *e.g.*, N=2, and hence there are three acoustic and three optical branches, each three comprising of one longitudinal and two transverse modes.

It is noted that in homopolar crystals (diamond, Si, Ge, and α-Sn), the fundamental vibration has no dipole moment and is IR inactive. In heteropolar crystals, as in GaAs, the first-order dipole moment gives rise to a very strong absorption band associated with optical modes having an of essentially zero $\boldsymbol{k}$ vector (i.e., long-wavelength optical phonons). A useful description of the theoretical and experimental aspects of this subject was given by Balkanski [7].

The complex dielectric permittivity $\varepsilon(\omega)$ in the long-wavelength limit can be generally explained by a single harmonic oscillator model:

$$\varepsilon(\omega) = \varepsilon_\infty \left(1 + \frac{\omega_{LO}^2 - \omega_{TO}^2}{\omega_{TO}^2 - \omega^2 - i\omega\gamma} \right) \qquad (2.7)$$

where γ is the phonon damping constant. This expression can be modified using the Lyddane–Sachs–Teller relation of Eq. (2.2) as

$$\varepsilon(\omega) = \varepsilon_\infty + \frac{\omega_{TO}^2 (\varepsilon_s - \varepsilon_\infty)}{\omega_{TO}^2 - \omega^2 - i\omega\gamma} \qquad (2.8)$$

The reststrahlen parameters, such as ε_s, ε_∞, ω_{TO}, ω_{LO}, and γ, can be determined from measurements of absorption, refractive-index dispersion, and reflection at normal and oblique incidence.

We show in Fig. 2.5 spectral variations of $\varepsilon_1(\omega)$ and $\varepsilon_2(\omega)$ for GaAs obtained from a numeric modeling of Eq. (2.7) [8]. Kachare *et al.* [9] measured the room-temperature IR reflectance spectra of undoped, high-resistivity GaAs. The samples used were chemically polished to remove all traces of surface damage. The reststrahlen parameters they obtained were ω_{TO}=269.2 cm^{-1}, ω_{LO}=292.77 cm^{-1}, γ=2.5 cm^{-1}, and ε_∞=10.90. Holm *et al.* [10] also measured the room-temperature reflectance of semiinsulating GaAs samples. They reported ω_{TO}=268.2 cm^{-1}, ω_{LO}=291.5 cm^{-1}, γ=2.3 cm^{-1}, and ε_∞=11.1. We averaged in Fig. 2.5 Kachare and Holm's parameters to obtain ω_{TO}=268.7 cm^{-1}, ω_{LO}=292.1 cm^{-1}, γ=2.4 cm^{-1}, and ε_∞=11.0.

The imaginary part of Eq. (2.7) can be modified as

$$\varepsilon_2(\omega) = \frac{S\omega_{TO}^2\omega\gamma}{(\omega_{TO}^2-\omega^2)^2+\omega^2\gamma^2} \tag{2.9}$$

with

$$S \equiv \frac{\varepsilon_\infty(\omega_{LO}^2-\omega_{TO}^2)}{\omega_{TO}^2} \tag{2.10}$$

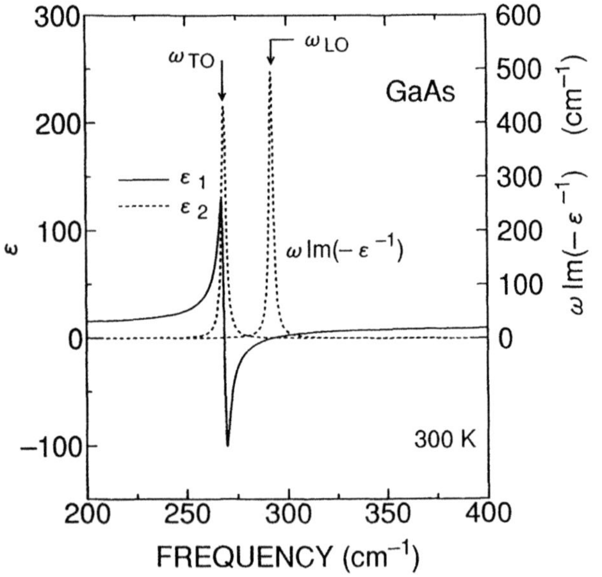

Figure 2.5 Spectral variation of $\varepsilon(\omega)$ obtained from a numeric modeling of Eq. (2.7) with the reststrahlen parameters of semiinsulating GaAs determined at room temperature (ω_{TO}=268.7 cm^{-1}, ω_{LO}=292.1 cm^{-1}, γ=2.4 cm^{-1}, and ε_∞=11.0). (From Adachi [8].) The spectral dependence of ωIm-$\varepsilon(\omega)^{-1}$ is also plotted.

Since the frequency ω_{TO} is usually much larger than γ, $\varepsilon_2(\omega)$ has a significant value only when ω is close to ω_{TO}. For such values of ω, it is a good approximation to replace ω by ω_{TO} everywhere in Eq. (2.9), except in the factor ω_{TO}-ω, which gives

$$\varepsilon_2(\omega) = \frac{S\omega_{TO}\gamma/4}{(\omega_{TO}-\omega)^2+(\gamma/2)^2} \quad (\gamma \ll \omega_{TO}) \qquad (2.11)$$

The $\varepsilon_2(\omega)$ thus reaches a maximum value $S\omega_{TO}/\gamma$ for frequency $\omega=\omega_{TO}$ (resonance frequency) and drops to half that peak value for frequencies $\omega=\omega_{TO}\pm\gamma/2$. Equation (2.11) is known as the Lorentzian line shape. The frequency region just above ω_{TO} is one in which ε_1 is driven strongly negative (see Fig. 2.5). Note that in homopolar crystals (Si, Ge, and α-Sn), $\omega_{TO}=\omega_{LO}$ then S=0 [$\varepsilon_s=\varepsilon_\infty$, see Eq. (2.2)].

We also show that one has the relation [7]

$$\mathrm{Im} - \varepsilon(\omega)^{-1} = \frac{\varepsilon_2(\omega)}{\varepsilon_1(\omega)^2+\varepsilon_2(\omega)^2} = \frac{(\varepsilon_\infty^{-1}-\varepsilon_s^{-1})\omega_{LO}^2\gamma\omega}{(\omega_{LO}^2-\omega^2)^2+\gamma^2\omega^2} \qquad (2.12)$$

which becomes for $\omega\sim\omega_{LO}$

$$\mathrm{Im} - \varepsilon(\omega)^{-1} = \frac{(\varepsilon_\infty^{-1}-\varepsilon_s^{-1})\omega_{LO}^2\gamma}{\omega[4(\omega_{LO}-\omega)^2+\gamma^2]} \qquad (2.13)$$

Hence the maximum of the curve $\omega\mathrm{Im}$-$\varepsilon(\omega)^{-1}$ determines the frequency ω_{LO}. This calculated result is shown in Fig. 2.5.

The $n(\omega)$ and $k(\omega)$ curves obtained for GaAs from Eqs. (1.19a) and (1.19b) are shown in Fig. 2.6 [8]. The absorption coefficient $\alpha(\omega)$ and normal-incidence reflectivity $R(\omega)$ as calculated from Eqs. (1.21) and (1.22) are also shown in Figs. 2.7 and 2.8, respectively. Both n and k (α) dispersion curves show a strong peak at the resonance frequency $\omega=\omega_{TO}$. The reflectivity R at ω close to ω_{LO} is found to be nearly zero.

The optical properties at the reststrahlen region of semiinsulating crystals can be adequately well interpreted by the above-mentioned single harmonic oscillator model. If free carriers (electrons) are present in the medium, we must take into account plasma contribution to the dielectric permittivity (see details in Sec. 5.3.1). When the frequency of free-carrier plasma excitations (*plasmons*) is close to the frequency of the LO phonons (ω_{LO}), the two excitations interact via their macroscopic electric fields. The dielectric permittivity $\varepsilon(\omega)$ in such case is given by [cf. Eq. (2.7)]

$$\varepsilon(\omega) = \varepsilon_\infty\left(1+\frac{\omega_{LO}^2-\omega_{TO}^2}{\omega_{TO}^2-\omega^2-i\omega\gamma} - \frac{\omega_p^2}{\omega(\omega+i\Gamma)}\right) \qquad (2.14)$$

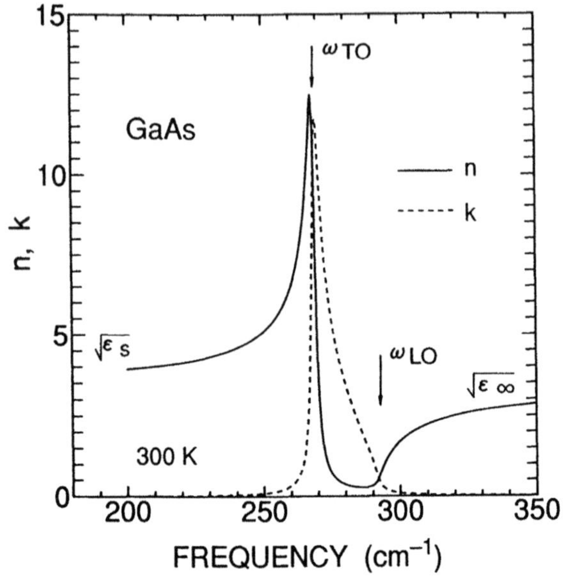

Figure 2.6 Numerically calculated spectral dependence of the real refractive index, $n(\omega)$, and extinction coefficient, $k(\omega)$, for GaAs using the same set of reststrahlen parameters as in Fig. 2.5. (From Adachi [8].)

where

$$\omega_p = \left(\frac{ne^2}{m_e \varepsilon_\infty \varepsilon_0}\right)^{1/2} \qquad (2.15)$$

is the plasma frequency, Γ is the plasmon damping constant, and n and m_e are, respectively, the concentration and effective mass ("*optical mass*") of the free electrons. The frequencies of the coupled plasmon–LO phonon modes are deduced in the limit of zero damping as the roots of the equation $\varepsilon(\omega)=0$ and are given by

$$\omega_\pm = \frac{\omega_{LO}^2+\omega_p^2}{2} \pm \left[\left(\frac{\omega_{LO}^2+\omega_p^2}{2}\right)^2 - \omega_p^2\omega_{TO}^2\right]^{1/2} \qquad (2.16)$$

As the damping values increase, however, this equation is no longer valid. The two sharp dips in the reflectivity minima get smeared out and the coupled-mode frequencies become complex.

In the case of III–V crystals and their alloys, we must take into consideration the effects of multivalley nature of the conduction-band structure. A three-valley

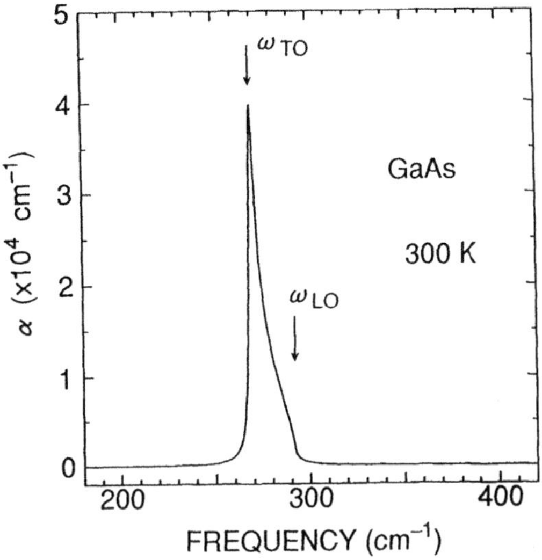

Figure 2.7 Numerically calculated spectral dependence of the absorption coefficient, $\alpha(\omega)$, for GaAs using the same set of reststrahlen parameters as in Fig. 2.5.

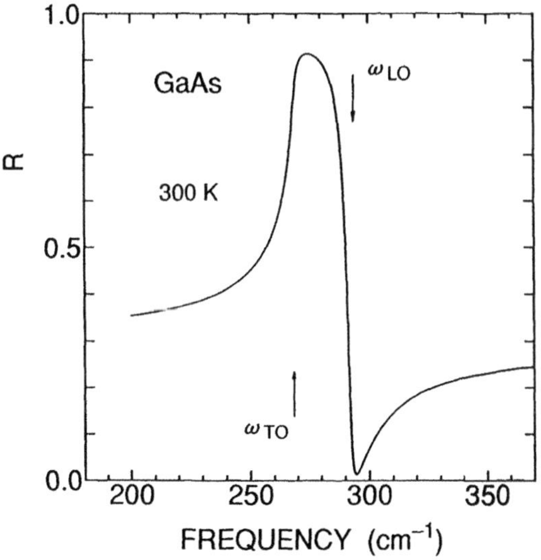

Figure 2.8 Numerically calculated spectral dependence of the normal-incidence reflectivity, $R(\omega)$, for GaAs using the same set of reststrahlen parameters as in Fig. 2.5. (From Adachi [8].)

model is used to account for the Γ, L, and X valleys in the conduction band of such materials. Considering the Γ minimum as the energy reference, the electron concentration in the Γ minimum and those in the L (X) minima relative to the Γ are given by [8]

$$n_\Gamma = 2\left(\frac{2\pi m_e^\Gamma kT}{h^2}\right)^{3/2} \exp\left(-\frac{E_F}{kT}\right) \quad (2.17a)$$

$$\frac{n_{L(X)}}{n_\Gamma} = \left(\frac{m_e^{L(X)}}{m_e^\Gamma}\right)^{3/2} \exp\left(-\frac{\Delta E_{\Gamma L(X)}}{kT}\right) \quad (2.17b)$$

where E_F is the Fermi energy and $\Delta E_{\Gamma L}$ ($\Delta E_{\Gamma X}$) is the energy gap between the Γ and L (X). The total free electron density can be given by $n=n_\Gamma+n_L+n_X$. Then, the plasmon term in Eq. (2.14) is modified as follows:

$$\frac{\omega_p^2}{\omega(\omega+i\Gamma)} \to \frac{\omega_{p\Gamma}^2}{\omega(\omega+i\Gamma_\Gamma)} + \frac{\omega_{pL}^2}{\omega(\omega+i\Gamma_L)} + \frac{\omega_{pX}^2}{\omega(\omega+i\Gamma_X)} \quad (2.18)$$

with

$$\omega_{p\Gamma} = \left(\frac{n_\Gamma e^2}{m_e^\Gamma \varepsilon_\infty \varepsilon_0}\right)^{1/2}, \quad \omega_{pL} = \left(\frac{n_L e^2}{m_e^L \varepsilon_\infty \varepsilon_0}\right)^{1/2}, \quad \omega_{pX} = \left(\frac{n_X e^2}{m_e^X \varepsilon_\infty \varepsilon_0}\right)^{1/2} \quad (2.19)$$

where m_e^Γ, m_e^L, and m_e^X are the effective masses at the Γ, L, and X conduction bands, respectively.

In the case of *p*-type semiconductors, we must consider the contributions from the HH and LH subbands (with the hole effective masses m_{HH}^* and m_{LH}^* and densities p_{HH} and p_{LH}, respectively). However, it should be noted that the sum of these contributions has the form similar to Eq. (2.18).

Equation (2.14) has been used as the basis for sorting out plasmon–LO phonon effects in a number of investigations [10–15]. Kukharskii [15], however, showed that this expression oversimplifies the coupling of plasmons and phonons and can lead, principally, to errors in deducing the damping constants γ and Γ. In Kukharskii's model (factorized form), ε(ω) is given by

$$\varepsilon(\omega) = \varepsilon_\infty \frac{(\omega^2+i\Gamma_-\omega-\omega_-^2)(\omega^2+i\Gamma_+\omega-\omega_+^2)}{\omega(\omega+i\Gamma)(\omega^2+i\gamma\omega-\omega_{TO}^2)} \quad (2.20)$$

In his model, two additional parameters are the damping constants $\Gamma_{\pm}$ of the coupled plasmon–phonon modes. The plasmon and single-oscillator models are related to each other through the equation

$$\varepsilon_{+}\varepsilon_{-} = \omega_p \omega_{TO} \qquad (2.21)$$

Chandrasekhar and Ramdas [14] measured the IR reflectivity spectra of *n*-type GaAs doped with Te in the concentration range from 1.8×10^{16} to 7.0×10^{18} cm^{-3}. We reproduce in Fig. 2.9 the reflectivity spectrum as a function of carrier concentration for *n*-GaAs [8]. These spectra were calculated using the reststrahlen parameters as determined by Chandrasekhar and Ramdas [14]. These authors used a least-squares minimization algorithm to fit the experimental data to the theoretical expression of Eq. (2.14). They also used the factorized form of the dielectric function, Eq. (2.20), to improve the fits and found to be unnecessary.

In Fig. 2.9, the reflectance minimum at about 300 cm^{-1} is due to the LO phonons. The position of the second minimum, called the plasma minimum, depends on the electron concentration. The plasma resonance minimum is very close to the TO-phonon line in the spectrum of the relatively lightly doped sample.

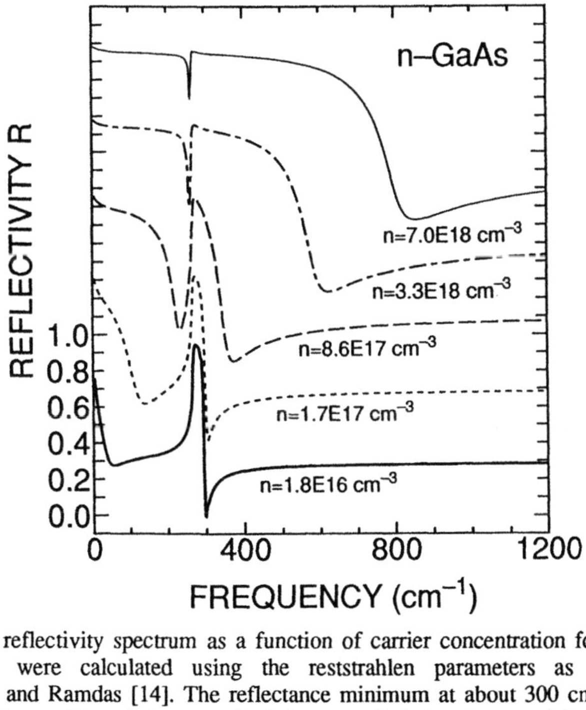

Figure 2.9 IR reflectivity spectrum as a function of carrier concentration for *n*-type GaAs. These spectra were calculated using the reststrahlen parameters as determined by Chandrasekhar and Ramdas [14]. The reflectance minimum at about 300 cm^{-1} is due to the LO phonons. The position of the second minimum, called the plasma minimum, depends on the electron concentration. (From Adachi [8].)

When the electron (carrier) concentration is larger, n=8.6×10^{17}–7.0×10^{18} cm^{-3}, the plasma minimum is shifted toward higher frequencies as seen in Fig. 2.9.

(b) Wurtzite-Type Crystals—At normal temperature and pressure CdS crystallizes in the wurtzite lattice with two molecular units in the primitive cell which gives rise to twelve phonon branches (N=4). At the center of the BZ (Γ) the acoustic modes comprise the representations

$$1A_1 + 1E_1 \qquad (2.22a)$$

and the remaining nine optical modes the representations

$$1A_1 + 2B_1 + 1E_1 + 2E_2 \qquad (2.22b)$$

Except the silent B_1 modes all the other optical modes are Raman active, and the A_1 and E_1 modes are IR active for incident radiation with the polarizations $\boldsymbol{E}\|c$ and $\boldsymbol{E}\perp c$, respectively, c being the hexagonal axis of the crystal. We reproduce in Fig. 2.10 the phonon frequency vs. reduced wave vector results for $\boldsymbol{q}$ along the principal symmetry directions (T, Σ, Δ) of ZnO, predicted by a shell model based on elastic constants and Raman spectrum by Hewat [16]. The dashed lines at region near $\boldsymbol{q}\to 0$, $\nu\to 0$ (Σ) represent the initial slopes of the acoustic modes estimated from the elastic constants (sound velocities).

In the uniaxial media, two different solutions of the wave equation exist, which describe the ordinary ($\boldsymbol{E}\perp c$) and extraordinary ($\boldsymbol{E}\|c$) rays [7]. Figure 2.11 shows spectral variations of $\varepsilon_1(\omega)$ and $\varepsilon_2(\omega)$ for CdS obtained from a numeric modeling of Eq. (2.8). The numeric values used in the calculation are as follows: ω_{TO}=240 cm^{-1}, γ=4.7 cm^{-1}, ε_s=8.4, and ε_∞=5.3 ($\boldsymbol{E}\perp c$); ω_{TO}=232 cm^{-1}, γ=6.3 cm^{-1}, ε_s=8.9, and ε_∞=5.4 ($\boldsymbol{E}\|c$). These values were determined from the reflectivity spectra for a crystal having its optical axis perpendicular to the surface and to the two polarizations of light at 290 K [7]. The LO phonon frequencies deduced from the curve $\omega\mathrm{Im}\text{-}\varepsilon(\omega)^{-1}$ are ω_{LO}=302 cm^{-1} ($\boldsymbol{E}\perp c$) and 298 cm^{-1} ($\boldsymbol{E}\|c$).

(c) Alloys—In the classical picture, the atoms of a crystal are visualized as joined by harmonic springs and the crystal dynamics can be analyzed in terms of a linear combination of 3N normal modes of vibration. In alloy crystals, the nature of the lattice optical spectrum depends on how large is the difference between the quantities representing the lattice vibronic properties of the components (i.e., atomic masses and force constants) [17]. If these quantities are similar, then the optical response of an alloy is similar to the response of a crystal with the quantities averaged over the composition ("*one-mode*" behavior). In one-mode system, such as most I–VII alloys, a single set of long-wavelength optical modes appears. The phonon spectrum is approximately constant in strength but varies in energy as the alloy composition is varied from one end member to the other.

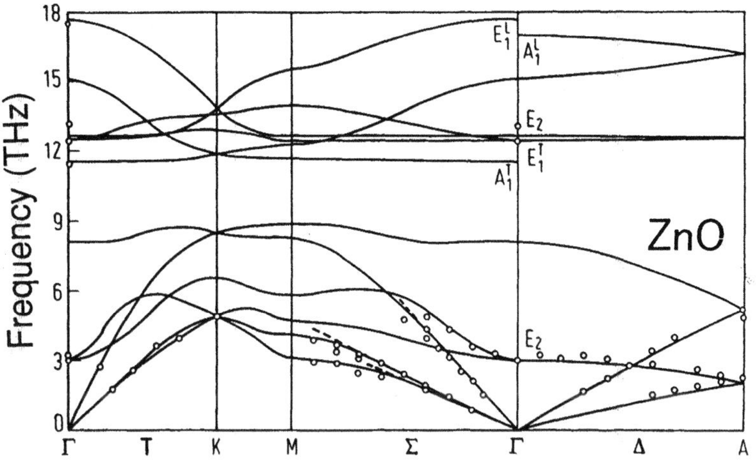

Figure 2.10 Phonon dispersion curves along the principal symmetry directions of ZnO, predicted by a shell model based on elastic constants and Raman spectrum. (From Hewat [16].) The dashed lines at region near $q \to 0$, $\nu \to 0$ (Σ) represent the initial slopes of the acoustic modes estimated from the elastic constants.

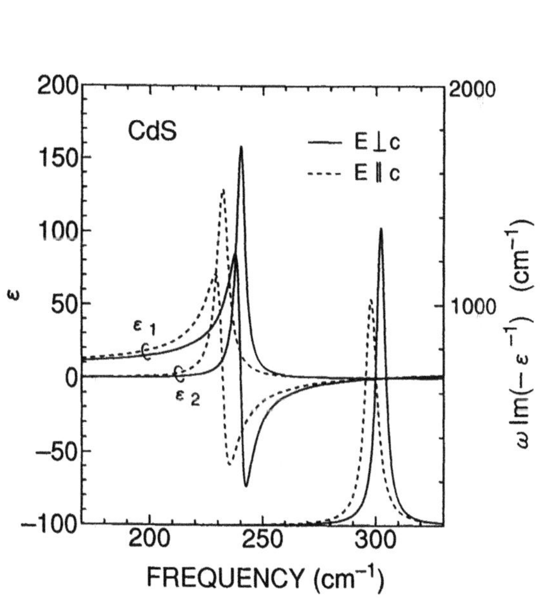

Figure 2.11 Spectral variations of ε_1 and ε_2 for CdS obtained from a numeric modeling of Eq. (2.7). The numeric values used in the calculation are as follows: ω_{TO}=240 cm^{-1}, γ=4.7 cm^{-1}, ε_0=8.4, and ε_∞=5.3 (***E***⊥*c*); ω_{TO}=232 cm^{-1}, γ=6.3 cm^{-1}, ε_0=8.9, and ε_∞=5.4 (***E***∥*c*). These values were determined from the reflectivity spectra for a crystal having its optical axis perpendicular to the surface and to the two polarizations of light at 290 K by Balkanski [7].

When the parameters differ strongly, the response of a system is more complex [17]; the spectrum contains a number of bands each of which corresponds to one of the components and has an intensity governed by its content in the alloy ("*multimode*" behavior). For example, two-mode system exhibits two distinct sets of optical modes with frequencies characteristic of each end member and strengths roughly proportional to the respective concentration. As we will see next, the long-wavelength optical phonons in III–V and II–VI ternary alloys exhibit either one-mode or *two-mode* behavior, or more rigorously, three different types of mode behavior, viz., one-, two-, and "*one-two-mode*" behaviors. The one-two-mode system exhibits a single mode only a part of the composition range, with two modes observed over the remaining range of composition.

An analysis of the optical spectra of multimode alloys shows that they are described satisfactorily by a set of oscillators and the number of such oscillators is equal to the number of different unit cells that occur in an alloy [17]. For example, in the case of $A_xB_{1-x}C$ alloy there are unit cells of two types—AC and BC, so that the corresponding optical spectrum can be described satisfactorily by a two-oscillator model. Tables 2.2 and 2.3 list behaviors of the long-wavelength optical modes in Si_xGe_{1-x}, some mixed III–V, II–VI, and IV–VI compounds obtained by IR optical and RS spectroscopies. Figure 2.12 shows typical examples of (a) one-mode and (b) two-mode behaviors of the long-wavelength optical modes observed in AlAsSb [18] and AlGaAs ternaries [8].

It is known that Si and Ge form a continuous range of alloys, Si_xGe_{1-x}, from x=0 (Ge) to x=1 (Si). The translational symmetry of the perfect diamond lattice forbids single phonon absorption of radiation because of a lack of a linear electric moment. Consequently, the absorption is probably due to the multiple-phonon processes which allow a second-order electric moment to be induced providing the coupling to the radiation field [19]. The loss of translational symmetry with alloying modifies the pure-crystal selection rules in IR optical spectra [20]. This also yields additional allowed RS from the entire density-of-phonon states as well as resonant or local modes [21,22]. It is found that the Si_xGe_{1-x} alloy has three types of cell and, consequently, shows multimode behavior consisting of three main peaks due to the vibrations of Si–Si, Ge–Ge, and Si–Ge pairs [22].

Most III–V ternary compounds were found to exhibit either two- or one-two-mode behavior [1]. However, it has recently been shown [18] that AlAsSb ternary displays one-mode behavior over the entire range of composition [$0 \leq x \leq 1$, see Fig. 2.12(a)]. GaInP is also the III–V ternary alloy for which one-mode behavior was currently assumed [23]. However, Jusserand and Slempkes [24] concluded from RS study that the alloy shows a somewhat modified two-mode behavior. Galtier *et al.* [25] considered from hydrostatic-pressure dependence of Raman spectra that the concept of a one-mode system is not appropriate for description of the physics of GaInP alloy.

In a quaternary alloy of the $A_xB_{1-x}C_yD_{1-y}$ type, there are four kinds of unit cells: AC, AD, BC, and BD. On the other hand, in the $A_xB_yC_zD$ ($AB_xC_yD_z$) type there

TABLE 2.2 *Behavior of the long-wavelength optical modes in mixed Si–Ge and III–V alloy crystals.*

System	Behavior	References
Si–Ge	three	*a*
AlGaN	one	*b*
AlGaP	two	*c*
AlGaAs	two	*d*
AlGaSb	two	*e*
AlInP	one-two	*f*
AlInAs	two	*g*
AlInSb	two	*h*
GaInN	one	*i*
GaInP	two	*j*
GaInAs	two	*k*
GaInSb	one-two	*l*
AlAsSb	one	*m*
GaNAs	two	*n*
GaPAs	two	*o*
GaPSb	two	*p*
GaAsSb	two	*q*
InPAs	two	*r*
InPSb	two	*p*
InAsSb	one-two	*s*
AlGaAsSb	three	*t*
GaInPSb	four	*u*
GaInPAs/InP	four	*v*
GaInPAs/GaAs	four	*w*
GaInAsSb	three	*x*
AlGaInP	three	*f*
AlGaInAs	three	*y*
InPAsSb	three	*z*

[a] T. Ishida *et al.*, *J. Phys. Soc. Jpn* **53**, 2584 (1984).
[b] K. Hayashi *et al.*, *Solid State Commun.* **77**, 115 (1991).
[c] G. Armelles *et al.*, *Solid State Commun.* **65**, 779 (1988).
[d] B. Jusserand and J. Sapriel, *Phys. Rev. B* **24**, 7194 (1981).
[e] G. Lucovsky *et al.*, *Phys. Rev. B* **12**, 4135 (1975).
[f] H. Asahi *et al.*, *J. Appl. Phys.* **65**, 5007 (1989).
[g] S. Emura *et al.*, *J. Appl. Phys.* **62**, 4632 (1987).
[h] V. P. Gnezdilov *et al.*, *J. Appl. Phys.* **74**, 6883 (1993).
[i] K. Osamune *et al.*, *J. Appl. Phys.* **46**, 3432 (1975).
[j] B. Jusserand and S. Slempkes, *Solid State Commun.* **49**, 95 (1984).
[k] G. Landa *et al.*, *Solid State Commun.* **86**, 351 (1993).
[l] Z. C. Feng *et al.*, *Can. J. Phys.* **69**, 386 (1991).
[m] I. Sela *et al.*, *Phys. Rev. B* **46**, 16142 (1992).
[n] T. Prokofyeva *et al.*, *Appl. Phys. Lett.* **73**, 1409 (1998).
[o] N. D. Strahm and A. L. McWhorter, in *Light Scattering Spectra of Solids*, edited by G. B. Wright (Springer, Berlin, 1969), p. 455.
[p] Y. T. Cherng *et al.*, *J. Appl. Phys.* **65**, 3285 (1989).
[q] T. C.McGlinn *et al.*, *Phys. Rev. B* **33**, 8396 (1986).
[r] N. P. Kekelidze *et al.*, *J. Phys. Chem. Solids* **34**, 2117 (1973).

[s] G. Lucovsky and M. F. Cohen, *Solid State Commun.* **8**, 1397 (1970).
[t] D. H. Jaw *et al.*, *J. Appl. Phys.* **69**, 2552 (1991).
[u] D. H. Jaw and G. B. Stringfellow, *J. Appl. Phys.* **72**, 4265 (1992).
[v] R. K. Soni *et al.*, *J. Appl. Phys.* **59**, 2184 (1986).
[w] T. Sugiura *et al.*, *Jpn. J. Appl. Phys.* **32**, 2718 (1993).
[x] C. Pickering, *J. Electron. Mater.* **15**, 51 (1986).
[y] R. Borroff *et al.*, *Appl. Phys. Lett.* **53**, 1652 (1988).
[z] A. G. Belov *et al.*, *Sov. Phys. Solid State* **22**, 839 (1980).

are three kinds of unit cells: AD, BD, and CD (AB, AC, and AD). We can, thus, once expect "*four-*" or "*three-mode*" behavior of the long-wavelength optical modes in such quaternary alloys. Indeed, the quaternary compounds GaInPSb and GaInPAs are reported to show four-mode behavior. Also, GaInAsSb, AlGaInP, AlGaInAs, and InPAsSb exhibit three-mode behavior (see Table 2.2). However, the GaInAsSb system showed a three-phonon mode with GaAs, InSb, and mixed InAs/GaAs characteristics [26]. It was considered that the behavior of this alloy system may be termed "*three-mode*" because of the degeneracy of the GaAs- and InAs-like modes.

Mixed II–VI ternaries grow either in the zinc-blende or wurtzite structure and exhibit one- or two-mode behavior (see Table 2.3). However, optical phonons in ZnMn-based diluted magnetic semiconductors (such as ZnMnSe and ZnMnTe) exhibit a variation with composition that is intermediate between the one- and two-mode situation. The II–VI quaternaries of the $A_xB_yC_zD$ or $AB_xC_yD_z$ type show a three-mode behavior, as can be expected from the number of different cells.

IR optical spectra of bulk alloy semiconductors consist of lattice absorption bands, the number of which is usually equal to the number of different unit cells. For example, reflectivity spectra of $Al_xGa_{1-x}As$ alloy consist of two lattice bands [two-mode behavior, see Table 2.2 and also Fig. 2.12 (b)]. The dielectric permittivity in such material can be successfully interpreted by the following additive multioscillator form:

$$\varepsilon(\omega) = \varepsilon_\infty + \sum_j \frac{S_j \omega_{\mathrm{TO}j}^2}{\omega_{\mathrm{TO}j}^2 - \omega^2 - i\omega\gamma_j} \tag{2.23}$$

or by the factorized form:

$$\varepsilon(\omega) = \varepsilon_\infty \prod_j \frac{\omega_{\mathrm{LO}j}^2 - \omega^2 - i\omega\gamma_j(\mathrm{LO})}{\omega_{\mathrm{TO}j}^2 - \omega^2 - i\omega\gamma_j(\mathrm{TO})} \tag{2.24}$$

where S_j $[=\varepsilon_\infty(\omega_{\mathrm{LO}j}^2-\omega_{\mathrm{TO}j}^2)]$ is the oscillator strength, $\omega_{\mathrm{TO}j}$ ($\omega_{\mathrm{LO}j}$) is the TO (LO) phonon frequency, and γ_j is the damping constant of the jth lattice oscillator.

Figure 2.13 shows the calculated $\varepsilon_1(\omega)$ and $\varepsilon_2(\omega)$ spectra for $Al_xGa_{1-x}As$ alloy

TABLE 2.3 *Behavior of the long-wavelength optical modes in mixed II–VI and IV–VI crystals.*

System	Behavior	References
MgZnS	one	*a*
MgZnSe	two	*b*
MgCdTe	two	*c*
ZnCdS	one	*d*
ZnCdSe	intermediate	*e*
ZnCdTe	two	*f*
ZnMnS	one	*g*
ZnMnSe	intermediate	*h*
ZnMnTe	intermediate	*i*
CdMnSe	two	*j*
CdMnTe	two	*k*
HgZnTe	two	*l*
HgCdTe	two	*m*
ZnSSe	two	*n*
ZnSTe	two	*o*
ZnSeTe	one	*p*
CdSSe	two	*n*
CdSTe	two	*q*
CdSeTe	two	*r*
MgZnSSe	three	*a*
MgCdMnTe	three	*c*
ZnCdMnTe	three	*s*
ZnCdHgTe	three	*t*
CdSSeTe	three	*u*
PbSeTe	two	*v*

[a] D. Wang *et al.*, *J. Appl. Phys.* **80**, 1248 (1996).
[b] D. Huang *et al.*, *Appl. Phys. Lett.* **67**, 3611 (1995).
[c] E. Oh *et al.*, *Phys. Rev. B* **48**, 15040 (1993).
[d] M. Ichimura *et al.*, *Phys. Rev. B* **46**, 4273 (1992).
[e] R. G. Alonso *et al.*, *Phys. Rev. B* **40**, 3720 (1989).
[f] D. N. Talwar *et al.*, *Phys. Rev. B* **48**, 17064 (1993).
[g] A. Anastassiadou *et al.*, *Solid State Commun.* **69**, 137 (1989).
[h] P. D. Lao *et al.*, *Phys. Rev. B* **48**, 11701 (1993).
[i] D. L. Peterson *et al.*, *Phys. Rev. B* **33**, 1160 (1986).
[j] R. G. Alonso *et al.*, *Phys. Rev. B* **43**, 9610 (1991).
[k] S. Venugopalan *et al.*, *Phys. Rev. B* **25**, 2681 (1982).
[l] M. P. Volz *et al.*, *Solid State Commun.* **75**, 943 (1990).
[m] S. P. Kozyrev *et al.*, *Sov. Phys. Solid State* **34**, 1984 (1992).
[n] I. F. Chang and S. S. Mitra, *Phys. Rev.* **172**, 924 (1968).
[o] P. Gupta *et al.*, *Thin Solid Films* **227**, 66 (1993).
[p] S. Nakashima *et al.*, *J. Phys. Soc. Jpn* **30**, 1508 (1971).
[q] R. Pal *et al.*, *J. Phys. D: Appl. Phys.* **26**, 704 (1993).
[r] Z. C. Feng *et al.*, *J. Cryst. Growth* **138**, 239 (1994).
[s] E. Oh *et al. Phys. Rev. B* **45**, 10934 (1992).
[t] J. Polit *et al. Phys. Status Solidi B* **208**, 21 (1998).
[u] V. M. Burlakov *et al.*, *Phys. Status Solidi B* **128**, 389 (1985).
[v] H. Finkenrath *et al.*, *Phys. Status Solidi B* **95**, 179 (1979).

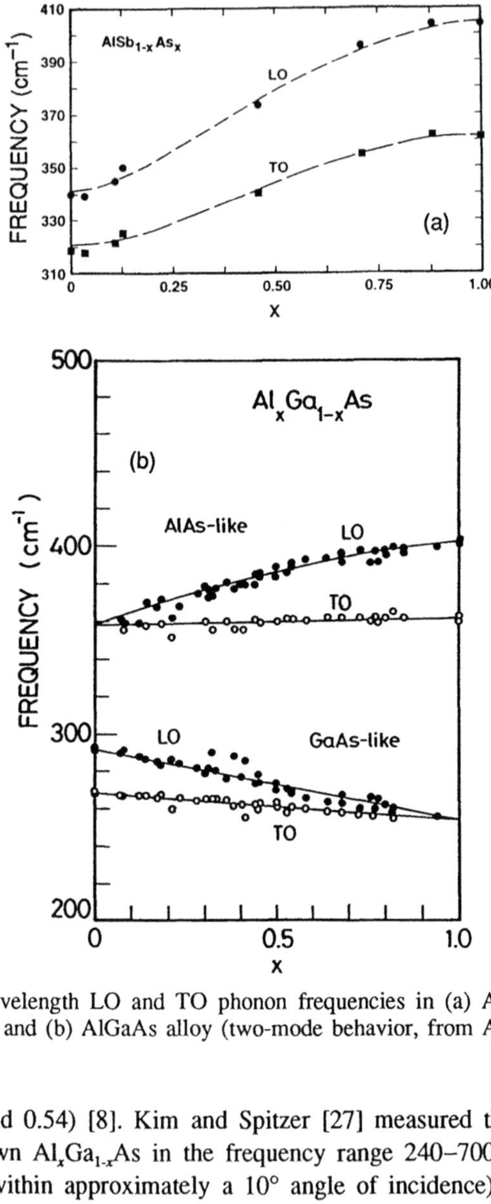

Figure 2.12 Long-wavelength LO and TO phonon frequencies in (a) AlAsSb (one-mode, from Sela *et al.* [18]) and (b) AlGaAs alloy (two-mode behavior, from Adachi [8]).

(x=0, 0.18, 0.36, and 0.54) [8]. Kim and Spitzer [27] measured the IR reflectivity spectra of LPE-grown $Al_xGa_{1-x}As$ in the frequency range 240–700 cm^{-1} at close to normal incidence (within approximately a 10° angle of incidence). The reflectivity data were analyzed by using the KK dispersion relations and by curve fitting to the two forms for the classical dielectric function, one the additive form of Eq. (2.23) and the other the factorized form of Eq. (2.24). In Fig. 2.13, the reststrahlen parameters used were taken from the factorized-form results of Kim and Spitzer. The numeric values of these parameters are listed in Table 2.4. The high-frequency dielectric constant, ε_∞, is simply obtained from the known values of GaAs and AlAs using the linear variation assumption of Eq. (2.3). The $\varepsilon(\omega)$ spectrum of AlAs is also calculated using the reststrahlen parameters of the additive form reported by Perkowitz *et al.* [28] and is shown in Fig. 2.13.

The corresponding $n(\omega)$ and $k(\omega)$ spectra are shown in Fig. 2.14 [8]. The two phonon bands, GaAs-like and AlAs-like, are clear and distinct in Figs. 2.13 and 2.14. No additional structure is seen in the reflectivity spectra in the frequency range 50–700 cm^{-1}.

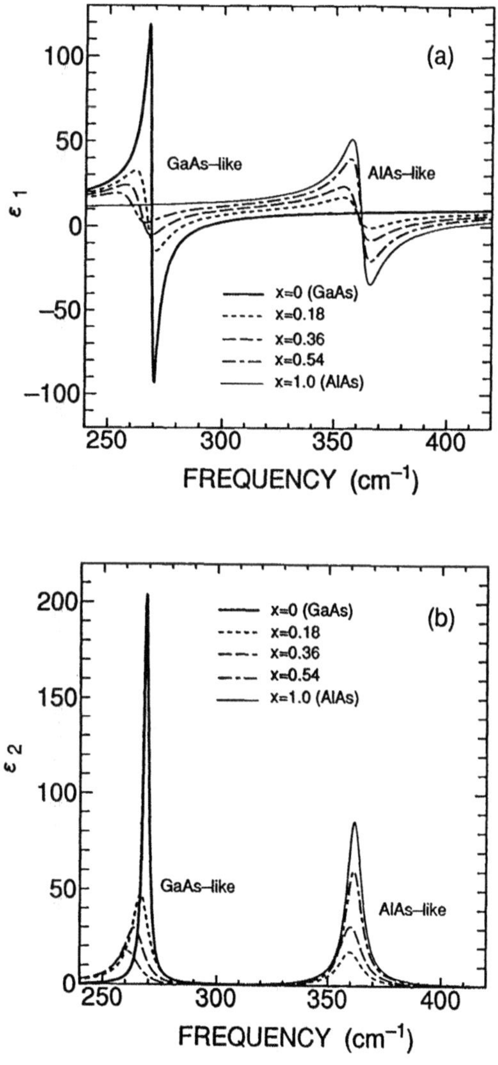

Figure 2.13 Numerically calculated spectral dependence of the dielectric function, $\varepsilon(\omega)=\varepsilon_1(\omega)+i\varepsilon_2(\omega)$, for $Al_xGa_{1-x}As$ alloy using the reststrahlen parameters as determined by Kim and Spitzer (x=0–0.54, see Table 2.4) [27] and by Perkowitz *et al.* (x=1.0) [28]. (From Adachi [8].)

TABLE 2.4 *Reststrahlen parameters for $Al_xGa_{1-x}As$ alloy obtained from the factorized form [O. K. Kim and W. G. Spitzer, J. Appl. Phys.* ***50****, 4362 (1979)].*

	GaAs-Like Mode				AlAs-Like Mode				
x	ω_{TO}	ω_{LO}	γ(TO)	γ(LO)	ω_{TO}	ω_{LO}	γ(TO)	γ(LO)	ε_∞
0	268.8	292.8	2.65	2.85					10.90
0.14	267.1	285.7	5.67	4.85	358.8	369.0	10.56	11.31	10.57
0.18	266.9	283.4	8.76	4.24	360.1	372.4	12.20	10.24	10.47
0.30	265.2	278.3	8.64	6.15	360.2	379.1	12.10	9.42	10.16
0.36	264.5	276.5	10.69	5.58	360.4	381.3	12.23	8.08	10.04
0.44	262.9	273.7	10.05	6.44	360.2	385.4	9.55	7.90	9.84
0.54	261.8	269.8	12.43	7.97	361.5	390.1	8.75	8.68	9.60

(d) Temperature Effect—It is noted that the phonon frequency usually decreases with increasing temperature. The static and high-frequency dielectric constants are functions of temperature; they are usually increase with increasing temperature (see Sec. 2.1.3). The damping constant γ in Eqs. (2.7)–(2.14) may also be strongly dependent on temperature (that will increase with increasing temperature). We can thus easily expect that the optical constants in the reststrahlen range of semiconductors are dependent on temperature.

Figure 2.15 shows, as examples, the reststrahlen refractive-index [(a)] and transmission spectra [(b)] for high-resistivity GaAs measured at 300 and 8 K by Johnson *et al.* [29]. These results show that the ionic contribution to the far-IR refractive index is different at 300 and 8 K, while the near-IR refractive index remains unchanged. The two transmission spectra shown in Fig. 2.15 (b) also suggest that cooling of this material has a marked effect on the optical absorption process.

2.2.2 Amorphous Materials

The simplest way to describe the structure of an amorphous material consists of considering that the short-range order is preserved whereas the extended atomic network is random (i.e., the lack of long-range order) [30]. If the short-range order in the amorphous form is similar to that in the corresponding crystal, one can expect the Raman intensity and IR absorption to show some similarity to the phonon DOS of the crystal: essentially, the DOS in the amorphous form should be a broadened version of the crystalline state [31,32].

The main features of the phonon DOS in amorphous solids can be extracted from neutron scattering, IR absorption, and RS spectroscopies. In IR absorption spectroscopy, the absorption coefficient $\alpha(\omega)$ is related to the phonon DOS $g(\omega)$ as

$$\alpha(\omega) \sim |M(\omega)|^2 g(\omega) \qquad (2.25)$$

where $M(\omega)$ is a frequency-dependent dipole matrix element. In a single crystal

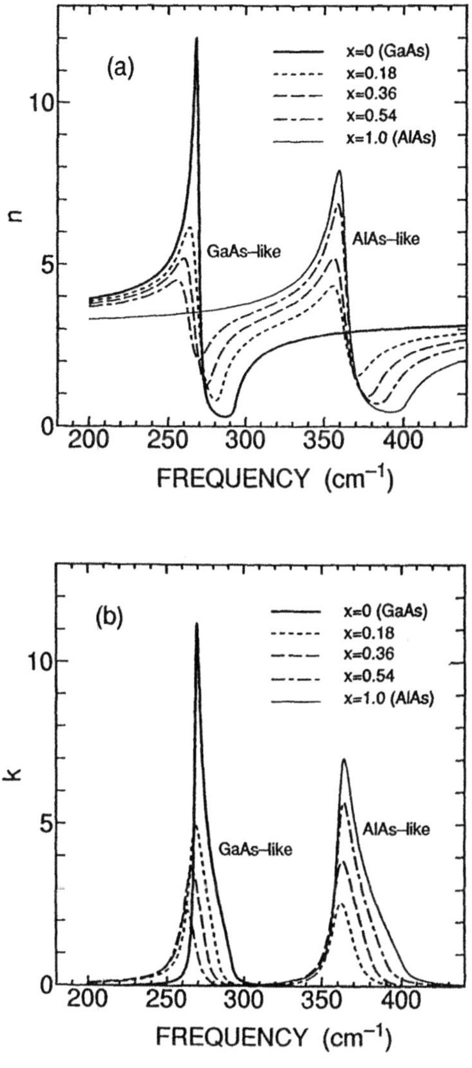

Figure 2.14 Numerically calculated spectral dependence of the real refractive index, $n(\omega)$, and extinction coefficient, $k(\omega)$, for $Al_xGa_{1-x}As$ alloy using the same sets of reststrahlen parameters as in Fig. 2.12. (From Adachi [8].)

with one optical active phonon at $\omega_{TO}(\Gamma)$, $|M(\omega)|^2$ is the Lorentzian-type function centered at $\omega_{TO}(\Gamma)$, whose width is given by the damping constant resulting from the anharmonicity [see Eq. (2.11)]. In this case, $|M(\omega)|^2$ projects a narrow line out of the phonon density of the crystal.

In an amorphous material, where the translational symmetry is relaxed, at least in principle, all phonons are optically active and $M(\omega)$ may be different from zero in the whole frequency range of the phonons [33].

Figure 2.16 (a) shows the room-temperature IR absorption coefficient $\alpha(\omega)$ for

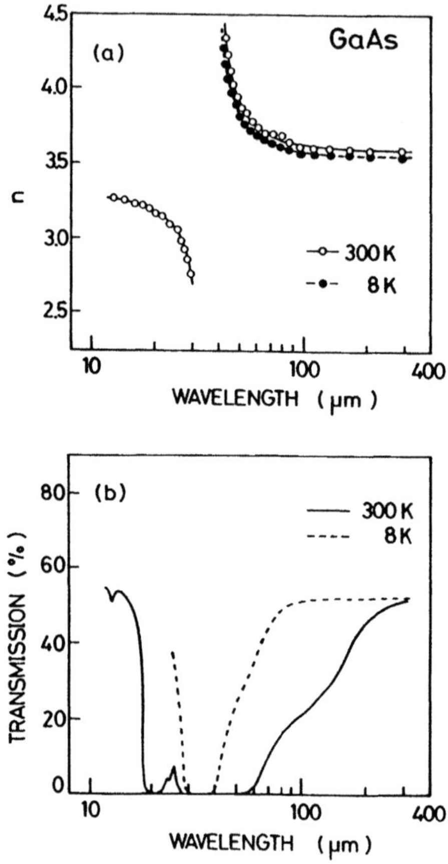

Figure 2.15 Far-IR real refractive-index [(a)] and transmission spectra [(b)] for high-resistivity GaAs at 300 and 8 K. (From Johnson *et al.* [29].)

a-Ge [34]. The lower part of the figure shows the phonon DOS for *c*-Ge as obtained by neutron inelastic scattering (dashed line) and a Gaussian-broadened version of the crystalline DOS (solid line). We represent in Fig. 2.17 the equivalent results for GaAs [33,35,36].

All first-order IR absorption is forbidden in *c*-Ge and the absorption band in the amorphous material [Fig. 2.16 (a)] arises from the lack of crystallinity. If the transition is IR active in the crystal, it is relatively strong in the spectrum of the amorphous counterpart but is broadened considerably [Fig. 2.17 (b)]. It is also apparent from Figs. 2.16 and 2.17 that there is some relations between the IR absorption and the vibrational DOS. As mentioned before, this is in contrast to the typical crystalline case where only a few special vibrations of suitable momenta and symmetry are IR active (see Fig. 2.7). All structures are also shifted to lower frequencies in the amorphous material relative to the crystal.

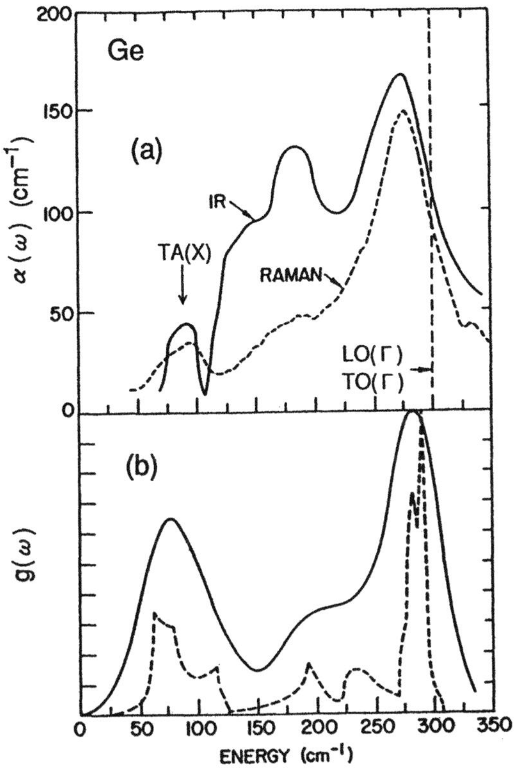

Figure 2.16 (a) Room-temperature IR absorption coefficient α(ω) vs. wave number for *a*-Ge. Also shown for comparison is the reduced Raman spectrum. (b) Phonon density of states of *c*-Ge (dashed line) and *a*-Ge (solid line). (From Brodsky and Lurio [34].)

If the matrix elements $M(\omega)$ were weakly dependent on ω, the absorption curve would very resemble the amorphous phonon DOS. A comparison in Figs. 2.16 and 2.17 of the broadened DOS with the experimental absorption spectra clearly shows that this is not the case: while $M(\omega)$ is nearly constant in the higher-frequency range, it must decrease rapidly at lower frequencies. A consequence of this fact is that the TA branch does not appear in the IR absorption (*a*-GaAs) or its strength is very weak (*a*-Ge). In the elastic quasi-continuous limit, Prettl *et al.* [33] have argued that the frequency dependence of $M(\omega)$ can be given by

$$|M(\omega)|^2 \propto \sin^4\left(\frac{\pi\omega}{4\omega_M}\right) \qquad (2.26)$$

where ω_M is the frequency at the maximum of the phonon DOS. Above ω_M they

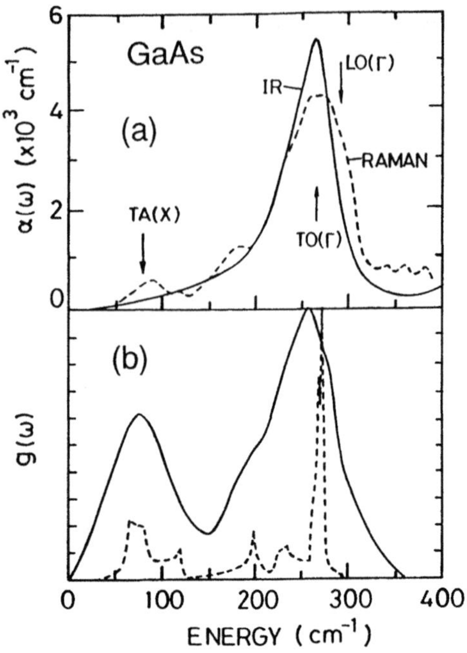

Figure 2.17 (a) Room-temperature IR absorption coefficient α(ω) vs. wave number for *a*-GaAs. Also shown for comparison is the reduced Raman spectrum. (From Prettl *et al.* [33].) (b) Phonon density of states of *c*-GaAs (dashed line) and *a*-GaAs (solid line) taken from Dolling and Cowley [35] and Smith *et al.* [36], respectively.

assumed the $|M(\omega)|^2$ to be independent of ω. The agreement with experiment was found to be excellent.

Also shown by the dashed lines in Figs. 2.16 (a) and 2.17 (a) are the room-temperature Raman spectra. The RS cross section σ_R of an amorphous material is given by

$$\sigma_R \propto \frac{1}{\omega} |C_{ij}|^2 f\left(\frac{\hbar\omega}{kT}\right) g(\omega) \qquad (2.27)$$

where $f(x)$ is the Bose–Einstein distribution factor and C_{ij} measures the coupling of the vibrational modes of frequency ω with the incident and scattered light. Considering the quasi-continuous limit, Prettl *et al.* [33] also obtained the dependence of $|C_{ij}(\omega)|^2$ on ω ($\omega<\omega_M$):

$$|C_{ij}(\omega)|^2 \propto \sin^2\left(\frac{\pi\omega}{4\omega_M}\right) \qquad (2.28)$$

Above ω_M it was assumed to be independent of ω. They found fair agreement between the IR absorption, RS, and neutron scattering DOS. For small ω Connell [37] simply assumed that $M(\omega)\propto\omega^2$ and $C_{ij}(\omega)\propto\omega$ so that

$$g(\omega) \propto \frac{\alpha(\omega)}{\omega^4} \propto \frac{\sigma_R'}{\omega^2} \qquad (2.29)$$

where $\sigma_R' \equiv \omega f^{-1}\sigma_R$ is the reduced Raman cross section (representing the density of phonon states).

Felty *et al.* [38] studied the IR absorption oands in charcogenide glass alloy $As_2S_xSe_{3-x}$. They observed two clear bands at frequencies approximately 300 and 215 cm^{-1} which did not vary in position but only in strength. The frequencies corresponded to the reststrahlen bands in pure As_2S_3 (~300 cm^{-1}) and As_2Se_3 (~215 cm^{-1}). On the other hand, Taylor *et al.* [39] observed a qualitatively different behavior in a $Tl_2Se_xTe_{1-x}As_2Se_yTe_{3-y}$ glass system. The intensity of the two observed IR bands did not change much with composition but their frequencies shifted. To our knowledge, however, no detailed experimental or theoretical data on the reststrahlen bands in tetrahedrally-coordinated amorphous semiconductor alloys has been reported so far.

REFERENCES

1. G. P. Srivastava, *The Physics of Phonons* (Adam Hilger, Bristol, 1990).
2. P. Giannozzi, S. de Gironcoli, P. Pavone, and S. Baroni, *Phys. Rev. B* **43**, 7231 (1991).
3. S. Adachi, *J. Appl. Phys.* **53**, 8775 (1982).
4. G. A. Samara, *Phys. Rev. B* **27**, 3494 (1983).
5. I. Strzalkowski, S. Joshi, and C. R. Crowell, *Appl. Phys. Lett.* **28**, 350 (1976).
6. Y. P. Varshni, *Physica* **34**, 149 (1967).
7. M. Balkanski, in *Optical Properties of Solids*, edited by F. Abelès (North-Holland, Amsterdam, 1972), p. 529.
8. S. Adachi, *GaAs and Related Materials: Bulk Semiconducting and Superlattice Properties* (World Scientific, Singapore, 1994).
9. A. H. Kachare, W. G. Spitzer, F. K. Euler, and A. Kahan, *J. Appl. Phys.* **45**, 2938 (1974).
10. R. T. Holm, J. W. Gibson, and E. D. Palik, *J. Appl. Phys.* **48**, 212 (1977).

11. A. Mooradian and G. B. Wright, *Phys. Rev. Lett.* **16**, 999 (1966).
12. J. D. Axe and R. Hammer, *Phys. Rev.* **162**, 700 (1967).
13. C. G. Olson and D. W. Lynch, *Phys. Rev.* **177**, 1231 (1969).
14. H. R. Chandrasekhar and A. K. Ramdas, *Phys. Rev. B* **21**, 1511 (1980).
15. A. A. Kukharskii, *Solid State Commun.* **13**, 1761 (1973).
16. A. W. Hewat, *Solid State Commun.* **8**, 187 (1970).
17. G. M. Zinger, I. P. Ipatova, and A. I. Ryskin, *Sov. Phys. Semicond.* **18**, 13 (1984).
18. I. Sela, C. R. Bolognesi, and H. Kroemer, *Phys. Rev. B* **46**, 16142 (1992).
19. R. Braunstein, *Phys. Rev.* **130**, 879 (1963).
20. S. C. Shen and M. Cardona, *Solid State Commun.* **36**, 327 (1980).
21. J. S. Lannin, *Phys. Rev. B* **16**, 1510 (1977).
22. T. Ishidate, S. Katagiri, K. Inoue, M. Shibuya, K. Tsuji, and S. Minomura, *J. Phys. Soc. Jpn* **53**, 2584 (1984).
23. R. Beserman, C. Hirlimann, and M. Balkanski, *Solid State Commun.* **20**, 485 (1976).
24. B. Jusserand and S. Slempkes, *Solid State Commun.* **49**, 95 (1984).
25. P. Galtier, J. Chevallier, M. Zigone, and G. Martinez, *Phys. Rev. B* **30**, 726 (1984).
26. C. Pickering, *J. Electron. Mater.* **15**, 51 (1986).
27. O. K. Kim and W. G. Spitzer, *J. Appl. Phys.* **50**, 4362 (1979).
28. S. Perkowitz, R. Sudharsanan, S. S. Yom, and T. J. Drummond, *Solid State Commun.* **62**, 645 (1987).
29. C. J. Johnson, G. H. Sherman, and R. Weil, *Appl. Opt.* **8**, 1667 (1969).
30. See, for example, J. Tauc, in *Amorphous and Liquid Semiconductors*, edited by J. Tauc (Plenum, London, 1974), p. 159.
31. J. E. Smith, Jr., M. H. Brodsky, B. L. Crowder, and M. I. Nathan, *Phys. Rev. Lett.* **26**, 642 (1971).
32. D. Weaire and R. Alben, *Phys. Rev. Lett.* **29**, 1505 (1972).
33. W. Prettl, N. J. Shevchik, and M. Cardona, *Phys. Status Solidi B* **59**, 241 (1973).
34. M. H. Brodsky and A. Lurio, *Phys. Rev. B* **9**, 1646 (1974).
35. G. Dolling and R. A. Cowley, *Proc. Phys. Soc.* **88**, 463 (1966).
36. J. E. Smith, M. H. Brodsky, B. L. Crowder, and M. I. Nathan, *Proc. 2nd Int. Conf. Light Scattering in Solids*, edited by M. Balkanski (Flammarion, Paris, 1971), p. 330.
37. G. A. N. Connell, *Phys. Status Solidi B* **69**, 9 (1975).
38. E. J. Felty, G. Lucovsky, and M. B. Myers, *Solid State Commun.* **5**, 555 (1967).
39. See, Reference 30, p. 165.

Chapter 3

THE INTERBAND TRANSITION REGION: CRYSTALLINE MATERIALS

3.1 OPTICAL DISPERSION MODEL

3.1.1 Critical Points and Dispersion Theory

The optical constants in the interband transition region of semiconductors depend fundamentally on the electronic energy-band structure of the semiconductors. The relation between the energy-band structure and $\varepsilon_2(E)$ can be given by [1]

$$\varepsilon_2(E) = \frac{4e^2\hbar^2}{\pi\mu^2E^2}\int d\boldsymbol{k}\,|P_{cv}(\boldsymbol{k})|^2\,\delta[E_c(\boldsymbol{k})-E_v(\boldsymbol{k})-E] \qquad (3.1)$$

where μ is the combined DOS mass, the Dirac δ function represents the spectral joint DOS between the valence- $[E_v(\boldsymbol{k})]$ and conduction-band $[E_c(\boldsymbol{k})]$ states differing by the energy $E=\hbar\omega$ of the incident light, $P_{cv}(\boldsymbol{k})$ is the momentum-matrix element between the valence- and conduction-band states, and the integration is performed over the first BZ. There have been several analytic models that can be used to explain $\varepsilon(E)$ spectra in the interband transition region of semiconductors. In the following, we briefly review such analytic models, namely HOA, SCP, and MDF, and show the analyzed results of $\varepsilon(E)$ for GaAs using these models.

(a) Harmonic Oscillator Approximation—In an HOA, the dielectric function of a solid is adequately reproduced as a sum over a finite number of the Lorentz oscillators, writing [2]

$$\varepsilon(E) = 1 + \sum_{i=1}^{N} A_i \left(\frac{1}{E+E_{ci}+i\Gamma_i} - \frac{1}{E-E_{ci}+i\Gamma_i} \right) \qquad (3.2)$$

where i is the index for a particular oscillator, A_i is the amplitude, E_{ci} is the center energy of the oscillator, and Γ_i is the oscillator broadening coefficient. Note that this expression is a simplified form of the more general one, namely Eq. (3.1) [3]. By using harmonic oscillators as a basis, the causality, linearity, reality, and KK requirements, compulsory properties of $\varepsilon(E)$, are satisfied automatically.

We have already shown in Fig. 1.11 the electronic energy-band structure of GaAs. The fitted results of HOA calculation to the experimental $\varepsilon(E)$ spectra for GaAs are shown in Figs. 3.1 and 3.2. The experimental $\varepsilon(E)$ data are taken from Aspnes *et al.* [4]. These were measured by SE at room temperature. The heavy solid lines show the calculated results using the four harmonic oscillators. Individual contributions to ε of these oscillators are also shown in the figures by the light solid lines. The center energies E_{ci} (i=1–4) of the oscillators correspond to the E_1-, $E_1+\Delta_1$-, E_0'-, and E_2-gap energies. It is clear from these figures that the four-oscillator HOA fit shows very poor agreement with the experimental data. To improve the quality of the HOA fit, we add three extra oscillators to the four OA. As shown by the dashed lines, the seven HOA results in dramatic improvement of the fit with the experimental data. It should be noted, however, that these extra oscillators have no physical origin (i.e., they do not correspond to any singularity or CP in the energy-band structure). The parameters obtained from these fits are summarized in Table 3.1.

The normal HOA assumes that all oscillators are in phase (i.e., all A_i's are *real* and *positive*). Terry [5] proposed a modified version of the HOA which allows each oscillator to have an independent phase [i.e., $A_i = |A_i| \exp(i\phi_i)$]. He found that this modification significantly improves the accuracy of the approximation. It was also demonstrated [5] that the nine harmonic oscillators would accurately fit the experimental $\varepsilon(E)$ spectra over the 1.5–5.0-eV range of $Al_xGa_{1-x}As$ (Ref. [4]). The mean square error for the fit was less than 0.03 for all cases. However, because of its phenomenological, nonphysical character, the HOA model completely fails to reproduce either the line shape of $\varepsilon_2(E)$ near the $E_0/(E_0+\Delta_0)$ edges or the line shape of the derivatives of $\varepsilon(E)$ near any CP. Also, it cannot be expected *a priori* to provide satisfactory fits to the dependence of $\varepsilon(E)$ on external perturbations, such as temperature, pressure, or electric field.

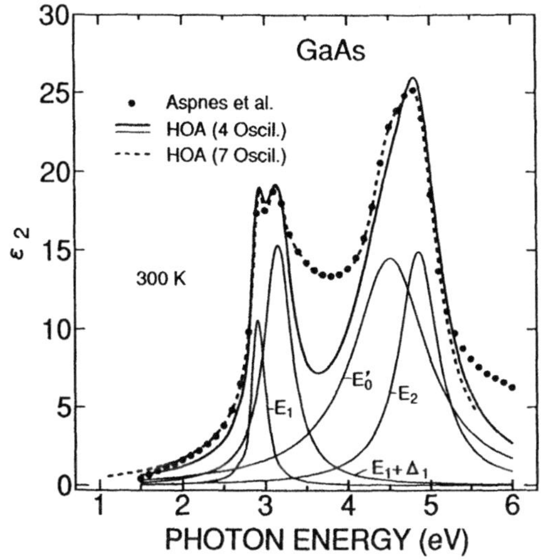

Figure 3.1 Fitted results of HOA model to the experimental $\varepsilon_2(E)$ spectrum for GaAs. The experimental data are taken from Aspnes *et al.* [4]. Individual contributions to ε_2 of the four harmonic oscillators are shown by the light solid lines. The center energies of these oscillators correspond to the E_1-, $E_1+\Delta_1$-, E_0'-, and E_2-gap energies. The parameter values obtained from these fits are summarized in Table 3.1.

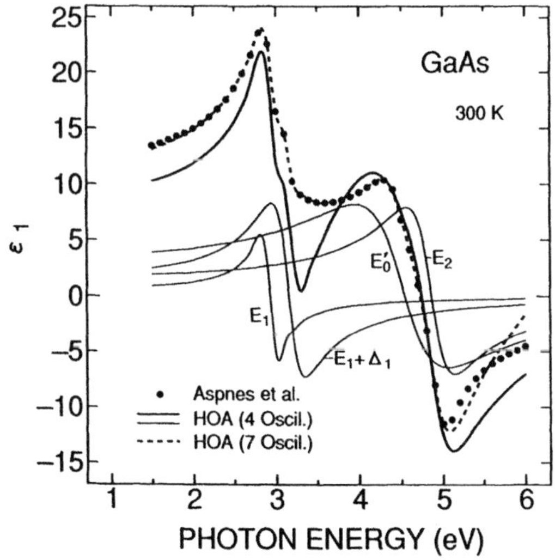

Figure 3.2 Fitted results of HOA model to the experimental $\varepsilon_1(E)$ spectrum for GaAs. The experimental data are taken from Aspnes *et al.* [4]. Individual contributions to ε_1 of the four harmonic oscillators are shown by the light solid lines. The center energies of these oscillators correspond to the E_1-, $E_1+\Delta_1$-, E_0'-, and E_2-gap energies. The parameter values obtained from these fits are summarized in Table 3.1.

TABLE 3.1 *Oscillator parameters for GaAs determined by fitting with HOA at 300 K.*

	HOA (Four Oscil.)		
i	E_{ci} (eV)	A_i	Γ_i (eV)
1	2.92	1.0	0.09
2	3.15	3.1	0.20
3	4.51	8.0	0.55
4	4.85	4.3	0.29
	HOA (Seven Oscil.)		
i	E_{ci} (eV)	A_i	Γ_i (eV)
1	2.921	0.8081	0.0905
2	3.124	1.2899	0.1565
3	3.364	4.2014	0.4753
4	3.930	2.3744	0.4792
5	4.500	4.0758	0.3270
6	4.853	4.7893	0.2873
7	7.145	6.0627	0.2472

(b) Standard Critical-Point Model—Expression (3.1) can be transformed into a surface integral

$$\varepsilon_2(E) = \frac{4e^2\hbar^2}{\pi\mu^2E^2}\int_S \frac{dS\,|P_{cv}(\boldsymbol{k})|^2}{|\nabla_k(E_c-E_v)|_{E_c-E_v=E}} \tag{3.3}$$

where the integration goes over the equal-energy-difference surface in $\boldsymbol{k}$ space defined by E_c-E_v=E. In most cases $|P_{cv}(\boldsymbol{k})|^2$ can safely be assumed to be a slowly varying function such that we can define

$$J_{cv}(E) = \int_S \frac{dS}{|\nabla_k(E_c-E_v)|_{E_c-E_v=E}} \tag{3.4}$$

the joint-DOS function measuring the product density of full and empty states of equal energy difference. The $\varepsilon_2(E)$ can, then, be simply written as

$$\varepsilon_2(E) = \frac{4e^2\hbar^2}{\pi\mu^2E^2}|P_{cv}(E)|^2J_{cv}(E) \tag{3.5}$$

The $J_{cv}(E)$ shows strong variations as a function of E for those frequencies at which

$$\nabla_k\,[E_c(\boldsymbol{k})-E_v(\boldsymbol{k})] = \nabla_kE_c(\boldsymbol{k})-\nabla_kE_v(\boldsymbol{k}) = 0 \tag{3.6}$$

for some value of k. These CPs defined by Eq. (3.6) are of paramount importance for the discussion of optical spectra.

The energy difference E_c-E_v can be expanded about a CP of energy $E_c(k_c)$ in a Taylor series [1]

$$E_c-E_v = E_c+\sum_{j=1}^{3}\left[\frac{d^2(E_c-E_v)}{dk_j^2}\right]_{k=k_c}(k_j-k_{cj})^2 \qquad (3.7)$$

The linear term in the expansion is identically zero at a CP because of Eq. (3.6) and in the parabolic approximation we retain only the quadratic term. Inserting Eq. (3.7) into Eq. (3.4) and performing the integral, one obtains the analytic behavior of the joint-DOS function J_{cv} at CPs. According to different combinations of the sign of the

$$a_j = \left[\frac{d^2(E_c-E_v)}{dk_j^2}\right]_{k=k_c} \qquad (3.8)$$

one has to distinguish between four types of CPs, M_0 to M_3. For example, an M_0-type CP is realized by having all a_j>0. We summarize in Table 3.2 the analytic behavior of J_{cv} at the four types of CPs (3D CPs). The different constants depend on the values of the a_j. We also summarize in this table the energy dependence of J_{cv} in the cases of 2D and 1D k spaces.

Figure 3.3 shows qualitatively the form of J_{cv} for all these nine cases (3D, 2D, and 1D). We note that there is a characteristic line shape for each CP. For example, one finds a step function for the 2D M_0- (a_j>0) or M_2-type (a_j<0) and a logarithmic singularity for the 2D M_1 saddle point (a_1>0, a_2<0) [see Fig. 3.3 (b)]. Such a feature gives us a valuable tool for identifying the structure in optical spectra which are experimentally observed.

In an SCP model, $\varepsilon(E)$ is approximated by a sum of several terms. Each of these terms is a function of energy and represents the contribution to Eq. (3.1) from the neighborhood of a CP in the joint DOS. These CPs are associated with electronic transitions in the energy-band structure at the energies labeled E_0, E_0+Δ_0, E_1, E_1+Δ_1, E_2, E_0', etc. The commonly used line shapes in the optical spectrum analysis are the SCP expressions given by [6,7]

$$\varepsilon(E) = \sum_{i=1}^{N}\left[C_i-A_i e^{i\phi_i}(E-E_{ci}+i\Gamma_i)^n\right] \qquad (3.9)$$

where the CP is described by the amplitude A_i, CP energy E_{ci}, broadening Γ_i, and exciton phase angle ϕ_i. The exponent n has the value -1/2 for the 1D, 0 [loga-

TABLE 3.2 *Joint-DOS function $J_{cv}(E)$ for different critical points.*

Three-Dimensional (3D)						
					J_{cv}	
Type of CP	Notation	a_1	a_2	a_3	$E<E_c$	$E>E_c$
Minimum	M_0	+	+	+	0	$C_0(E-E_c)^{1/2}$
Saddle Point	M_1	+	+	–	$C_1-C_1'(E_c-E)^{1/2}$	C_1
Saddle Point	M_2	+	–	–	C_2	$C_2-C_2'(E-E_c)^{1/2}$
Maximum	M_3	–	–	–	$C_3(E_c-E)^{1/2}$	0

Two-Dimensional (2D)					
				J_{cv}	
Type of CP	Notation	a_1	a_2	$E<E_c$	$E>E_c$
Minimum	M_0	+	+	0	B_1
Saddle Point	M_1	+	–	$(B_1/\pi)(B_2-\ln\lvert E_c-E\rvert)$	$(B_1/\pi)(B_2-\ln\lvert E_c-E\rvert)$
Maximum	M_2	–	–	B_1	0

One-Dimensional (1D)				
			J_{cv}	
Type of CP	Notation	a_1	$E<E_c$	$E>E_c$
Minimum	M_0	+	0	$A(E-E_c)^{-1/2}$
Maximum	M_1	–	$A(E_c-E)^{-1/2}$	0

rithmic, i.e., $\ln(E-E_{ci}+i\Gamma_i)$] for the 2D, and 1/2 for the 3D CP. Discrete excitons are represented by n=-1. A phase angle of $\phi\neq 0$ with n=-1 corresponds to a Fano profile, i.e., the line shape which results from the interaction with a continuous background.

We show in Fig. 3.4, as an example, the line shapes of the 2D CP (n=0) with five different phase angles ϕ=0°, 45°, 90°, 135°, and 180°. We can recognize from this figure that ϕ=0°, 90°, and 180° correspond to a minimum (2D-M_0 CP), saddle point (2D-M_1 CP), and maximum (2D-M_2 CP), respectively. Similarly, in Eq. (3.9), taking A_i>0 and n=1/2 (3D CP), ϕ=0°, 90°, 180°, and 270° correspond to M_0, M_1, M_2, M_3, and M_0 CPs, respectively.

Figures 3.5 and 3.6 show the best-fitted results of SCP model to the experimental $\varepsilon(E)$ spectra for GaAs [4]. The heavy solid lines show the calculated results of the SCP model (ΣC_i=2.50). Individual contributions to ε of the E_0, $E_0+\Delta_0$, E_1, $E_1+\Delta_1$, E_0', and E_2 transitions are also shown in the figures by the light solid lines. The CP parameters determined from this fit are listed in Table 3.3. In the case of the $E_0/(E_0+\Delta_0)$ CPs, the best fit was obtained with a 3D-M_0 line shape. The $E_1/(E_1+\Delta_1)$ CPs were fitted with a 2D line shape, whereas a 1D line shape yielded the best representation for the E_0' and E_2 CPs. The SCP model, however, showed remarkable disagreement with the experimental data for photon energies below 3

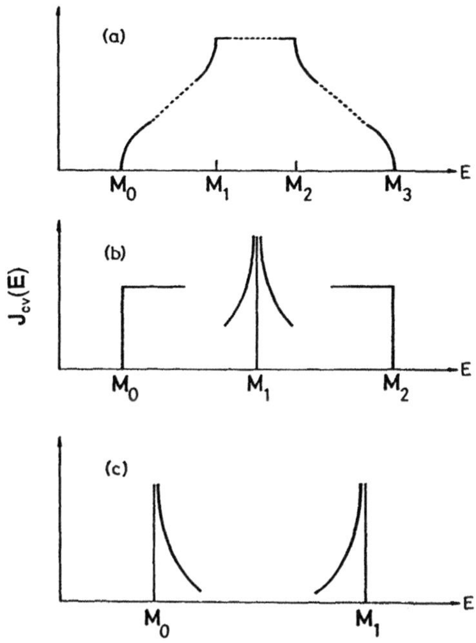

Figure 3.3 Joint-DOS function $J_{cv}(E)$ plotted against energy E for the 3D (a), 2D (b), and 1D CPs (c).

eV. Despite this fact, the SCP model can successfully explain the dielectric-function derivative spectra, such as $d\varepsilon/dE$ and $d^2\varepsilon/dE^2$ [6,7].

(c) Model Dielectric Function

Let us now present a new model of the dielectric function $\varepsilon(E)$, named the MDF, which covers the optical response of semiconductors over a wide range of photon energies. The model is based on the KK transformation [see Eq. (1.16)] and is strongly connected with the electronic energy-band structure of the medium. The model predicts distinct structures at energies of the various CPs.

In Figs. 3.7 and 3.8, we show the MDF $\varepsilon_2(E)$ and $\varepsilon_1(E)$ spectra of GaAs at 300 K with the experimental data [4]. As mentioned before, the joint-DOS function $J_{cv}(E)$ is mainly determining the interband contribution to $\varepsilon_2(E)$ and thus to the optical constants of solids. The MDF representation of $\varepsilon_2(E)$ allows for the joint DOS of the interband transitions from states in the valence band to the conduction band in the BZ. Agreement with the experimental data is found to be very good (see Figs. 3.7 and 3.8). The dielectric function $\varepsilon_2(E)$ can be related to $J_{cv}(E)$ via Eq. (3.5). Analytic behaviors of $J_{cv}(E)$ at the various types of CP have already been summarized in Table 3.2. The KK relations assure that $\varepsilon_1(E)$ can be calculated at

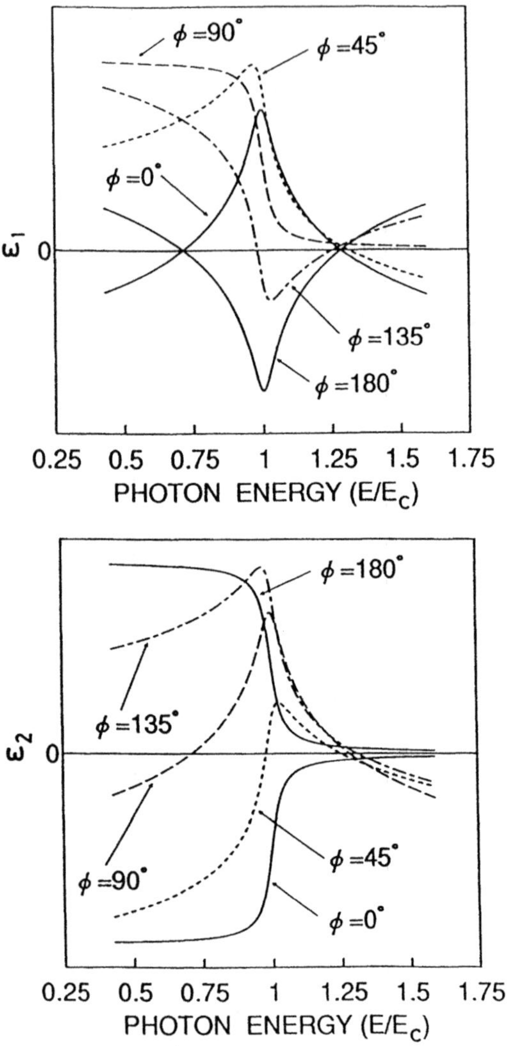

Figure 3.4 Line shape of the 2D CP (n=0) with five different phase angles ϕ=0°, 45°, 90°, 135°, and 180° (SCP).

each photon energy if $\varepsilon_2(E)$ is known explicitly over the entire photon-energy range, and vice versa. In the following, we summarize the MDF for CP of each energy gap (E_0, E_0+Δ_0, E_1, E_1+Δ_1, E_0', E_2, and indirect-band gap), for III–V, zinc-blende-type materials [8,9]. Combining all these contributions, we obtain the spectral dependence of $\varepsilon(E)$ in the entire range of photon energies. It is noted that the MDF can also be easily applied to the diamond-, wurtzite-, and rocksalt-type

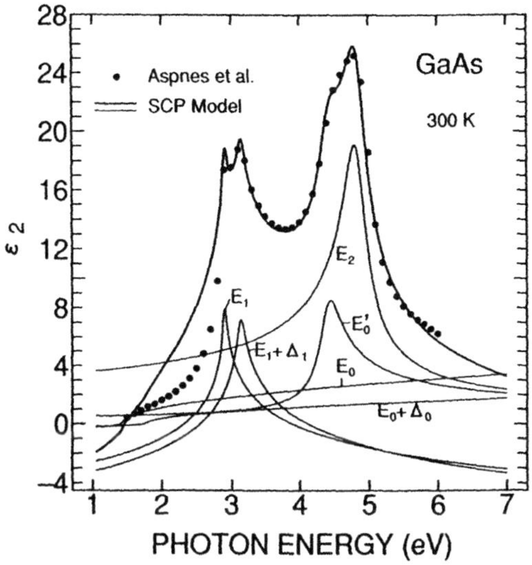

Figure 3.5 Fitted result of SCP model to the experimental $\varepsilon_2(E)$ spectrum for GaAs. The experimental data are taken from Aspnes *et al.* [4]. Individual contributions to ε_2 of the $E_0/(E_0+\Delta_0)$, $E_1/(E_1+\Delta_1)$, E_0', and E_2 transitions are shown by the light solid lines. The CP parameters determined from this fit are listed in Table 3.3.

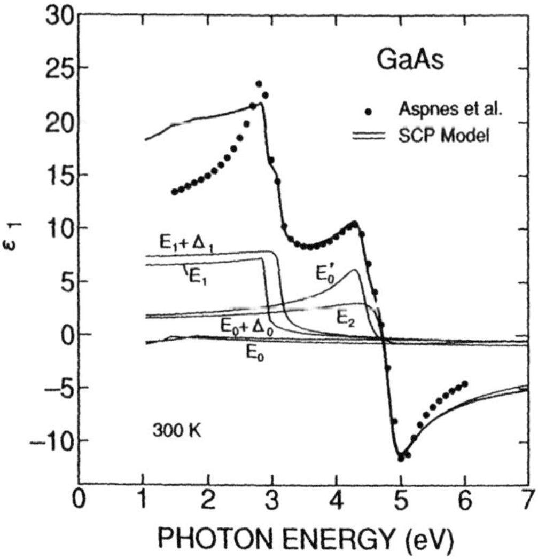

Figure 3.6 Fitted result of SCP model to the experimental $\varepsilon_1(E)$ spectrum for GaAs. The experimental data are taken from Aspnes *et al.* [4]. Individual contributions to ε_1 of the $E_0/(E_0+\Delta_0)$, $E_1/(E_1+\Delta_1)$, E_0', and E_2 transitions are shown by the light solid lines. The CP parameters determined from this fit are listed in Table 3.3.

TABLE 3.3 *CP parameters for GaAs determined by fitting with SCP model at 300 K.*

	Parameter				
CP	E_{ci} (eV)	n	A_i	ϕ_i (deg)	Γ_i (eV)
E_0	1.42	0.5	1.5	285	0.01
$E_0+\Delta_0$	1.77	0.5	0.8	285	0.01
E_1	2.90	0	2.2	80	0.02
$E_1+\Delta_1$	3.13	0	2.5	80	0.04
E_0'	4.40	-0.5	3.5	-75	0.14
E_2	4.83	-0.5	7.7	335	0.15

semiconductors.

1. E_0 and $E_0+\Delta_0$ transitions—The E_0 and $E_0+\Delta_0$ transitions in the zinc- blend-type semiconductors occur at the center of the BZ. These transitions are of the 3D M_0 type. Assuming that the bands are parabolic and using the KK relations, we obtain the contribution f these gaps to $\varepsilon(E)$:

$$\varepsilon(E) = AE_0^{-3/2}\left[f(\chi_0)+\frac{1}{2}\left(\frac{E_0}{E_0+\Delta_0}\right)^{3/2} f(\chi_{so})\right] \qquad (3.10)$$

with

$$A = \frac{4}{3}\left(\frac{3}{2}\mu\right)^{3/2}P^2 \qquad (3.11)$$

$$f(\chi_0) = \chi_0^{-2}[2-(1+\chi_0)^{1/2}-(1-\chi_0)^{1/2}] \qquad (3.12a)$$

$$f(\chi_{so}) = \chi_{so}^{-2}[2-(1+\chi_{so})^{1/2}-(1-\chi_{so})^{1/2}] \qquad (3.12b)$$

$$\chi_0 = \frac{E+i\Gamma}{E_0} \qquad (3.13a)$$

$$\chi_{so} = \frac{E+i\Gamma}{E_0+\Delta_0} \qquad (3.13b)$$

where P^2 is the squared momentum-matrix element and Γ is the broadening constant of the $E_0/(E_0+\Delta_0)$ transitions. Here, for simplicity, the reduced masses μ for the three valence bands are assumed to be equal.

The calculated $\varepsilon_2(E)$ [Im $\varepsilon(E)$] spectrum of Eq. (3.10) in the limit $\Gamma\to 0$ eV provides the well-known line shape of

$$\varepsilon_2(E) = \frac{A}{E^2}\left[(E-E_0)^{1/2}H(\chi_0-1) + \frac{1}{2}(E-E_0-\Delta_0)^{1/2}H(\chi_{so}-1)\right] \qquad (3.14)$$

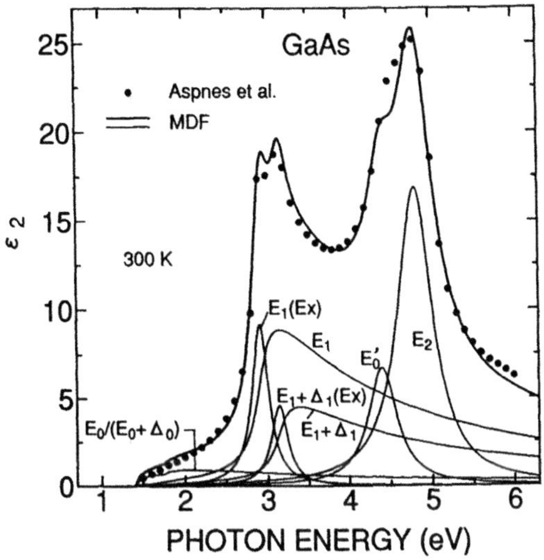

Figure 3.7 Fitted result of MDF calculation to the experimental $\varepsilon_2(E)$ spectrum for GaAs. The experimental data are taken from Aspnes *et al.* [4]. Individual contributions to ε_2 of the $E_0/(E_0+\Delta_0)$ [Eq. (3.10)], $E_1/(E_1+\Delta_1)$ [Eqs. (3.20) and (3.23)], E_0' [Eq. (3.26)], and E_2 transitions [Eq. (3.26)] are shown by the light solid lines. The CP parameters determined from this fit are listed in Table 3.4.

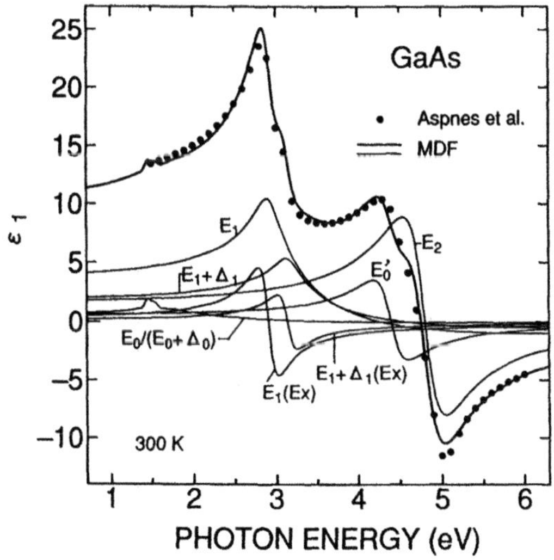

Figure 3.8 Fitted result of MDF calculation to the experimental $\varepsilon_1(E)$ spectrum for GaAs. The experimental data are taken from Aspnes *et al.* [4]. Individual contributions to ε_1 of the $E_0/(E_0+\Delta_0)$ [Eq. (3.10)], $E_1/(E_1+\Delta_1)$ [Eqs. (3.20) and (3.23)], E_0' [Eq. (3.26)], and E_2 transitions [Eq. (3.26)] are shown by the light solid lines. The CP parameters determined from this fit are listed in Table 3.4.

where H's are functions defined by

$$H(z) = \begin{cases} 1 & \text{for } z \geq 0 \\ 0 & \text{for } z < 0 \end{cases} \qquad (3.15)$$

The 3D-exciton effect may profoundly modify the CP singularity structure and may thus take an important part in the analysis of the optical spectra of semiconductors. However, its effect in materials of interest here (i.e., III–V semiconductors) is present at low temperatures only. We will discuss this effect in more detail in Sec. 5.1.2.

2. *E_1 and $E_1+\Delta_1$ transitions*—The E_1 and $E_1+\Delta_1$ transitions in the zinc- blend-type semiconductors may occur along the <111> direction or at L point in the BZ. These transitions are of the 3D M_1 type. The contributions to ε_2 of this type of CPs are given by

$$\varepsilon_2(E) = \begin{cases} \pi\chi_1^{-2}[B_1 - B_{11}(E_1 - E)^{1/2}] & \text{for } E < E_1 \\ \pi B_1 \chi_1^{-2} & \text{for } E \geq E_1 \end{cases} \qquad (3.16a)$$

for the E_1 transitions, and

$$\varepsilon_2(E) = \begin{cases} \pi\chi_{1s}^{-2}[B_2 - B_{21}(E_1 + \Delta_1 - E)^{1/2}] & \text{for } E < E_1 + \Delta_1 \\ \pi B_2 \chi_{1s}^{-2} & \text{for } E \geq E_1 + \Delta_1 \end{cases} \qquad (3.16b)$$

for the $E_1+\Delta_1$ transitions, where

$$\chi_1 = \frac{E}{E_1} \qquad (3.17a)$$

$$\chi_{1s} = \frac{E}{E_1 + \Delta_1} \qquad (3.17b)$$

The B's in Eq. (3.16) are the strength parameters of the $E_1/(E_1+\Delta_1)$ transitions.

Since the M_1-CP longitudinal effective mass m_L is much larger than its transverse counterparts, the m_T's, one can treat these 3D M_1 CPs as 2D minima, 2D M_0. The contribution to ε_2 of this type of 2D minima is given by

$$\varepsilon_2(E) = \pi [B_1 \chi_1^{-2} H(\chi_1 - 1) + B_2 \chi_{1s}^{-2} H(\chi_{1s} - 1) \qquad (3.18)$$

where H's are the Heaviside functions defined by Eq. (3.15).

The KK transformation of Eq. (3.18) gives

$$\varepsilon_1(E) = -B_1\chi_1^{-2}\ln(1-\chi_1^2) - B_2\chi_{1s}^{-2}\ln(1-\chi_{1s}^2) \qquad (3.19)$$

Calculation of Eqs. (3.18) and (3.19) satisfactorily interprets experimental ε spectra of various semiconductors in these CP regions. However, it is worth noting that these expressions do not account for any damping effect, that is, for a lifetime broadening. We thus introduce in these expressions such an effect in a phenomenological manner by replacing E by E+iΓ. [It is noted that the same manner as here has been used in Eq. (3.13).] The contribution to $\varepsilon(E)$ of the E_1 and $E_1+\Delta_1$ transitions can, then, be finally written as

$$\varepsilon(E) = -B_1\chi_{1d}^{-2}\ln(1-\chi_{1d}^2) - B_2\chi_{1sd}^{-2}\ln(1-\chi_{1sd}^2) \qquad (3.20)$$

with

$$\chi_{1d} = \frac{E+i\Gamma}{E_1} \qquad (3.21a)$$

$$\chi_{1sd} = \frac{E+i\Gamma}{E_1+\Delta_1} \qquad (3.21b)$$

Equation (3.20) with Γ=0 eV exhibits a divergence in ε_1 [Re $\varepsilon(E)$] at the $E_1/(E_1+\Delta_1)$ CPs. The damping effect can successfully decrease ε_1 strength of the $E_1/(E_1+\Delta_1)$ structures and lead to agreement with experiment [10–13]. In the limit $\Gamma\rightarrow0$ eV, Eq. (3.20) [Im $\varepsilon(E)$] exactly agrees with Eq. (3.18).

There may be only two analytic expressions that enable us to treat the exciton effects at the $E_1/(E_1+\Delta_1)$ spectral region: (1) the effective-mass approximation (EMA) [14,15] and (2) the Koster-Slater method (KSM) [16,17].

In the case of the 3D M_1 CPs (i.e., the saddle-point excitons or hyperbolic excitons), the EMA equation is much more difficult to solve. However, in the limit $m_L^{-1}\rightarrow0$ the equation predicts a series of the 2D Wannier-type excitons [14]

$$E_{x1}^{2D} = E_1 - \frac{G_1^{2D}}{(n-\frac{1}{2})^2} \qquad (3.22a)$$

$$E_{x\Delta}^{2D} = E_1 + \Delta_1 - \frac{G_\Delta^{2D}}{(n-\frac{1}{2})^2} \qquad (3.22b)$$

where G_1^{2D} (G_Δ^{2D}) is the 2D-exciton Rydberg energy. The contribution of the 2D-exciton transitions to $\varepsilon(E)$ is written, with a Lorentzian line shape, as

$$\varepsilon(E) = \sum_{n=1}^{\infty} [B_{1x}^n(E_{x1}^{2D}-E-i\Gamma)^{-1} + B_{2x}^n(E_{x\Delta}^{2D}-E-i\Gamma)^{-1}] \qquad (3.23)$$

where B_{1x}^n (B_{2x}^n) is the exciton strength parameter proportional to the envelope function of the nth exciton. The envelope function can now be given by

$$|\phi_n^{2D}(0)|^2 = \frac{16V_0}{\pi(a_B^{2D})^3(2n-1)^3} \qquad (3.24)$$

In Eq. (3.24) a_B^{2D} is the 2D-exciton Bohr radius and V_0 is the volume of the unit cell. The 2D EMA also gives the continuum part of the exciton states [14]. One can, however, consider that the contribution of this part may be similar to that of the one-electron approximation [i.e., Eq. (3.20)].

In the KSM, the electron-hole interaction (Coulomb potential) is evaluated in the tight-binding approximation and its attraction is approximated by an on-site contact interaction. An approximate expression of $\varepsilon(E)$ based on this model is written as [16,17]

$$\varepsilon(E) - 1 = \frac{\mathcal{E}(E)-1}{1-g[\mathcal{E}(E)-1]}, \qquad g>0 \qquad (3.25)$$

where $\mathcal{E}$ is the one-electron dielectric function and the parameter g is proportional to the depth of the assumed square-well potential. This expression was successfully used to show the presence of the exciton effects in the E_1 structure of low-temperature ER data of Ge [16].

In Fig. 3.9, we show the calculated line shapes of the E_1-gap contribution to ε_2 of GaAs (T=22 K) [13]. The solid line corresponds to the sum of the discrete-exciton [Eq. (3.23), dashed line] and one-electron term [Eq. (3.20), dashed line]. Here, in Eq. (3.23), we considered only the n=1 ground-state exciton, since the strengths of the n≥2 excited states are extremely weaker than that of the n=1 state [see Eq. (3.24)]. The dotted line shows the result of Eq. (3.25) with g=0.065. A difficulty arises in obtaining the one-electron dielectric function $\mathcal{E}$ to be inserted into Eq. (3.25), since this function is not directly measurable. We, therefore, assumed the dielectric function of Eq. (3.20) to be $\mathcal{E}(E)$.

Optical spectra in the $E_1/(E_1+\Delta_1)$-structure region of GaAs become sharp when the temperature is lowered. Such spectral change cannot be explained within the framework of the one-electron approximation with lifetime-broadening correction [i.e., Eq. (3.20)]. This fact is clearly suggestive of the contribution of exciton effect to the $E_1/(E_1+\Delta_1)$ transitions. As seen in Fig. 3.9, not only the EMA (solid line)

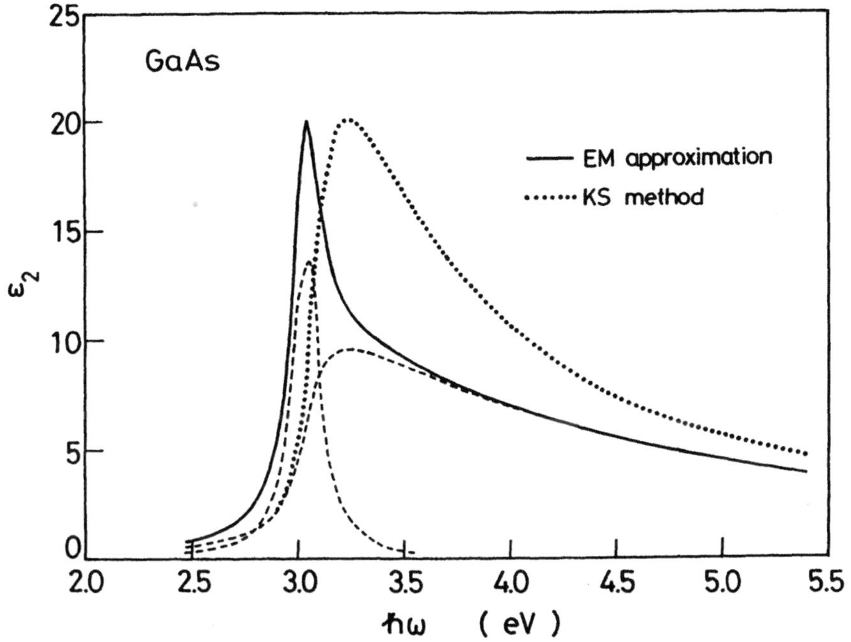

Figure 3.9 Line shape of the E_1-gap contribution to ε_2 of GaAs (T=22 K). The solid line corresponds to the sum of the discrete-exciton [Eq. (3.23), dashed line] and one-electron term [Eq. (3.20), dashed line]. The dotted line shows the result of Eq. (3.25) with g=0.065.

but also the KSM (dotted line) dramatically modify and sharpens the $E_1/(E_1+\Delta_1)$ CP structure. However, the degree of sharpness is larger for the EMA than for the KSM. We find better agreement with the experiment using the EMA expression [13].

3. E_0' and E_2 transitions—As seen in Fig. 3.7, the more pronounced structure found in GaAs in the region higher in energy than E_1 is labeled E_2 (E_0'). The nature of the E_2 (E_0') transitions is more complicated. They do not correspond to a single, well-defined CP (see Table 1.3). If the E_0' transitions occur at Γ (Δ), then the CP should be a 3D M_0 (3D M_1) type. However, the fit shows that neither the 3D M_0 nor the 3D M_1 model represents the peculiar line shapes of ε_1 and ε_2 in the E_0'–E_2 spectral region. The best fit is obtained with a DHO model.

The DHO gives

$$\varepsilon(E) = \frac{C}{(1-\chi_2^2) - i\chi_2\gamma} \qquad (3.26)$$

with

$$\chi_2 = \frac{E}{E_2} \quad \left(\chi_2 = \frac{E}{E_0{}'}\right) \qquad (3.27)$$

where C is the nondimensional strength parameter and γ is the nondimensional broadening parameter. Note that in the limit $\gamma \to 0$ the ε_1 spectrum of the DHO is equivalent to the classical Drude–Lorentz formula: namely

$$\varepsilon_1(E) = \frac{C}{1-\chi_2^2} \qquad (3.28)$$

The DHO model gives a Lorentzian-like line shape, and in the limit $\gamma \to 0$ $\varepsilon_2(E)$ spectrum exhibits a divergence at $E=E_2$. It is possible to consider the DHO as the CP of 2D M_1 type. The contribution to $\varepsilon_2(E)$ of this type is (see Table 3.2)

$$\varepsilon_2(E) = \pi \chi_2^{-2} (C_1 \ln|1-\chi_2| - C_2) \qquad (3.29)$$

where C's are the strength parameters. Equation (3.29) exhibits a divergence at $\chi_2=1.0$ ($E=E_2$), and its spectrum resembles that of the DHO when we take into account the damping effect in the model. Thus, the DHO may be of the 2D M_1 type in good approximation.

Many-particle effects at the E_2-structure region of semiconductors have been treated with their detailed electronic-energy band structures [18,19]. Results have shown that the absorption at the E_2 CP is markedly weakened, with no drastic change in its shape, by introducing the exciton interaction. This fact supports the hypothesis that the DHO model is a good representation for the E_2 CP both with and without the presence of the exciton interaction.

The E_2 structures in the III–V semiconductors are well fitted by the DHO model [Eq. (3.26)]. However, the structures in the group-IV semiconductors (Si, Ge, and α-Sn) could not be successfully fitted only by the DHO [20–2]. We find that the best result can be obtained with a mixture of the DHO and a 2D maximum (M_2).

The contribution of the 2D maximum to $\varepsilon(E)$ can now be given by [20–2]

$$\varepsilon(E) = -F \chi_{2m}^{-2} \ln \frac{1-\chi_{cl}^2}{1-\chi_{2m}^2} \qquad (3.30)$$

with

$$\chi_{2m} = \frac{E+i\Gamma}{E_2} \qquad (3.31a)$$

$$\chi_{cl} = \frac{E+i\Gamma}{E_{cl}} \qquad (3.31b)$$

where F and Γ are the strength and broadening parameters of the E_2 (2D-M_2) transitions, respectively, and E_{cl} is a low-energy cutoff assumed to be E_{cl}=E_1. This contribution will be taken into consideration in the MDF calculation of μc-Si (see Sec. 4.2.3).

4. *Indirect-band-gap transitions*—It is well known that not only the direct transitions at the CPs but also the indirect transitions at the indirect band gap influence the optical dispersion relations of semiconductors. The indirect transitions in indirect-band-gap materials (*e.g.*, GaP) take part at energies below the onset of the direct transitions, and vice versa above the onset of the direct transitions in direct-band-gap materials (*e.g.*, GaAs). Because the indirect transitions are higher order in the perturbation process than the direct ones, their strength is usually very weak. As a result, one can expect to observe them only in spectrum below the direct threshold as a tail of the direct absorption edge in indirect-band-gap materials.

The optical transition mechanism at the indirect band gap, E_g^{ID}, is expressed by a second-order process in the perturbation. This perturbation gives the contribution of the indirect transitions to ε_2 as (see also Sec. 5.1.1)

$$\varepsilon_2(E) = \frac{D}{E^2}[n_q(E-E_g^{ID}+\hbar\omega_q)^2+(n_q+1)(E-E_q^{ID}-\hbar\omega_q)^2]H(1-\chi_g) \quad (3.32)$$

with

$$n_q = \frac{1}{e^{\hbar\omega_q/kT}-1} \quad (3.33a)$$

$$\chi_g = \frac{E_g^{ID}\pm\hbar\omega_q}{E} \quad (3.33b)$$

where D is the strength parameter, $\hbar\omega_q$ is the energy of the phonons involved in the indirect transitions, and H is the Heaviside function [see Eq. (3.15)].

The parabolic bands extending to infinite energies implied by Eq. (3.32) should be nonphysical. One must, thus, modify the model by taking into account a cutoff at energy E_c. This modification provides

$$\varepsilon_2(E) = \frac{D}{E^2}[n_q(E-E_g^{ID}+\hbar\omega_q)^2+(n_q+1)(E-E_q^{ID}-\hbar\omega_q)^2]H(1-\chi_g)H(1-\chi_c) \quad (3.34)$$

with

$$\chi_c = \frac{E}{E_c} \quad (3.35)$$

Assuming that the strength parameter D is independent of the photon energy [also $(E-E_g^{ID}\pm\hbar\omega_q)\rightarrow(E-E_g^{ID})$], the KK transformation of Eq. (3.34) gives

$$\varepsilon(E) = \frac{2D}{\pi}(2n_q+1)\Big[-\frac{(E_g^{ID})^2}{(E+i\Gamma)^2}\ln\Big(\frac{E_c}{E_g^{ID}}\Big)+\frac{1}{2}\Big(1+\frac{E_g^{ID}}{E+i\Gamma}\Big)^2\ln\frac{E+i\Gamma+E_c}{E+i\Gamma+E_g^{ID}}$$

$$+\frac{1}{2}\Big(1-\frac{E_g^{ID}}{E+i\Gamma}\Big)^2\ln\frac{E+i\Gamma-E_c}{E+i\Gamma-E_g^{ID}}\Big] \qquad (3.36)$$

The damping effect is taken into account in Eq. (3.36). As shown in Fig. 3.10, thus, the ε_1 spectrum of Eq. (3.36) [Re $\varepsilon(E)$] exhibits no divergence at E_c. It is also noted that in the limit $\Gamma\rightarrow 0$ eV, the calculated ε_2 spectrum [Im $\varepsilon(E)$] agrees exactly with that predicted by Eq. (3.34).

In Figs. 3.7 and 3.8, the heavy solid lines are obtained from the sum of Eqs. (3.10), (3.20), (3.23), and (3.26). Individual contributions to ε of each CP are also shown in the figures by the light solid lines. The parameters of the fit are listed

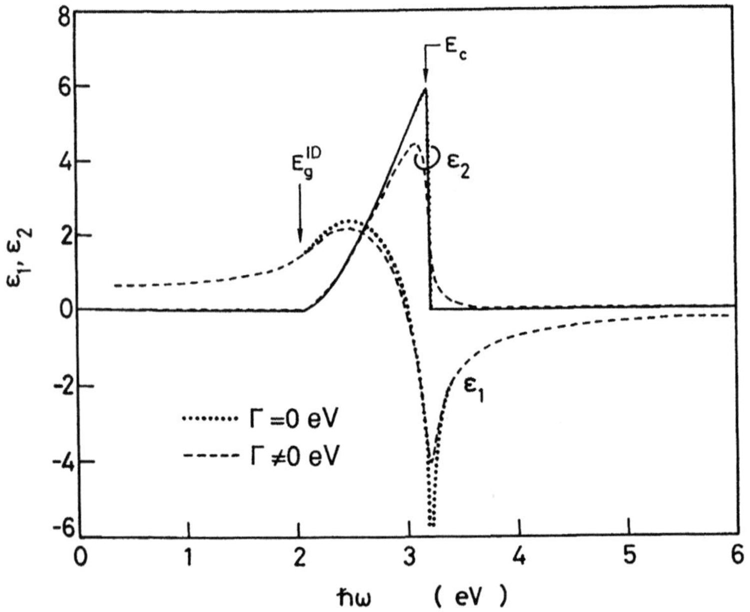

Figure 3.10 Line shape of the indirect-band-gap (E_g^{ID}) contribution to ε with two different broadening parameters [Eq. (3.36), $\Gamma=0$ eV (dotted line) and $\Gamma\neq 0$ eV (dashed line)]. The solid line represents the dependence of ε_2 on E with high-energy cutoff correction in Eq. (3.34).

TABLE 3.4 *CP parameters for GaAs determined by fitting with MDF at 300 K.*

Parameter	Numeric Value
E_0 (eV)	1.42
$E_0+\Delta_0$ (eV)	1.76
A (eV$^{1.5}$)	3.6
Γ (meV)	25
E_1 (eV)	2.91
$E_1+\Delta_1$ (eV)	3.14
B_1	4.0
B_2	2.0
$B_{1x}{}^1$ (eV)	1.1
$B_{2x}{}^1$ (eV)	0.55
Γ (eV)	0.12
E_0' (eV)	4.45
C	0.64
γ	0.10
E_2 (eV)	4.77
C	1.77
γ	0.11
$\varepsilon_{1\infty}$	1.2

in Table 3.4. The experimental data of $\varepsilon_1(E)$ are found to be usually somewhat larger than those of the MDF calculation. In order to improve the fit, therefore, we considered an additional term, $\varepsilon_{1\infty}$, to $\varepsilon_1(E)$, in the calculation. This term is assumed to be nondispersive (i.e., constant) and may arise from other higher-lying, interband transitions (E_1', $E_1'+\Delta_1'$, etc.). Excellent agreement can then be achieved between the MDF calculation (heavy solid lines) and experimental data (solid circles) over the entire range of photon energies.

It is easy to obtain the E-derivative spectra $d^n\varepsilon(E)/dE^n$ (n=1,2,...) using the MDF in pure analytic form. It has been shown that these analytic functions successfully explain the experimental $d^n\varepsilon(E)/dE^n$ spectra [23–5] and also the dependence of $\varepsilon(E)$ on some external perturbation [26–9].

In the MDF, the matrix element in Eq. (3.5) is taken to be constant over the BZ. Jenkins [30] modified Adachi's MDF to take the matrix element to be a specific function of energy for the direct terms labeled E_1 and $E_1+\Delta_1$ and the indirect term $E_g{}^{ID}$. Rakić and Majewski [31] augmented the MDF model by introducing Gaussian-like broadening function instead of Lorentzian broadening. Kim *et al.* [32] have also carried out modeling the optical dielectric function of semiconductors by extending the CP parabolic-band approximation.

3.1.2 Plasma and *d*-Band Effects

We have seen that the fundamental optical spectra of materials such as the III–V zinc-blende semiconductors are dominated by strong interband transitions occurring between the fundamental absorption edge and about 8 eV. Around 20 eV further weak structure is observed due to the onset of real transitions from atomic *d*-band levels into the conduction band. Between these two regions the optical spectra are characterized by a rapid and smooth decrease in the reflectivity or absorption which is similar to the behavior seen in certain metals in the UV region [1]. In this so-called *plasma region* the valence electrons can behave like free particles and take part in collective oscillations. The reflectivity curve for GaAs illustrates this behavior well, as shown in Fig. 3.11 (b) [33]. Because Si has no *d* electrons, we cannot observe *d*-band transitions in the optical spectra of this material [Fig. 3.11 (a)].

Optical properties in the plasma and *d*-band region of semiconductors have been extensively discussed by Philipp and Ehrenreich [33] starting from the frequency-dependent dielectric function obtained in the random-phase approximation [3]. The simplest system is represented when a single group of the valence bands $|v>$ is well separated from the core states. Then, at frequencies sufficiently high so that the oscillator strengths of transitions from $|v>$ to higher bands is exhausted, $\varepsilon(\omega)$ can be described by the Drude model modified by a damping term with the relaxation time τ_v:

$$\varepsilon(\omega) = 1 - \frac{\omega_{pv}^2}{(\omega + i\tau_v^{-1})^2} \qquad (3.37)$$

where the valence-electron plasma frequency ω_{pv} is given by (N_v=the density of valence electrons)

$$\omega_{pv} = \left(\frac{N_v e^2}{m_e \varepsilon_\infty \varepsilon_0} \right) \qquad (3.38)$$

This situation is met with Si where no *d* bands exist and the valence band lies approximately 100 eV above the highest 2*p* core states ($L_{2,3}$ shell). The optical constants of Si above ~6 eV can thus be described well by Eq. (3.37) with a value of four electrons per atom for N_v.

When *d* bands are present, as in Ge and in the zinc-blende-type semiconductors, Eq. (3.37) is modified by the oscillator strength coupling between the valence and *d*-band electrons and can now be written as

$$\varepsilon(\omega) = (1 + \delta\varepsilon_0) \left[1 - \frac{\Omega_{pv}^2}{(\omega + i\tau_v^{-1})^2} \right] \qquad (3.39)$$

where the contribution of the *d*-band electrons to the static dielectric constant is

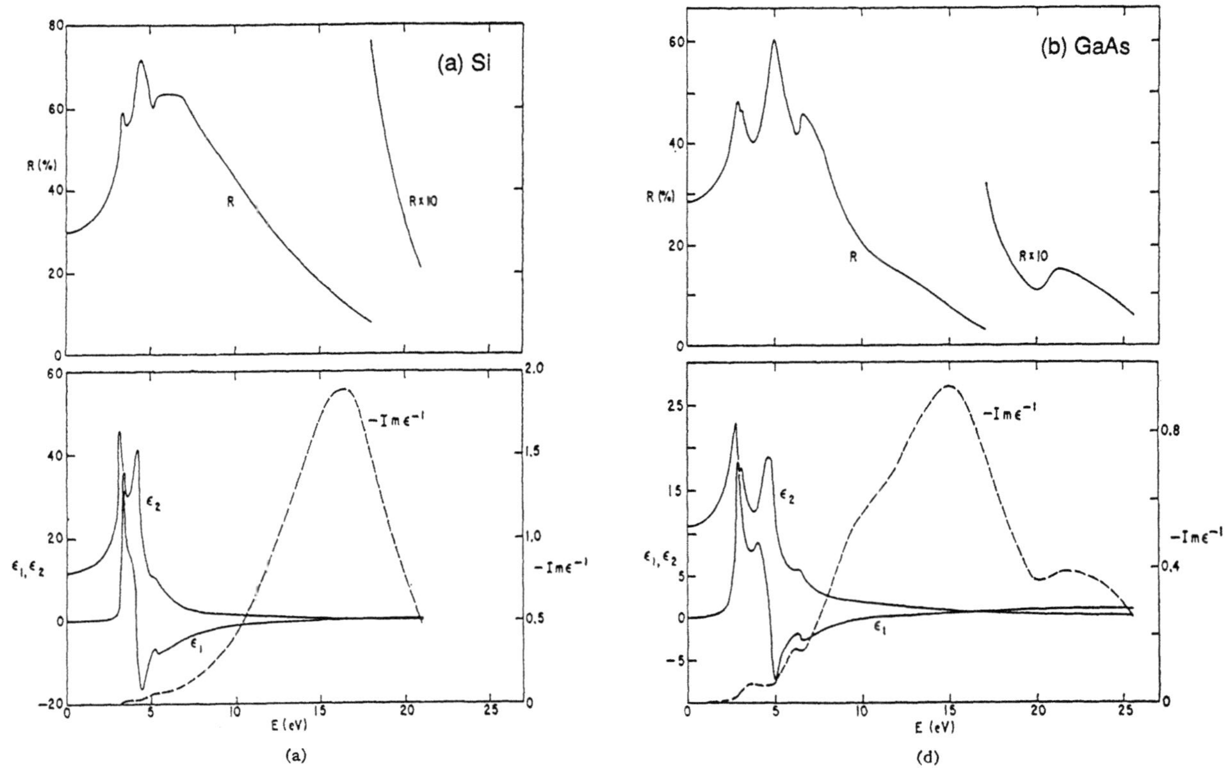

Figure 3.11 Spectral dependence of $R(E)$, $\varepsilon_1(E)$, $\varepsilon_2(E)$, and energy-loss function $-\mathrm{Im}\varepsilon(E)^{-1}$ for (a) Si and (b) GaAs. (From Philipp and Ehrenreich [33].)

described by $\delta\varepsilon_0$ and Ω_{pv} is an effective plasma frequency different from ω_{pv} due to the influence of the *d*-band electrons. The relaxation time τ_v in Eq. (3.39) [Eq. (3.37)] is found to lie in the range 1.4 to 1.8×10^{-16} s for the materials of diamond- and zinc-blende types [34]. This scattering time is several orders of magnitude smaller than that for lattice or impurity scattering. It is thus considered that the electron-electron scattering is the dominant scattering mechanism in the plasma region.

3.2 TEMPERATURE, PRESSURE, AND DOPING EFFECTS

3.2.1 Temperature Effect

Temperature can cause a shift of the band-gap (CP) energies of semiconductors at constant pressure. Two effects contribute to shift of the band-gap energy with temperature. The first is due to the thermal expansion of the lattice coupled with the change in the electron energy with volume. The second contribution is the direct renormalization of the band-gap energy by electron–phonon interactions. The total shift can then be written by the derivatives as

$$\left(\frac{\partial E_g}{\partial T}\right)_P = \left(\frac{\partial E_g}{\partial T}\right)_V + \left(\frac{\partial E_g}{\partial T}\right)_T \qquad (3.40)$$

with

$$\left(\frac{\partial E_g}{\partial T}\right)_V = \left(\frac{\partial \ln V}{\partial T}\right)_P \left(\frac{\partial p}{\partial \ln V}\right)_T \left(\frac{\partial E_g}{\partial p}\right)_T = -3\alpha B\left(\frac{\partial E_g}{\partial p}\right)_T \qquad (3.41)$$

where α is the thermal expansion coefficient and $B=-V(\partial p/\partial V)_T$ is the bulk modulus. Hence the magnitude of the differential thermal-expansion shift can be calculated by means of experimental values for α, B, and $(\partial E_g/\partial p)_T$, where care of the fact that α is strongly dependent on T, and even reverses sign at low T [9], must be taken. Of course, it is possible to separate these two terms from measurements of both the temperature and pressure coefficients.

The temperature dependence of the band-gap energy E_g can be commonly given in terms of the α and β coefficients of the Varshni equation [35]

$$E_g(T) = E_g(0) - \frac{\alpha T^2}{T+\beta} \qquad (3.42)$$

where $E_g(0)$ is the band-gap energy at T=0 K, α is in electronvolts per degree kelvin (not the thermal expansion coefficient), and β is proportional to the Debye temperature (in kelvins).

Figure 3.12 shows, as an example, the temperature dependence of the interband CP energies, E_0, $E_0+\Delta_0$, E_1, E_0', and E_2, for GaAs obtained from the line-shape analysis of the SE data by Lautenschlager *et al.* [7]. The solid lines represent the best fit results with Eq. (3.42). The values of the Varshni parameters, $E_g(0)$ α, and β, for GaAs obtained by Lautenschlager *et al.* [7] and also by some other authors are summarized in Table 3.5.

The effect of temperature provides not only on the sift of the CP energies but also on the broadening of the CP structures. The line width of the CP vs. temperature is usually described with an expression

$$\Gamma(T) = \Gamma_0 + \left(\frac{\Gamma_1}{e^{\theta/T} - 1} \right) \qquad (3.43)$$

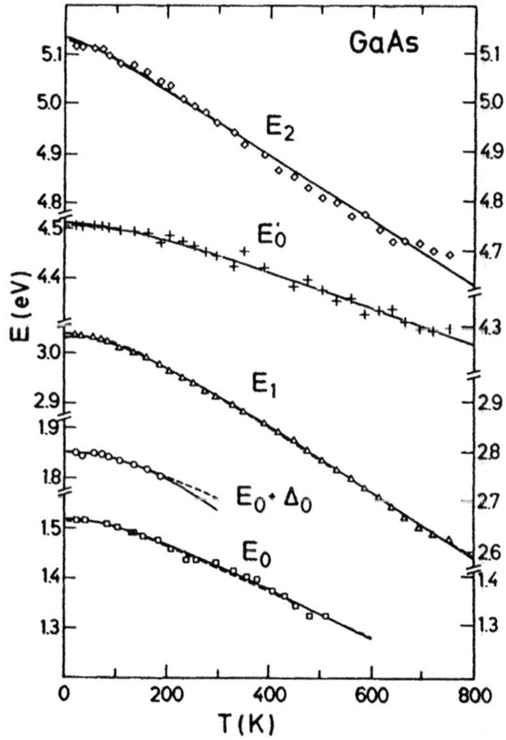

Figure 3.12 Temperature dependence of the interband CP energies, E_0, $E_0+\Delta_0$, E_1, E_0', and E_2, for GaAs obtained from the line-shape analysis of the SE data. The solid lines represent the best fit results with Eq. (3.42). (From Lautenschlager *et al.* [7])

TABLE 3.5 *Values of $E_g(0)$, α, and β for GaAs obtained by fitting the CP energies vs. temperature with Eq. (3.42).*

CP	$E_g(0)$ (eV)	α (10^{-4} eV/K)	β (K)	Ref
E_0	1.522	5.8	300	*a*
	1.517	5.5	225	*b*
	1.512	5.1	190	*c*
	1.5194	10.6	671	*d*
$E_0+\Delta_0$	1.851	3.5	225	*b*
E_1	3.041	7.2	205	*b*
	3.044	6.7	188	*e*
E_0'	4.509	4.0	241	*b*
E_2	5.133	6.6	43	*b*

[a]B. Panish and H. C. Casey, Jr., *J. Appl. Phys.* **40**, 163 (1969).
[b]P. Lautenschlager *et al.*, *Phys. Rev. B* **35**, 9174 (1987).
[c]H. Shen *et al.*, *Appl. Phys. Lett.* **53**, 1080 (1988).
[d]E. Grilli *et al.*, *Phys. Rev. B* **45**, 1638 (1992).
[e]E. G. Seebauer, *J. Appl. Phys.* **66**, 4963 (1989).

where Γ_0 represents the broadening due to temperature-independent mechanisms (such as crystalline imperfections, Auger process, electron–electron interaction, and surface scattering) and θ is a temperature which describes the average phonon frequency for the lattice. Equation (3.43) is based on the assumption of broadening by interband scattering via phonon absorption and holds only for the lowest gap (E_0). For the higher-lying gaps, one has to replace $n_B \equiv (e^{\theta/T}-1)^{-1}$ in the above expression with $2n_B+1$ since now both phonon absorption and emission processes can take place.

Let us consider the temperature dependence of the dielectric function $\varepsilon(E)$ of GaAs. SE measurements for GaAs as a function of temperature were carried out by Lautenschlager *et al.* [7] from 22 to 753 K in the 1.3–5.5-eV photon- energy range and by Yao *et al.* [36] from 295 to 884 K in the range of 1.6–4.45 eV. We reproduce in Fig. 3.13 the real and imaginary parts of the dielectric function, $\varepsilon(E)=\varepsilon_1(E)+i\varepsilon_2(E)$, of GaAs measured for several temperatures by Lautenschlager *et al.* [7]. The main structures are due to the E_1, $E_1+\Delta_1$, E_0', and E_2 transitions which are clearly observed from 22 up to 750 K. Other weak transitions, such as E_0 and $E_0+\Delta_0$, and some others in the 4.2–5.5-eV region, are observed only at lower temperatures and are partly difficult to distinguish in these figures.

The fits with the MDF calculation to the experimental data of GaAs at 22 K are shown in Figs. 3.14 (ε_1) and 3.15 (ε_2) [13]. The experimental data are taken from Refs. [7] (solid circles) and [37] (open circles). The solid lines represent the sum of Eqs. (3.10), (3.20), (3.23), (3.26), and $\varepsilon_{1\infty}=1.4$. The dashed line in Fig. 3.15 shows the calculated result without taking account of the saddle-point exciton contribution [i.e., Eq. (3.23)].

We recognize in Figs. 3.14 and 3.15 a considerable deviation of the MDF model from the experimental data in the energy region of ~4.8 eV. This is nonessential.

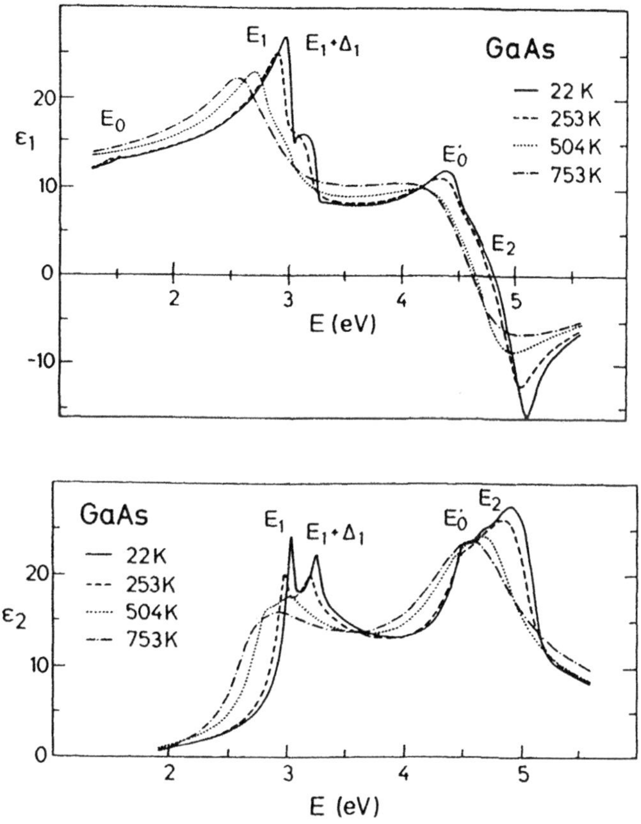

Figure 3.13 (a) Real and (b) imaginary parts of the dielectric function of GaAs for several temperatures (T=22–753 K). (From Lautenschlager *et al.* [7].)

As mentioned before, there is an accumulation of the interband CPs in the 4.2–5.5-eV spectral region (E_0' and E_2 multiplets). For simplicity, we considered only two DHOs (E_0'=4.5 eV and E_2=4.95 eV) in this spectral region. If we consider additional oscillators in this spectral region, the discrepancy may be lost or a better fit may be achieved, both for ε_1 and ε_2.

It is clear from Fig. 3.15 that the sharp peaks appearing at the $E_1/(E_1+\Delta_1)$ edge region cannot be explained only by the one-electron approximation [Eq. (3.20), dashed line]. The saddle-point exciton model of Eq. (3.23) can successfully improve the fit in this spectral region. The best-fitted exciton parameters obtained are $B_{1x}{}^1$=1.1 eV, $B_{2x}{}^1$=0.55 eV, and Γ=0.08 eV.

The fits with the MDF model to the experimental $\varepsilon(E)$ of GaAs at 504 K are shown in Figs. 3.16 and 3.17 [13]. The dashed line in Fig. 3.17 represents the calculated result without taking account of the saddle-point exciton contribution

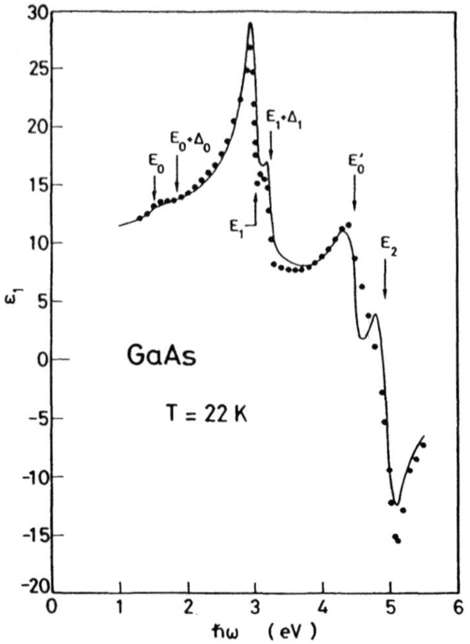

Figure 3.14 ε_1 spectrum of GaAs at 22 K. The experimental data are taken from Ref. [7]. The solid line is obtained from the sum of Eqs. (3.10), (3.20), (3.23), (3.26), and $\varepsilon_{1\infty}$. (From Adachi [13].)

[Eq. (3.23)]. As seen in Fig. 3.17, the calculated $\varepsilon_2(E)$ spectrum at 504 K becomes quite satisfactory when the saddle-point exciton contribution is taken into consideration (solid line). However, the exciton-induced sharp peak disappeared at 504 K both in the calculated and experimental spectra. This is a consequence of a relatively strong exciton-phonon interaction at such high temperature. The best-fitted Γ value at 504 K is 0.14 eV, which is considerably larger than the value determined at 22 K (Γ=0.08 eV). We can also see that the E_2 (E_0') structure at 504 K becomes much broader and its spectral region is successfully explained only by the two DHOs (i.e., E_0'=4.3 eV and E_2=4.70 eV), in contrast to the case at 22 K where a few more oscillators should be required to achieve complete agreement with the experimental spectra.

Individual contributions to ε_2 of the $E_0/(E_0+\Delta_0)$ [Eq. (3.10)], $E_1/(E_1+\Delta_1)$ [Eqs. (3.20) and (3.23)], and E_2 (E_0') [Eq. (3.26)] for GaAs at 22 and 504 K are shown in Figs. 3.18 (a) and (b), respectively [13]. The exciton contribution at the $E_1/(E_1+\Delta_1)$ edges at low temperature (22 K) gives sharp ε_2 peaks in this spectral region, in good agreement with the experimental data. At higher temperature (504 K), these peaks are strongly influenced by the exciton-phonon interaction and,

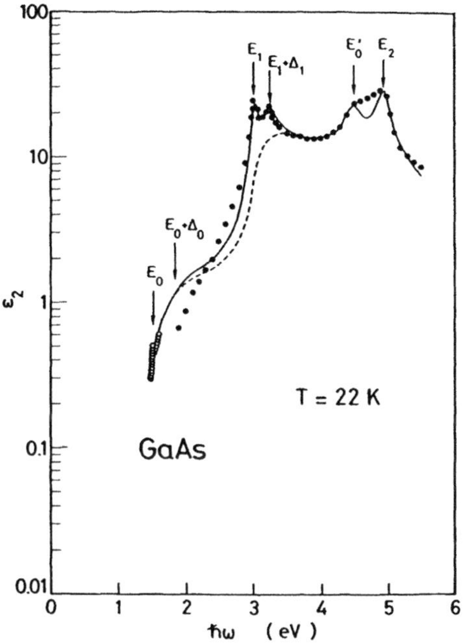

Figure 3.15 ε_2 spectrum of GaAs at 22 K. The experimental data are taken from Refs. [7] (solid circles) and [37] (open circles). The solid line is obtained from the sum of Eqs. (3.10), (3.20), (3.23), and (3.26). The dashed line shows the calculated result without taking account of the saddle-point exciton contribution [i.e., Eq. (3.23)]. (From Adachi [13].)

consequently, they become considerably broader.

The E_2 structure can be characterized by the DHO well. The DHO model of $\varepsilon_2(E)$ gives the Lorentzian-type line shape and ensures that the higher the temperature, the broader the spectrum. Many-particle effects on CPs in the interband continuum of semiconductors have been treated with their detailed electronic energy-band structures [18,19]. Results have shown that the absorption at the E_2 CP is markedly weakened, with no drastic change in its shape, by introducing the exciton interaction. Unfortunately, no analytic line shape suitable to fit the exciton-influenced E_2 line shape has been reported so far. We now find that the DHO model is a good representation for the E_2 CP both with and without the exciton interaction. In principle, the strength parameter C in Eq. (3.26) may be not dependent on the temperature. Thus, the change in C value may account for the exciton interaction (i.e., its value reflects the strength of the exciton interaction).

The strength of the $E_1/(E_1+\Delta_1)$ transitions is represented by B_1 (E_1) and B_2 ($E_1+\Delta_1$). In Fig. 3.19, we show the variations of B_1, B_2, and Γ as a function of temperature (T=22–754 K) for GaAs [13]. The strength of the $E_1/(E_1+\Delta_1)$ transitions in the zinc-blende-type semiconductors can be theoretically estimated with the

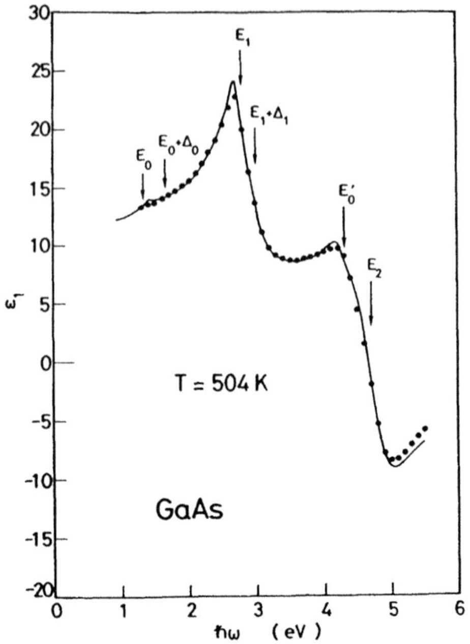

Figure 3.16 ε_1 spectrum of GaAs at 504 K. The experimental data are taken from Ref. [7]. The solid line is obtained from the sum of Eqs. (3.10), (3.20), (3.23), (3.26), and $\varepsilon_{1\infty}$. (From Adachi [13].)

following simple expression [7]:

$$B_1 = 44\frac{E_1+(\Delta_1/3)}{a_0E_1^2} \qquad (3.44a)$$

$$B_2 = 44\frac{E_1+(2\Delta_1/3)}{a_0(E_1+\Delta_1)^2} \qquad (3.44b)$$

where a_0 is the lattice constant in Å and E_1 and Δ_1 are in eV. The calculated results suggest that both B_1 and B_2 are only slowly increasing function of temperature from 2.6 (22 K) to 3.1 (754 K) for B_1 and from 2.3 (22 K) to 2.7 (754 K) for B_2, while the experimental B_1 and B_2 did not depend on the temperature within our fitting accuracy. The sum of the calculated B_1 and B_2 values is between 4.9 (22 K) and 5.8 (754 K), while the experimental sum is 6.0 over the whole temperature range. The agreement is seemed to be good in view of the crudeness of the theory used.

The experimental 2D-exciton strengths determined here are $B_{1x}{}^1$=1.1 eV and $B_{2x}{}^1$=0.55 eV over the whole temperature range. The corresponding ratio

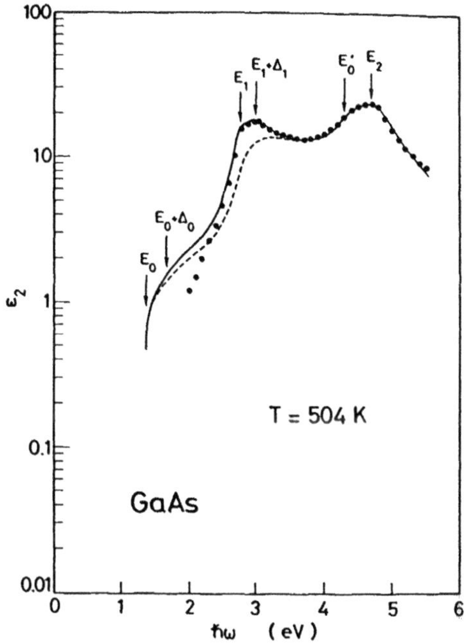

Figure 3.17 ε_2 spectrum of GaAs at 504 K. The experimental data are taken from Ref. [7] (solid circles). The solid line is obtained from the sum of Eqs. (3.10), (3.20), (3.23), and (3.26). The dashed line shows the calculated result without taking account of the saddle-point exciton contribution [i.e., Eq. (3.23)]. (From Adachi [13].)

$B_{1x}{}^1/B_{2x}{}^1$=2.0, is in exact coincidence with the experimental ratio of the one-electron transitions B_1/B_2.

As mentioned before, the effect of temperature provides not only the shift of the CP energies but also the sharpness (*broadening*) of the CP structures. Equation (3.43) ensures a relatively constant Γ value from low T up to, in many cases, T~50–200 K, at which point the temperature-dependent component becomes discernible, and increases nearly proportionally to T for high T. The experimental $\Gamma(T)$ plotted in Fig. 3.19 faithfully follows this behavior.

The variations of C and γ (E_2 transitions) as a function of temperature for GaAs are shown in Fig. 3.20 [13]. The saddle-point (2D-M_0) excitons profoundly modify the CP structures. We can find, however, no clear change in the E_2-CP spectral shape due to the exciton interaction. In a broad class of group-IV elemental, III–V, and II–VI semiconductors, the calculated $\varepsilon(E)$ spectrum within the one-electron approximation yielded an E_1 peak with less strength than measured, but elicited an E_2 peak with more strength than the measured one. These discrepancies have prompted studies of exciton effects on the optical spectra of these semiconductors, and the inclusion of the exciton effect has led to an improvement over the

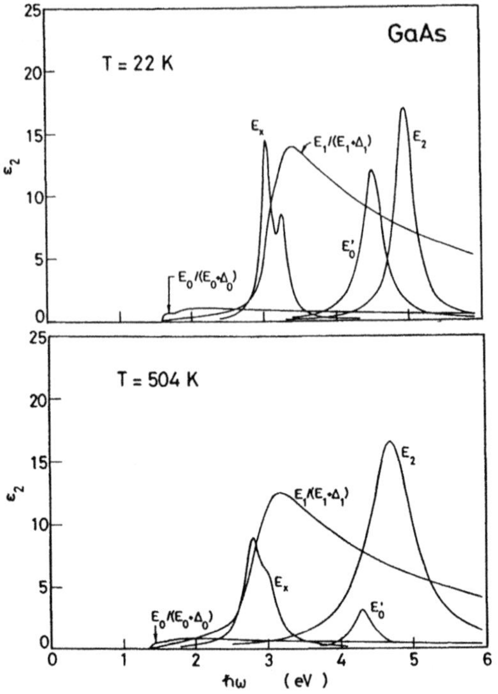

Figure 3.18 Individual contributions to ε_2 of the $E_0/(E_0+\Delta_0)$ [Eq. (3.10)], $E_1/(E_1+\Delta_1)$ [Eqs. (3.20) and (3.23)], and E_2 (E_0') [Eq. (3.26)] for GaAs at 22 [(a)] and 504 K [(b)]. (From Adachi [13].)

one-electron approximation in the sense that the E_1 peak becomes stronger and the E_2 peak weaker [14,18,19]. These studies also indicated that if one takes into account the exciton effect, the E_1 structure becomes relatively sharp, but the E_2 structure shows no rigorous change in its spectral shape. The experimental variation of C suggests that the lower the temperature, the smaller the C value. This temperature variation is in reasonable agreement with the theoretical anticipation (i.e., the smaller C values at lower T may be due to the exciton modification of the E_2 strength). The temperature dependence of γ is essentially the same as that observed for the $E_1/(E_1+\Delta_1)$ transitions (see Fig. 3.19).

The strength of the $E_0/(E_0+\Delta_0)$ transitions (A) as a function of temperature for GaAs is plotted in Fig. 3.21 [13]. The strength parameter A may, in principle, be independent of temperature. The plots, however, suggest that A decreases with almost linearly with increasing temperature. The relatively large values of A at lower temperatures may be the result of the exciton modification at the 3D M_0 CP. Unfortunately, however, it is difficult to further proceed with detailed discussion on this problem because of the poor accuracy of the experimental ε for small ε (and also the poor fit of our analyzed results, see Figs. 3.15 and 3.17) in this spectral

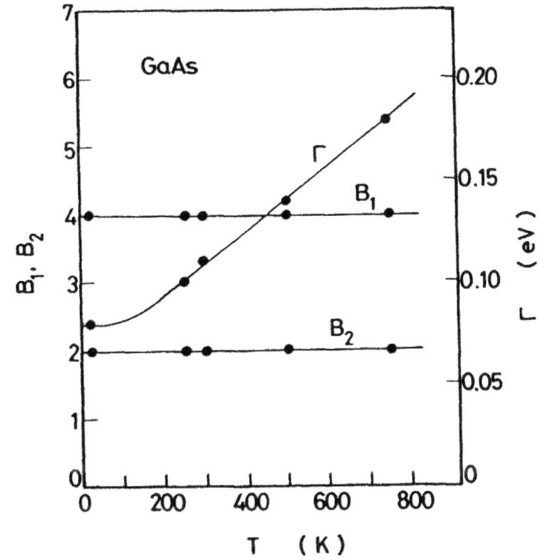

Figure 3.19 Strength parameters B_1 and B_2 and broadening factor Γ [$E_1/(E_1+\Delta_1)$ transitions] as a function of temperature T for GaAs. (From Adachi [13].)

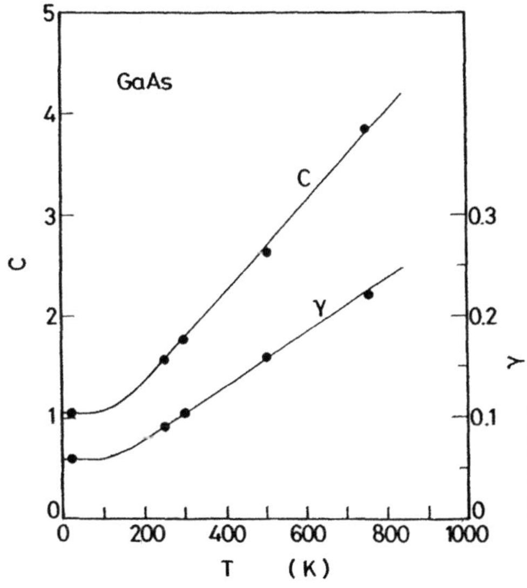

Figure 3.20 Strength parameter C and broadening factor γ (E_2 transitions) as a function of temperature T for GaAs. (From Adachi [13].)

region. The behavior of the 3D-M_0 excitons will be presented in Sec. 5.1.2.

The SE data of $\varepsilon(E)$ reported by Lautenschlager *et al.* [7] were obtained by mathematically removing the effect of surface oxide. The oxide thickness and its optical constants were assumed to remain constant throughout the entire tem-

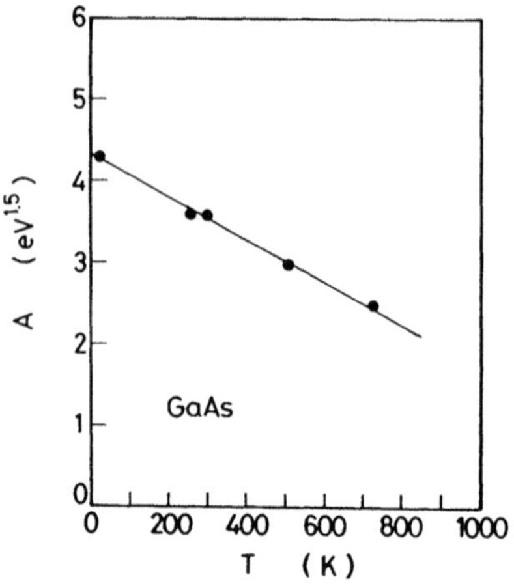

Figure 3.21 Strength parameter A [$E_0/(E_0+\Delta_0)$ transitions] as a function of temperature T for GaAs. (From Adachi [13].)

perature range. However, this latter assumption is found to be incorrect at higher temperatures [38], where the oxide reacts with the substrate. Yao *et al.* [36], therefore, performed to measure an oxide-free GaAs surface, to minimize errors due to surface overlayers and their evolution with temperature. A very clean, smooth surface was obtained by first growing an epitaxial layer of GaAs on a GaAs substrate and immediately capping it with a protective layer of arsenic. The cap prevented surface oxidation during transport to the measurement chamber, where it was evaporated under UHV at T~350°C. The room-temperature SE results were comparable with those reported for the best wet-chemically processed surfaces [4,7]. Because there was no As overpressure in the chamber, some surface roughening occurred at temperatures above 550°C, possibly due to congruent evaporation of Ga and As. The SE data at these temperatures were successfully fitted with a model which quantitatively accounted for the surface roughness (see Sec. 3.4 below).

Yao *et al.* [36] have determined the temperature sensitivity of ε_2 at various temperatures, by taking the numeric derivative of ε_2 with respect to temperature T, as shown in Fig. 3.22. The spectrum of $|d\varepsilon_2/dT|$ at a fixed temperature represents the relative sensitivity of ε_2, over the entire spectral range, to changes of temperature. From this figure, the highest sharp peak in each derivative spectrum falls into a spectroscopic window of photon energy ranging from ~2.4 to ~2.85 eV. The position of each peak in photon energy decreases by about the same amount as the E_1 CP, as the temperature increases, indicating that the temperature sensitivity is mostly associated with the temperature dependence of the E_1 CP. They then developed an algorithm, using the measured data at a number of fixed temperatures, to compute the $\varepsilon(E)$ spectrum based on the seven-oscillator HOA model at an arbitrary temperature (T=22–610°C). Thus, an unknown surface temperature can be determined by fitting the measured data, with temperature as an adjustable

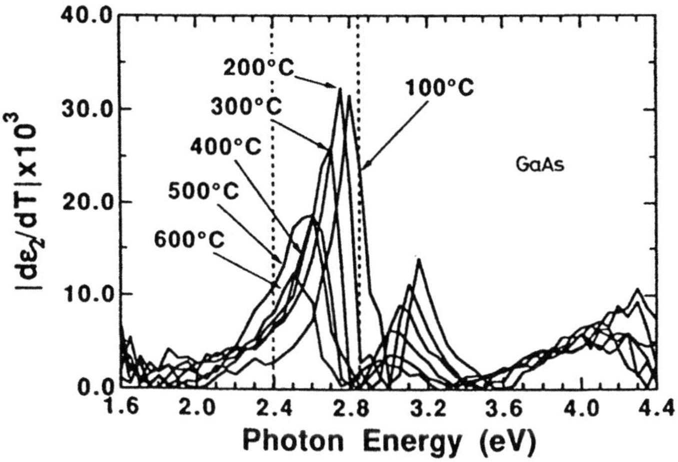

Figure 3.22 Temperature sensitivity of ε_2 at different photon energies. The two vertical dashed lines indicate a spectroscopic window containing the highest sensitivity peaks from each fixed temperature. (From Yao *et al.* [36].)

parameter. The SE can, therefore, be utilized as an optical thermometer to determine the sample-surface temperature in a growth chamber.

3.2.2 Pressure Effect

The lattice parameter is dependent relatively largely on the pressure. Thus, not only temperature but also pressure cause a shift in the CP energies of semiconductors. Figure 3.23 shows, as an example, the pressure dependence of the E_1 and $E_1+\Delta_1$ gaps in GaAs measured by Hanfland *et al.* [39] using the PR technique. The solid lines correspond to the results of least-squares fits using the following quadratic expression:

$$E_g(V) = E_g(V_0) + a\left(\frac{\Delta V}{V_0}\right) + b\left(\frac{\Delta V}{V_0}\right)^2 \qquad (3.45)$$

where V_0 is the volume at the normal pressure. Note that the volume change is related to pressure p with Murnaghan's equation:

$$p = \frac{B_u}{B_u'}\left[\left(\frac{a_0(0)}{a_0(p)}\right)^{3B_u'} - 1\right] \qquad (3.46)$$

where B_u' is the pressure derivative of the bulk modulus [$B_u'=(dC_{11}/dp)+2/3(dC_{12}/dp)$, C_{ij}=elastic stiffness constant], and $a_0(0)$ and $a_0(p)$ are the lattice parameters at pressures p=0 and p≠0, respectively. In Fig. 3.23, the E_1 gap tends to show the sublinear characteristic while the $E_1+\Delta_1$ gap keeps the linear characteristic up to $(\Delta V/V_0)=-1.2\times10^{-1}$.

The complex dielectric functions of Ge, GaAs, InSb, and CdTe have been studied by Alouani *et al.* [40] based on the band-structure calculation. They calculated the band structures and optical matrix elements from the relativistic self-consistent LMTO scheme. They also investigated the dependence on pressure of the dielectric functions of Ge and GaAs up to V/V_0=0.92 (Ge) and 0.859 (GaAs). All CP peaks in $\varepsilon_2(E)$ moved towards higher energies as the pressure was increased. Since their origins in $\boldsymbol{k}$ space remain unchanged under hydrostatic pressure, the shifts of the peaks to higher energies are considered to be due to the increase of the CP energies. To our knowledge, however, no experimental data has been reported on the pressure-dependent optical constants in the interband transition regions of such semiconductors [except near or at the fundamental absorption edges (see Chapter 5)].

It is known that at high pressures the group-IV elemental semiconductors (Si and Ge) show structural phase transitions in a sequence from cubic (diamond)→tetragonal (β-Sn)→simple hexagonal→intermediate structure→hexagonal closed-packed [41–3]. Similarly, the III–V and II–VI compounds exhibit a variety of crystal structures at high pressures. For instance, two orthorhombic phases and a tetragonal phase have been observed in crystalline InSb [44]. The NaCl (rocksalt), β-Sn, and orthorhombic (*Pmm2*) structures have also been identified in CdTe as the high-pressure phases [45].

Hanfland *et al.* [46] reported on an experimental and theoretical investigation of the optical response of the high-pressure phase Si. They used the gasketed-diamond-widow-cell technique in combination with a micro-optical system for the

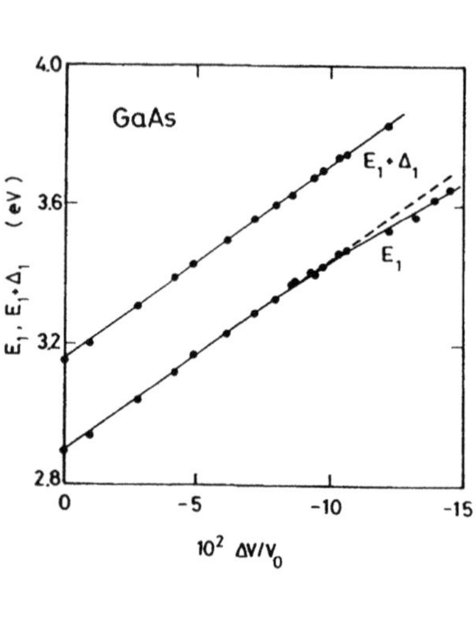

Figure 3.23 E_1 and $E_1+\Delta_1$ gaps of GaAs as a function of volume change $\Delta V/V$. (From Hanfland *et al.* [39].) The volume change is related to pressure p with Murnaghan's equation of Eq. (3.46).

spectral range from 0.5 to ~4 eV. Pressures were determined from the red shift of the ruby luminescence. Optical reflectivities were measured in the β-Sn, primitive hexagonal, and hexagonal-closed-packed phases of Si at pressures up to 44 GPa.

We reproduce in Fig. 3.24 (a) the reflectivity spectra of Si at four different pressures corresponding to the diamond, β-Sn, primitive hexagonal, and hexagonal closed-packed phases [46]. The three metallic phases exhibit a high reflectivity in the near-IR spectral range followed by a decrease towards the visible range, which they attributed to interband absorption. The visible interband absorption is obviously strongest in the β-Sn and primitive hexagonal phases and decreases in the hexagonal closed-packed phase, but shows structure for all three phases.

Figure 3.24 (b) shows the measured interband part of $\varepsilon(\omega)$ of Si at 13.2 GPa in the β-Sn and at two pressures in the primitive hexagonal phase [46]. The dominant absorption near 2.5 eV has a low-energy shoulder which is resolved only at pressures above 24 GPa. A weaker absorption is seen at the high-energy limit of the experiment near 4 eV.

Hanfland *et al.* [46] also performed band-structure calculations for the primitive hexagonal cell by using the self-consistent scalar-relativistic LMTO method. The resultant interband dielectric functions for the light polarization parallel and perpendicular to the *c* axis showed some differences which, however, were not essential in the present discussion. Figures 3.24 (c) and (d) show, respectively, the calculated polarization-averaged $\varepsilon_2(\omega)$ spectrum and energy-band structure of primitive hexagonal Si at V/V_0=0.65 [46]. The decomposition of $\varepsilon_2(\omega)$ into contributions from various pairs of bands is also shown in Fig. 3.24 (c). The dominant contribution near 2.6 eV is mainly due to transitions between the bands 2 and 3 and corresponds to the (101) splitting, characteristic of a polyvalent, nearly-free-electron metal. The second weaker contribution near this energy is thought to be due to transitions between the bands 3 and 4. The energy, however, does not correspond to the (001) splitting, because the absorption mainly involves states along the L–H direction, where the band splittings are a result of mixing of states from the (001), (100), and (101) planes. The peak in ε_2 near 3.7 eV, which arises from transitions between the bands 2 and 4, is also due to parallel bands near the L–H direction.

The uniaxial or biaxial stress also causes a change in the CP energies and, thus, in the optical spectra of semiconductors. The stress-induced change in the optical constants of solids is known as the *elastooptic* (*photoelastic*) effect [9]. Figure 3.25 shows the behaviors of $\varepsilon_2(\|)$ for the uniaxial stress $X\|[100]$ at three different compressive stresses around the $E_1/(E_1+\Delta_1)$ transitions of GaAs [47]. For a fixed photon energy, the evolution of the real and imaginary parts of $\varepsilon(E)$ was found to be linear up to ~600 MPa, except close to the $E_1/(E_1+\Delta_1)$ CP energies. An example of the nonlinearity in $\varepsilon_2(E)$ observed close to the E_1-gap energy is shown in the inset of Fig. 3.25. Using fits to the linear portions of the $\varepsilon(X)$ dependence, $\Delta\varepsilon_{ij}(X)=P_{ijkl}X_{kl}$, Etchegoin *et al.* [47] obtained three independent components of the photoelastic coefficient P_{ijkl} (P_{11}, P_{12}, and P_{44}) and their liner combinations (P_{11}-P_{12} and $P_{11}+2P_{12}$). We reproduce in Fig. 3.26 their obtained three independent com-

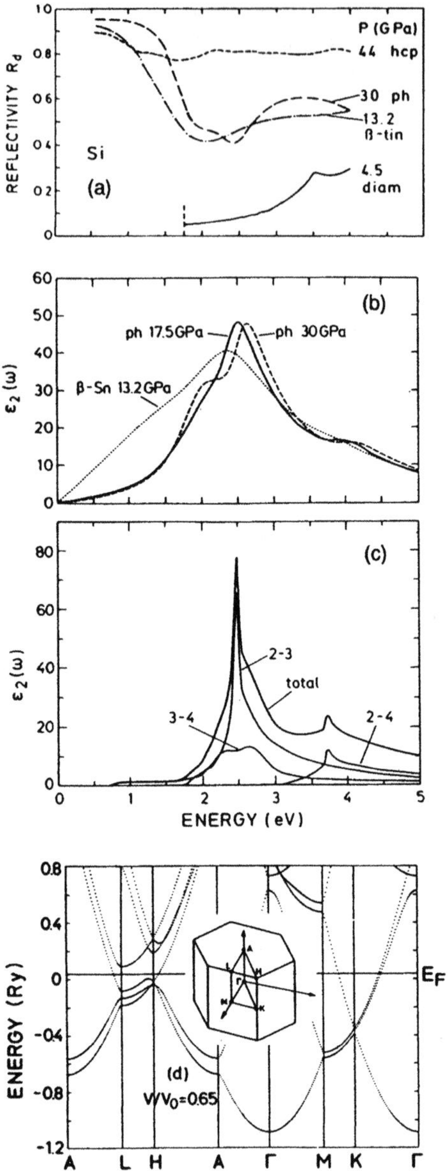

Figure 3.24 (a) Reflectivity spectra of Si in the diamond, β-Sn, primitive hexagonal, and hexagonal closed-packed phases at 4.5, 13.2, 30, and 44 GPa, respectively. (b) Interband dielectric function $\varepsilon_2(E)$ of Si at different pressures as derived from the reflectivity spectra. (c) Calculated polarization-averaged interband dielectric function $\varepsilon_2(E)$ of primitive hexagonal Si at V/V_0=0.65 (22.2 GPa) with decomposition of $\varepsilon_2(E)$ into various interband contributions. The numbers are enumerated in the sequence of increasing energy. (d) LMTO band structure of primitive hexagonal Si at V/V_0=0.65. (From Hanfland *et al.* [46].)

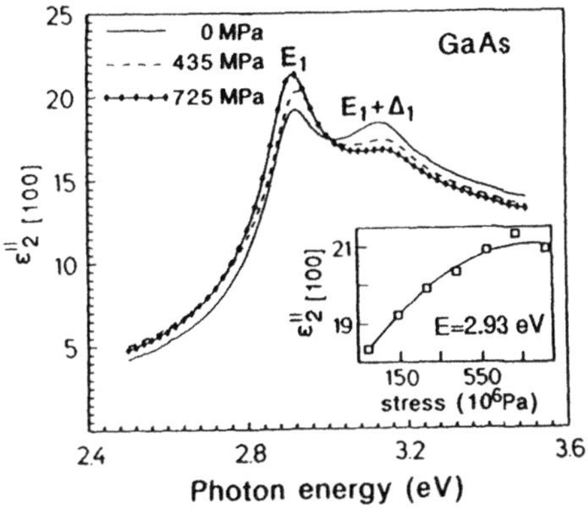

Figure 3.25 Imaginary part of the dielectric function, $\varepsilon_2(E)$, as a function of compressive stress $X\|[100]$ for GaAs in a region near the $E_1/(E_1+\Delta_1)$ transitions at room temperature. The inset shows the evolution of $\varepsilon_2(E)$ for a fixed photon energy (E=2.93 eV) as a function of stress. (From Etchegoin *et al.* [47].)

ponents, P_{11}, P_{12}, and P_{44}, in the ~1.6–5.4-eV photon-energy range [47]. The real part of each component is plotted in Fig. 3.26, together with the corresponding curve obtained by the KK transformation of the experimental imaginary part. This curve was obtained numerically from [48]

$$\mathrm{Re}\,[P_{ij}(E)] = \left(\frac{2}{\pi}\right) \mathcal{P}\int_{E_i}^{E_f} \frac{\Omega\,\mathrm{Im}\,[P_{ij}(\Omega)]}{(E^2-\Omega^2)}\,d\Omega + C' \qquad (3.47)$$

where E_i and E_f are the initial and final experimental energies, $\mathcal{P}$ means "Cauchy principal part", and C' is an *ad hoc* constant fixed so as to obtain the best fit with the directly determined Re$[P_{ij}(E)]$. The insets of Figs. 3.26 (a)–(c) also show the imaginary parts of each $P_{ij}(E)$ obtained in arbitrary units from piezoreflectance measurements. We can recognize from these figures that both the KK self-consistency and the qualitative comparison with the piezoreflectance data are in excellent agreement with the experimental results.

The distribution of strains in multilayered material systems is a subject of perennial, great interest since internal strains arise normally in thin epitaxial films during the preparations of the films by heteroepitaxial growth. Internal strains or stresses are then a consequence of the mismatch of lattice constants and/or the difference in the thermal expansion coefficients of the individual layers. Such strains have an important influence on the physical properties of semiconductor films because they change the electronic states, which result in an electronic energy-band structure via the DP interaction. It is thus easy to expect that the optical properties of layered semiconductor films are profoundly modified by this

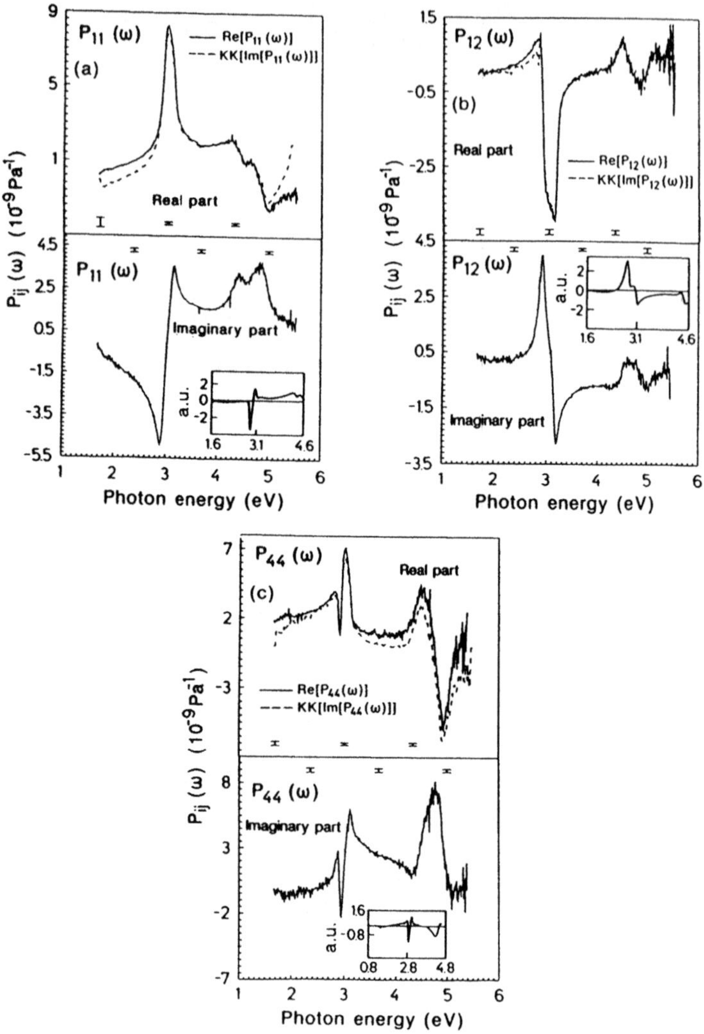

Figure 3.26 (a) P_{11}, (b) P_{12}, and (c) P_{44} for GaAs in the ~1.6–5.4-eV photon-energy range at room temperature. On top of each figure the real part of the component is shown together with the KK consistency check calculated with Eq. (3.47) from the imaginary part obtained in the experiment. The imaginary part is shown together with the corresponding qualitative result of piezoreflectance measurements (inset). (From Etchegoin *et al.* [47].)

effect.

$In_{1-x}Ga_xAs/GaAs$ is the well-known strained heterostructure system. Pickering *et al.* [49] have measured the dielectric-function spectra of strained $In_{1-x}Ga_xAs$ ($x \geq 0.75$) epilayers grown by MBE on GaAs(001) using SE. We reproduce in Fig. 3.27 their measured $\varepsilon_2(\omega)$ spectra and corresponding second-derivative spectra $d^2\varepsilon_2(\omega)/d(\hbar\omega)^2$ $[d^2\varepsilon_2(E)/d(E)^2]$ of both strained and relaxed $In_{0.24}Ga_{0.76}As$ epilayers. The positions of the E_1 and $E_1+\Delta_1$ CPs are shown by the vertical arrows. These are obtained from line-shape fitting of the SCP expression [see Eq. (3.9)]:

$$\frac{d^2\varepsilon(E)}{dE^2} = \sum_{i=1}^{N} n(n-1)A_i e^{i\phi_i}(E-E_{ci}+i\Gamma_i)^{n-2} \quad (3.48)$$

In perfect coherent (001) epitaxy, the lattice mismatch may be accommodated by the tetragonal deformation of the layer. The in-plane strain in this case can be given by

$$e_{xx} = e_{yy} = \frac{a_e - a_s}{a_s} \equiv -\delta \quad (3.49)$$

In Eq. (3.49) a_s and a_e are the unstrained lattice parameters of substrate and epitaxial layer, respectively. The strain along the growth direction is given, from elasticity theory, by

$$e_{zz} = 2\left(\frac{C_{12}}{C_{11}}\right)\delta \quad (3.50)$$

The above two equations show that the biaxial in-plane tensile (compressive) strain corresponds to the compressive (tensile) strain along the (001) growth direction (since C_{11}, $C_{12}>0$).

In Fig. 3.27 the shift in the CP energies with in-plane strain can be clearly seen in the second-derivative spectra. This shift is caused by a hydrostatic component (E_H) and a biaxial shear component (E_S) of the DP [49]:

$$\Delta E_1 = \left(\frac{\Delta_1}{2}\right) + E_H - \frac{1}{2}\left(\Delta_1^2 + 4E_S^2\right)^{1/2} \quad (3.51a)$$

$$\Delta(E_1+\Delta_1) = \left(\frac{-\Delta_1}{2}\right) + E_H + \frac{1}{2}\left(\Delta_1^2 + 4E_S^2\right)^{1/2} \quad (3.51b)$$

with

$$E_H = 2D_1^1\left(\frac{C_{11}-C_{12}}{C_{11}}\right)\delta \quad (3.52a)$$

$$E_S = \left(\frac{2}{3}\right)^{1/2} D_3^3\left(\frac{C_{11}+2C_{12}}{C_{11}}\right)\delta \quad (3.52b)$$

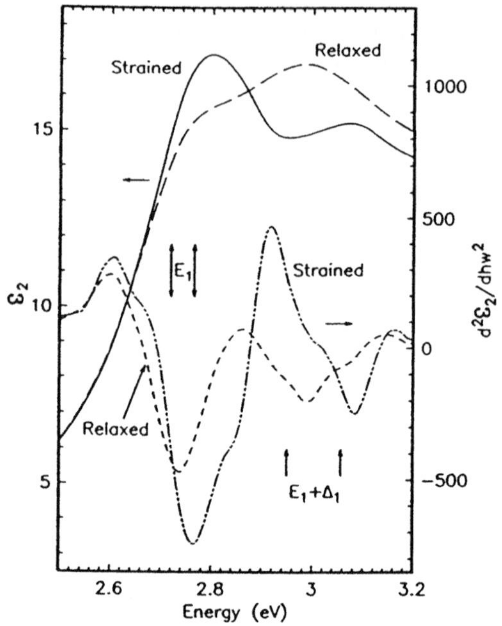

Figure 3.27 $\varepsilon_2(\omega)$ and $d^2\varepsilon_2(\omega)/d(\hbar\omega)^2$ spectra for strained (solid and dash-dotted lines) and relaxed (dashed lines) $In_{0.24}Ga_{0.76}As$ epilayers on GaAs(001). The vertical arrows indicate positions of the E_1 and $E_1+\Delta_1$ CPs obtained from line-shape fitting of Eq. (3.48). (From Pickering *et al.* [49].)

where $D_1^{\ 1}$ is the hydrostatic interband ($\Lambda_3 \rightarrow \Lambda_1$) DP and $D_3^{\ 3}$ is the interband DP for the Λ_3 valence band under the [001] uniaxial strain [9].

3.2.3 Doping Effect

It is well known that heavy doping of impurities influences considerably the physical properties of semiconductors, in particular, transport and optical response (see, *e.g.*, Ref. [9]). At moderate low doping levels, localized impurity states develop close to the band edges [50]. With increasing dopant concentration the localized states overlap, producing an impurity band. This impurity band is separated from the host conduction (valence) band for the case of donors (acceptors) by the so-called Mott gap. At still higher doping concentrations, the impurity and host bands become mixed to build a new continuum, so that there is no longer a clear distinction between band and impurity states. In this way the degenerate semiconductor resembles a metal [50].

The effects of doping on the fundamental absorption edge have been studied with electrical transport measurements as well as various optical techniques [9,50]. The main effects of doping on these edges are the formation of band tails because of the random potential of the doped impurities, the Burstein–Moss shift due to the filling of the conduction (valence) band by electrons (holes), and the shrinkage of the band gap by many-body effects where correlation as well as exchange play an important role. These absorption edges are also suitable for studying the Mott transitions [50].

The higher edges, above the fundamental absorption ones, have been less studied. Only optical techniques, such as reflectivity, ER, PL, SE, and resonant RS, have been used [50]. It should be noted that at higher edges the exchange term plays only a small role due to the fact that carriers are located in a small region of $\boldsymbol{k}$ space different from the region where the transitions occur. The Mott transition is also irrelevant to this problem [50].

SE has been successfully used to study the effects of doping for Si [51,52], Ge [50], and GaAs [53]. All these studies revealed that the dielectric-function spectrum of heavily doped sample is very similar to that of the undoped material, except that the E_1 and E_2 structures are broadened and shifted to lower energies. The effects of heavy doping at lower photon energies can be described by the Drude free-carrier model of Eq. (3.37) with lifetimes determined by scattering from lattice vibrations [51].

We show in Fig. 3.28 the pseudodielectric-function spectra $\varepsilon(E)$ for the five selected Ge samples [50]. The dotted lines represent the pseudodielectric function of undoped Ge. The effect of doping is clearly seen in this figure. There is an increasing lifetime broadening and red shift in the E_1, $E_1+\Delta_1$, E_2', and E_2. The dominant mechanism responsible for the red shift is due to the effect of the random impurities on the crystal potential [50–2].

In order to obtain the CP energy shift ΔE_i and the change in the Lorentzian broadening $\Delta\Gamma_i$, a line-shape analysis of the data was performed by fitting the numerically derived SE $d^n\varepsilon_2(E)/d(E)^n$ (n=3) spectra with the analytic expression of Eq. (3.48) [50,52,53]. Figure 3.29 shows the red shifts of the CPs, ΔE_1 and ΔE_2, plotted against carrier concentrations in Ge in a log–log scale [50]. The solid and dashed lines represent the theoretical and experimental fits to a N_I^{α} law. It is seen in the figure that the effect of doping becomes clearly noticeable for concentrations above 10^{19} cm^{-3}, as in the case of Si [52]. It is also found that the impurity-induced $\Delta\Gamma_i$ is much larger for E_1 than for E_2, indicating that a metastable excitonic resonance is well developed for the former but not for the latter [50–2].

When comparing the results obtained for Si and Ge [50–2] with those for GaAs, there appear some common features but also some noticeable differences [53]. The common features concern the qualitatively similar dependence of ΔE_i and $\Delta\Gamma_i$ on doping concentration and the possibility of expressing both effects with the same semiempirical relations, namely, $\Delta E_i \sim p^{\alpha}$ or n^{α}, $\Delta\Gamma_i \sim p^{\alpha}$ or n^{α} [53]. The most significant difference is that in GaAs the experimental values of both ΔE_i and $\Delta\Gamma_i$ differ considerably for n- and p-type samples. The ΔE_i and $\Delta\Gamma_i$ of n-type GaAs at $n=1\times10^{19}$ cm^{-3} is found to be about 3–4 times larger than those for p-type GaAs ($p=1\times10^{19}$ cm^{-3}). These results are compared with second-order perturbation-theory

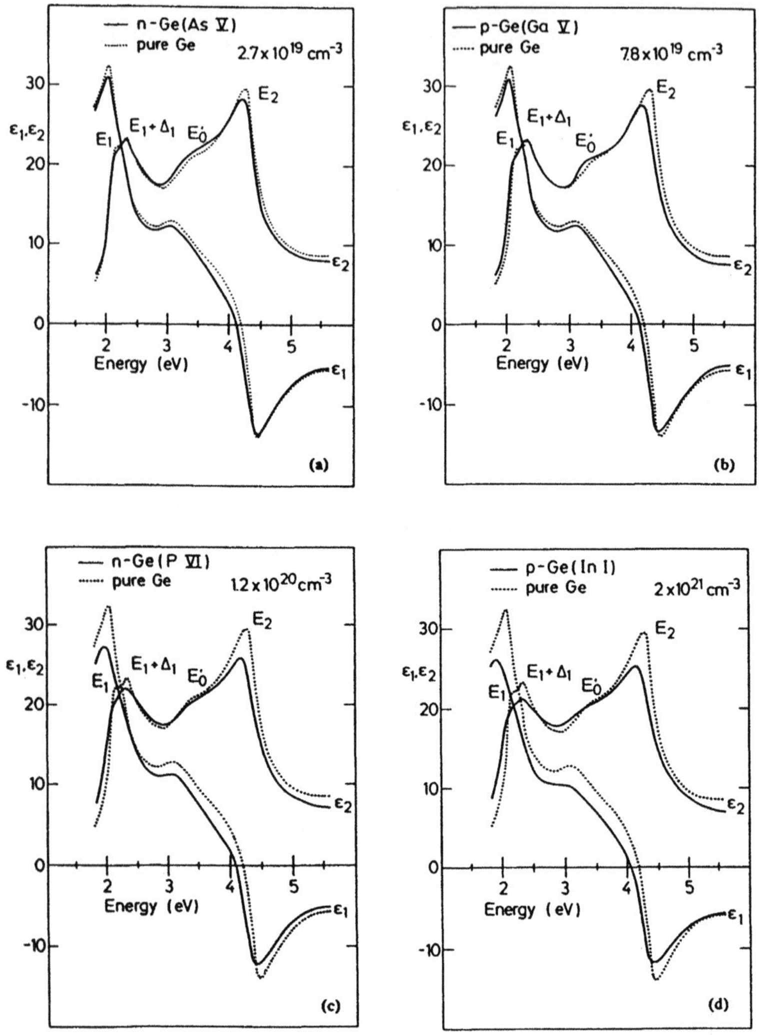

Figure 3.28 Pseudodielectric function of (a) As-doped (n=2.7×10^{19} cm^{-3}) Ge, (b) Ga-doped (p=7.8×10^{19} cm^{-3}) Ge, (c) P$^+$-implanted, laser-annealed (n=1.2×10^{20} cm^{-3}) Ge, and (d) In$^+$-implanted, laser-annealed (p=2×10^{21} cm^{-3}) Ge measured at room temperature. For the purpose of comparison, the pseudodielectric function of pure Ge (n<10^{13} cm^{-3}) is shown by the solid lines in (a)–(d). (From Viña and Cardona [50].)

calculations of the effects of substitutional impurities on the energy-band structure of GaAs [53].

There has been much interest in the subject of hydrogenation of semiconductors because it can passivate a variety of deep and shallow impurities [54]. The pas-

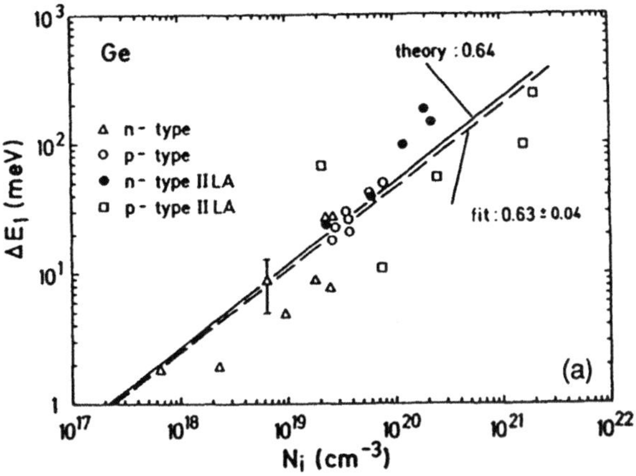

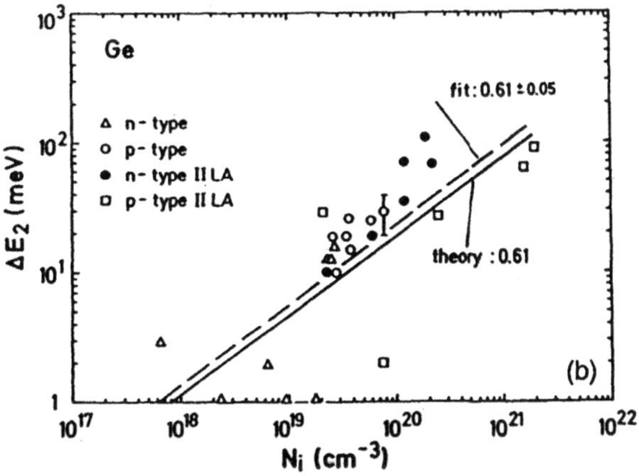

Figure 3.29 Dependence on doping concentration of the CP energies for Ge at room temperature: (a) red shift (ΔE_1) for the E_1 singularity; (b) red shift (ΔE_2) for the E_2 singularity. The shifts are measured with respect to the undoped sample. The dashed lines, best fits to a N_I^{α} law, yielding α=0.63±0.04 for E_1 and α=0.61±0.05 for E_2. The solid lines represent the results of second-order perturbation-theory calculation. (From Viña and Cardona [50].)

sivation of Zn acceptors by atomic deuterium in heavily doped *p*-GaAs has been studied by de Mierry and Stutzmann [55] using PL and IR reflectance techniques. These authors reported $\varepsilon(E)$ spectra of *p*-GaAs (p=7×10^{19} cm^{-3}) before and after ^{2}D plasma exposure (170°C, 20 h) and found that the E_1 and E_2 peaks in ε_1 and ε_2 are shifted to higher energies by about 0.1 eV after plasma exposure, definitely demonstrating the passivation of Zn dopants (free carriers and Zn random potentials).

Although it may not fall under the category of "*doping*," the matrix-element isotope shifts of the band gaps and its influence on the optical properties of semiconductors have been discussed both theoretically and experimentally [56,57].

3.3 Fresnel Reflection Coefficient, Ellipsometry Definition, and Surface Oxide Effect

3.3.1 Fresnel Reflection Coefficient

(a) Isotropic Medium—Ellipsometry involves the reflection of light from a surface. We consider the oblique reflection and transmission of an optical plane wave at the planar interface between two semi-infinite homogeneous optically isotropic media 1 and 2 with complex indices of refraction n_1^* ($=n_1+ik_1$) and n_2^* ($=n_2+ik_2$), respectively, as illustrated in Fig. 3.30. The wave incident from medium 1 gives rise to a reflected wave in the same medium and a transmitted (or refracted) wave in medium 2. We refer to plane polarized waves that are in the plane of incidence as *p waves* and those polarized perpendicular to the plane of incidence as *s waves*. The amplitudes of the electromagnetic waves can be determined by the boundary conditions requiring that the tangential components ***E*** and ***H*** (a solution of Maxwell's equations) must be the same on both sides. For a wave incident at an angle ϕ, the boundary conditions lead to *Fresnel's formulae* [58,59]

$$\frac{n_1^*}{n_2^*} = \frac{\sin\phi_2}{\sin\phi_1} \tag{3.53}$$

$$r_{12}^{p} = \frac{E_{1p}}{E_p} = \frac{n_2^* \cos\phi_1 - n_1^* \cos\phi_2}{n_2^* \cos\phi_1 + n_1^* \cos\phi_2} \tag{3.54a}$$

$$r_{12}^{s} = \frac{E_{1s}}{E_s} = \frac{n_1^* \cos\phi_1 - n_2^* \cos\phi_2}{n_1^* \cos\phi_1 + n_2^* \cos\phi_2} \tag{3.54b}$$

$$t_{12}^{p} = \frac{E_{2p}}{E_p} = \frac{2n_1^* \cos\phi_1}{n_2^* \cos\phi_1 + n_1^* \cos\phi_2} \tag{3.55a}$$

$$t_{12}^{s} = \frac{E_{2s}}{E_s} = \frac{2n_1^* \cos\phi_1}{n_1^* \cos\phi_1 + n_2^* \cos\phi_2} \tag{3.55b}$$

where the superscript refers to waves parallel (p) or perpendicular (s) to the plane of incidence and the subscript refers to either medium 1 or 2. Equation (3.53) is known as *Snell's law*.

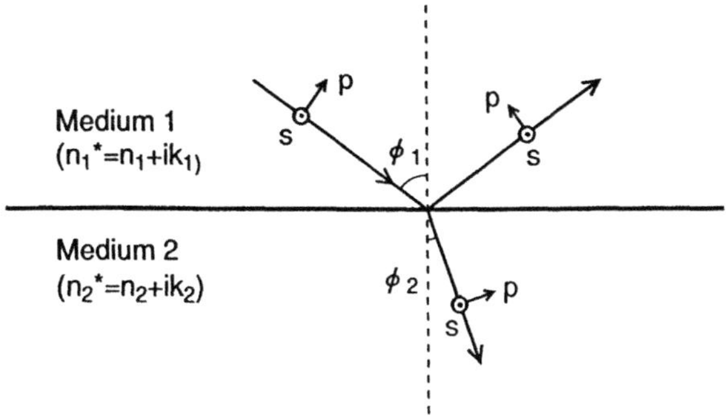

Figure 3.30 Oblique reflection and transmission of a plane wave by an ambient/substrate (medium 1/medium 2) system. ϕ_1 is the angle of incidence in the ambient and ϕ_2 is the angle of refraction in the substrate.

The intensity of the light is proportional to the square of amplitude. We can thus write the reflection coefficients, the ratios of the reflected to the incident intensities, as

$$R^p = |r^p|^2 \quad , \qquad R^s = |r^s|^2 \qquad (3.56)$$

An example of the plots of R^p and R^s vs. angle of incidence ϕ_1 is shown in Fig. 3.31. These curves are obtained assuming that medium 1 is air (n^*=1.0+i0.0) and medium 2 is GaP with n^*=3.334+i0.000 (E=2.0 eV, solid lines) and n^*=3.790 +i2.171 (E=4.0 eV, dashed lines). The n^* values for GaP are taken from Ref. [60]. As can be seen in the figure, the distinction between parallel and perpendicular disappears at ϕ_1=0° (normal incidence). Also, R^s is always larger than R^p.

If both media are dielectric in nature (i.e., $k_1=k_2=0$) then R^p becomes zero at a specific angle of incidence ϕ_B:

$$\phi_B = \tan^{-1}\left(\frac{n_2}{n_1}\right) \qquad (3.57)$$

This can be readily proved from Eqs. (3.53) and (3.54a) ($r_{12}^p=0$ at $\phi_1=\phi_B$). The significance of this is that when unpolarized light makes a reflection with this specific angle of incidence, the light polarized parallel to the plane of incidence is never reflected and thus it is all transmitted. The resulting reflected light is polarized perpendicular to the plane of incidence. This specific angle is called various names, such as the *Brewster angle*, *polarizing angle*, and *principal angle* [58,59]. For the example shown in Fig. 3.31, with n_2=3.334+i0.000, the Brewster angle is about 73.30°. This value is marked by the arrow.

When the reflecting surface is not a dielectric (i.e., $k_2 \neq 0$), the situation becomes more complex. The Fresnel reflection coefficients are complex numbers, so the reflectance does not go to zero at any ϕ_1. The angle for which R^p is minimum is called the *principal angle of incidence*. In Fig. 3.31, with n_2^*=3.790+i2.171, the principal angle is about 76.93°.

If medium 1 is air and $k_2 \neq 0$, we can easily obtain from Eq. (3.54) the normal-incidence reflectivity (ϕ_1=0° and, from Snell's law, ϕ_2=0°)

$$R = R^p = R^s = \frac{(n_2-1)^2 + k_2^2}{(n_2+1)^2 + k_2^2} \qquad (3.58)$$

It should be noted that this expression is the same as that given in Sec. 1.3.2 [Eq. (1.22)].

The electric field of the transmitted light ($\phi_{1,2}$=0°), as a function of position x in medium 2, can now be written as

$$E_{2p(s)}(x,t) = |E^0_{2p(s)}|\, e^{i(\frac{2\pi n_2^*}{\lambda}x - \omega t)} \qquad (3.59)$$

where ω is the angular frequency and λ is the light wavelength in vacuum. The corresponding light intensity is

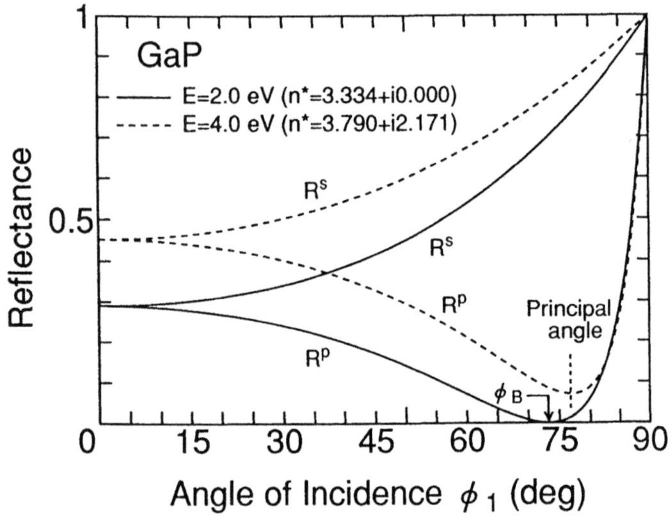

Figure 3.31 Plots of R^p and R^s vs. angle of incidence ϕ_1. These curves are obtained assuming that medium 1 is air (n^*=1.0+i0.0) and medium 2 is GaP with n^*=3.334+i0.000 (E=2.0 eV, solid lines) and n^*=3.790+i2.171 (E=4.0 eV, dashed lines). The n^* values for GaP are taken from Ref. [60].

$$E_{2p(s)}(x,t)^2 = |E^0_{2p(s)}|^2 e^{i(\frac{4\pi n_2^*}{\lambda}x - 2\omega t)} = I_0 e^{i(\frac{4\pi n_2}{\lambda}x - 2\omega t)} \qquad (3.60)$$

with

$$I_0(E) = |E^0_{2p(s)}|^2 e^{-\frac{4\pi k_2(E)}{\lambda}x} \qquad (3.61)$$

From a definition, we obtain the absorption coefficient

$$\alpha(E) = \frac{-1}{I_0(E)} \frac{dI_0(E)}{dx} = \frac{4\pi}{\lambda} k_2(E) \qquad (3.62)$$

(b) Anisotropic Medium—Let us assume that the light is incident at an angle ϕ from ambient (air) onto a uniaxially anisotropic crystal with the optic axis c parallel and perpendicular to the plane of incidence, as shown in Fig. 3.32. Fresnel's formulas then allow us to calculate the two complex reflection coefficients, r_{12}^{p} and r_{12}^{s}, for light polarized parallel and perpendicular to the plane of incidence, respectively. If the angle θ between the optic axis c and the plane of incidence has two extreme values (i.e., θ=0° and 90°, see Fig. 3.32), Fresnel reflection coefficients can be simply written as [61]

$$r_{12}^{p} = \frac{n_\perp^* n_\parallel^* \cos\phi - X_m}{n_\perp^* n_\parallel^* \cos\phi + X_m} \qquad (\theta=0°) \qquad (3.63a)$$

$$r_{12}^{s} = \frac{\cos\phi - X_m}{\cos\phi + X_m} \qquad (\theta=0°) \qquad (3.63b)$$

and

$$r_{12}^{p} = \frac{n_\perp^{*2} \cos\phi - X_m}{n_\perp^{*2} \cos\phi + X_m} \qquad (\theta=90°) \qquad (3.64a)$$

$$r_{12}^{s} = \frac{\cos\phi - X_n}{\cos\phi + X_n} \qquad (\theta=90°) \qquad (3.64b)$$

with

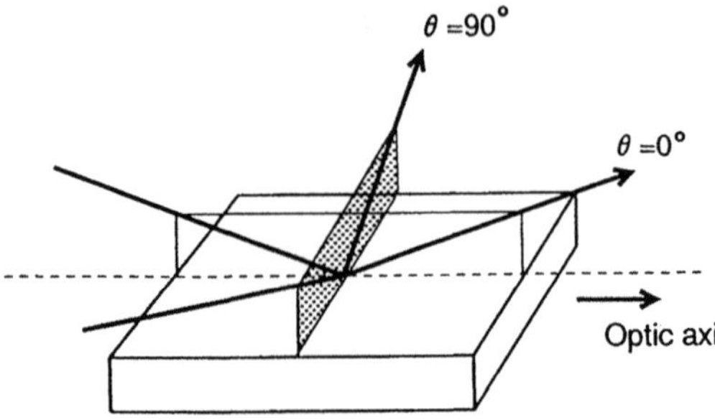

Figure 3.32 Reflection of a light beam from a surface. The light beam is incident at an angle ϕ from the ambient (air) onto a uniaxially anisotropic crystal with the optic axis c parallel (θ=0°) and perpendicular (θ=90°) to the plane of incidence.

$$X_m = \sqrt{n_\perp^{*2} - \sin^2\phi} \qquad (3.65a)$$

$$X_n = \sqrt{n_\parallel^{*2} - \sin^2\phi} \qquad (3.65b)$$

where $n_\perp^*$ and $n_{||}^*$ are the complex refractive indices ($n^*=n+ik$) for the ordinary ($\boldsymbol{E}\perp c$) and extraordinary ($\boldsymbol{E}\| c$) rays, respectively. Corresponding equations exist for transmission, but these are not presented here.

Figure 3.33 shows the calculated R^p ($=|r_{12}^p|^2$) and R^s ($=|r_{12}^s|^2$) as a function of angle of incidence ϕ with θ=0° (solid lines) and 90° (dashed lines). These curves correspond to those for CdS with $n_\perp^*$=2.49+i0.00, $n_{||}^*$=2.51+i0.00 [E=2.0 eV, (a)] and $n_\perp^*$=2.06+i1.55, $n_{||}^*$=2.34+i1.84 [E=5.5 eV, (b)]. The numeric n^* values are taken from Ref. [62].

Since the optical anisotropy for CdS at E=2.0 eV is very small, we can recognize no clear difference of R^p (and R^s) between θ=0° and 90°. In a case of uniaxial material, the normal-incidence (ϕ=0°) reflectivity can be written as

$$R_{||} = \frac{(n_{2\|}-1)^2 + k_{2\|}^2}{(n_{2\|}+1)^2 + k_{2\|}^2} \qquad (3.66a)$$

$$R_\perp = \frac{(n_{2\perp}-1)^2 + k_{2\perp}^2}{(n_{2\perp}+1)^2 + k_{2\perp}^2} \qquad (3.66b)$$

From these equations, we get nearly the same R values for p and s waves (ϕ=0°),

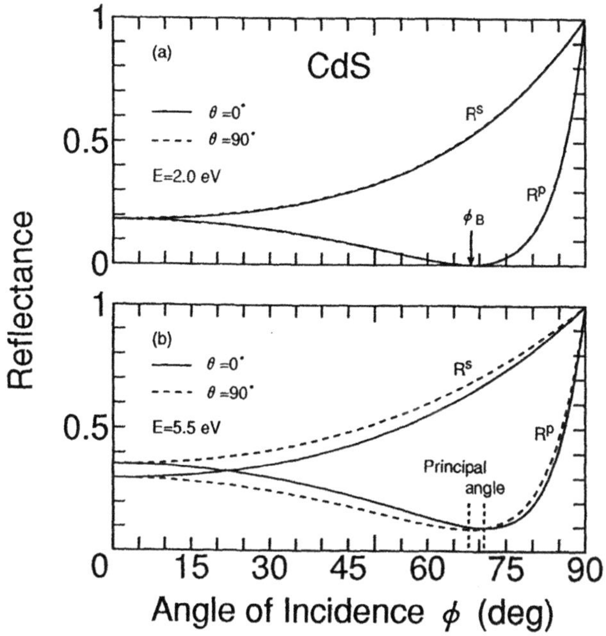

Figure 3.33 Plots of R^p $(=|r_{12}{}^p|^2)$ and R^s $(=|r_{12}{}^s|^2)$ vs. angle of incidence ϕ with θ=0° (solid lines) and 90° (dashed lines). These curves correspond to those for CdS with $n_\perp$*=2.49+i0.00, $n_{||}$*=2.51+i0.00 [E=2.0 eV, (a)] and $n_\perp$*=2.06+i1.55, $n_{||}$*=2.34+i1.84 [E=5.5 eV, (b)]. The numeric n values are taken from Ref. [62].

i.e., 0.185 [$R^p(\theta=0°)$ and $R^s(\theta=90°)$] and 0.182 [$R^s(\theta=0°)$ and $R^p(\theta=90°)$]. It should be noted that for ϕ=0° the following relations always hold in uniaxial materials: $R^p(\theta=0°)=R^s(\theta=90°)$, $R^s(\theta=0°)=R^p(\theta=90°)$. The Brewster angles ϕ_B shown in Fig. 3.33 (a) are 68.3° and 68.1° for θ=0° and 90°, respectively.

At E=5.5 eV, distinct anisotropic nature of R^p and R^s appears between θ=0° and 90° [Fig. 3.33 (b)]. R^s is found to be smaller than R^p at angles 0°≤ϕ≤22° (θ=0°). The principal angles, shown by the dashed lines, are 70.9° (θ=0°) and 67.8° (θ=90°), respectively.

(c) Reflection at Multiple Interfaces—The above discussion concerned a single interface. We thus disregarded the light transmitted through the interface. This may be not true in many real situations. A case of considerable importance in ellipsometry is that in which polarized light is reflected from, or transmitted by, a substrate covered by a single film, as suggested by Fig. 3.34.

In Fig. 3.34, the ambient (medium 1), film (medium 2, thickness d), and substrate (medium 3) are simply assumed to be all homogeneous and optically isotropic, with complex refractive indices n_1*, n_2*, and n_3*, respectively. The

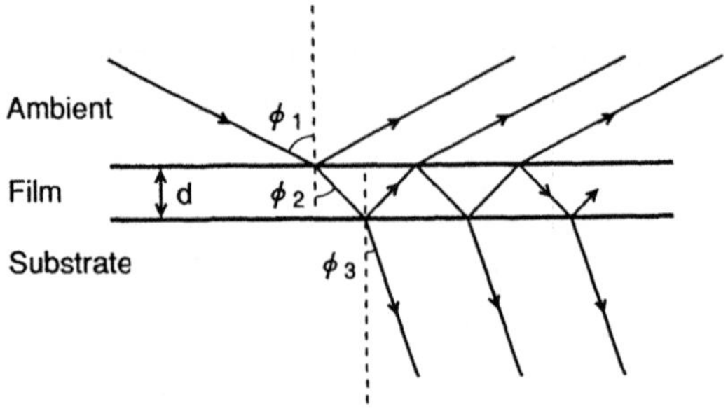

Figure 3.34 Oblique reflection and transmission of a plane wave by an ambient (air)/film (effective-medium layer)/substrate system with parallel-plane boundaries. d is the film thickness.

reflected wave returning to medium 1 is made up of the light reflected directly from the first interface plus all of the transmissions from the light approaching the first interface from medium 2. The addition of this infinite series of partial waves leads to the resultant wave. The total reflected amplitude r_m is then [58,59]

$$r_m = r_{12} + t_{12}t_{21}r_{23}\,e^{i2\beta} + t_{12}t_{21}r_{21}r_{23}^2\,e^{i4\beta} + t_{12}t_{21}r_{21}^2r_{23}^3\,e^{i6\beta} + \ldots \tag{3.67}$$

whose summation gives

$$r_m = r_{12} + \frac{t_{12}t_{21}r_{23}\,e^{i2\beta}}{1 - r_{21}r_{23}\,e^{i2\beta}} = \frac{r_{12} + r_{23}\,e^{i2\beta}}{1 + r_{12}r_{23}\,e^{i2\beta}} \tag{3.68}$$

where β is the film phase thickness and is given by

$$\beta = 2\pi\left(\frac{d}{\lambda}\right)n_2^*\cos\phi_2 \tag{3.69}$$

In obtaining Eq. (3.68), we used the following relations [see Eq. (3.59)]:

$$r_{21} = -r_{12}\ , \qquad t_{12}t_{21} = 1 - r_{12}^2 \tag{3.70}$$

These relations are obtained by interchanging n_1* and n_2* in Eqs. (3.54) and (3.55) [i.e., equating that the direction of propagation of the refracted wave in medium 2 is reversed (see Fig. 3.30)].

Equations (3.67) and (3.68) are valid when the incident wave is linearly polarized either parallel (p) or perpendicular (s) to the plane of incidence. We thus obtain

$$r_m^p = \frac{r_{12}^p + r_{23}^p e^{i2\beta}}{1 + r_{12}^p r_{23}^p e^{i2\beta}} \qquad (3.71a)$$

$$r_m^s = \frac{r_{12}^s + r_{23}^s e^{i2\beta}}{1 + r_{12}^s r_{23}^s e^{i2\beta}} \qquad (3.71b)$$

Then, the reflectance R is given by

$$R^p = |r_m^p|^2 \quad , \qquad R^s = |r_m^s|^2 \qquad (3.72)$$

3.3.2 Ellipsometry Definition and Optical Constants

In order to examine the effect of reflection on the amplitude and phase of the wave separately, we write down the complex Fresnel coefficients in the form [58,59]

$$r_{12}^p = |r_{12}^p| \, e^{i\delta_r^p} \qquad (3.73a)$$

$$r_{12}^s = |r_{12}^s| \, e^{i\delta_r^s} \qquad (3.73b)$$

Reflection ellipsometry is a technique based on measurements of the states of polarization of the incident and reflected light waves, leading to the determination of the ratio ρ of the complex Fresnel reflection coefficients [58,59]:

$$\rho = \frac{r^p}{r^s} \qquad (3.74)$$

It is often convenient to write ρ in the form

$$\rho = \tan\Psi \, e^{i\Delta} \qquad (3.75)$$

with

$$\tan\Psi = \frac{|r_{12}^p|}{|r_{12}^s|} \qquad (3.76a)$$

$$\Delta = \delta_{12}^r - \delta_{12}^s \qquad (3.76b)$$

where Ψ and Δ are the quantities measured by an ellipsometry.

The pseudodielectric function is a quantity derived from ellipsometry by means of the two-phase (air/substrate) model. By substituting for r^p and r^s in Eq. (3.74) their values from Eq. (3.54) and making use of Snell's law [Eq. (3.53)], we obtain an equation that can be solved for n_2^*/n_1^* in terms of ρ and ϕ_1 only, as follows [58,59]

$$\frac{n_2^*}{n_1^*} = \tan\phi_1 \sqrt{1 - \frac{4\rho \sin^2\phi_1}{(\rho + 1)^2}} \qquad (3.77)$$

If medium 1 is air (n_1^*=1.0+i0.0), Eq. (3.77) can be simply given by [58,59]

$$n_2^* = \tan\phi_1 \sqrt{1 - \frac{4\rho \sin^2\phi_1}{(\rho + 1)^2}} \qquad (3.78)$$

This expression can be separated into the following equations [63]:

$$\varepsilon_1 = n_2^2 - k_2^2 = \sin^2\phi_1 \left[1 + \frac{\tan^2\phi_1(\cos^2 2\Psi - \sin^2\Delta \sin^2 2\Psi)}{(1 + \sin 2\Psi \cos\Delta)^2}\right] \qquad (3.79a)$$

$$\varepsilon_2 = 2n_2k_2 = \frac{\sin^2\phi_1 \tan^2\phi_1 \sin 4\Psi \sin\Delta}{(1 + \sin 2\Psi \cos\Delta)^2} \qquad (3.79b)$$

Figure 3.35 plots (a) (n, k) and (b) (Ψ, Δ) vs. photon energy $E=\hbar\omega$ for a ambient (air)/substrate system. The substrate corresponds to GaP, and its (n, k) are taken from Ref. [60]. The ellipsometric variables (Ψ, Δ) are obtained by computing Eqs. (3.53), (3.54), (3.73), and (3.76). The (Ψ, Δ) for substrates always fall in the ranges (0–45°)/(0–180°), respectively. We note that Δ=0° or 180° occurs when the substrate is a dielectric (k=0). The phase shift Δ from a reflection from a dielectric is either 180° if it is less than the Brewster angle or zero if it is more than the Brewster angle. The k for GaP are nearly zero in the range E<2.5 eV. The Brewster angles will be 72.5°≤ϕ_B≤74.5° in this spectral range. In Fig. 3.35, thus, we obtain Δ=180° (ϕ_1=70°) and zero (ϕ_1=75°) for E<2.0 eV.

3.3.3 Effect of Surface Oxide

The pseudodielectric function is the ellipsometry-derived quantity. It is exactly equal to the true bulk dielectric function of a given sample if its surface is perfectly abrupt and film-free. It should be noted that such an ideal condition has never been achieved in practice.

When an etch-cleaned semiconductor is exposed to air, the thickness of surface

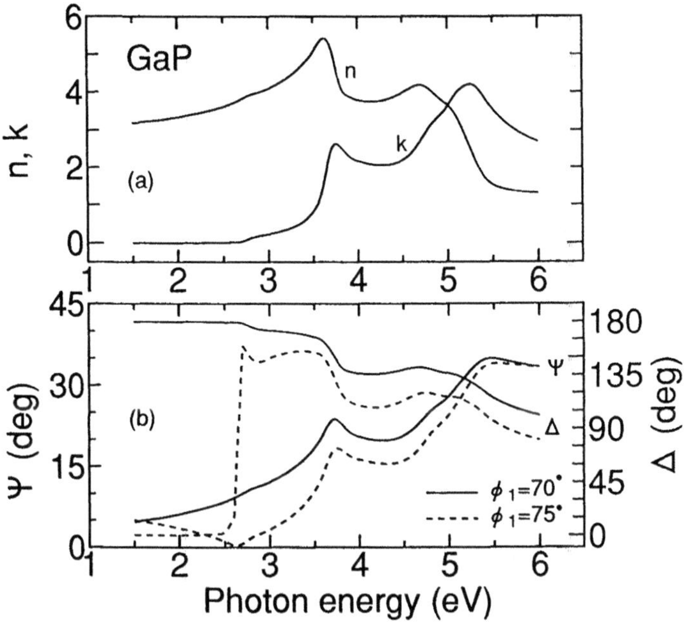

Figure 3.35 Plots of (a) (n, k) and (b) (Ψ, Δ) vs. photon energy $E=\hbar\omega$ for an ambient (air)/substrate system. The substrate corresponds to GaP, and its (n, k) are taken from Ref. [60] (T=300 K). The ellipsometric variables (Ψ, Δ) are obtained by computing Eqs. (3.53), (3.54), (3.73), and (3.76).

native oxide increases with time so that these thicknesses may be strongly dependent on time between the surface cleaning and ellipsometric measurement. As an example, we show in Fig. 3.36 the pseudodielectric-function data $\varepsilon_2(E)$ for GaAs etch-cleaned and exposed to room air for less than 30 min (solid circles) [64]. The experimental data for clean GaAs measured by Aspnes and Studna [60] are also shown by the solid line.

As seen in the figure, the most prominent air-exposure effect occurs near the absolute maximum in ε_2, i.e., at the energy of the E_2 peak in ε_2. Measurement of the $\varepsilon_2(E)$ spectra can, thus, yield direct information about the relative quality of semiconductor surfaces. In fact, Aspnes and Studna [65] reported chemical etching and cleaning procedures, producing the most abrupt dielectric discontinuities between the bulk and ambient (cleanest and/or smoothest surfaces) for several semiconductors (such as Si, Ge, and some III–V compounds) by monitoring SE $\varepsilon_2(E)$ values.

Equations (3.53), (3.54), (3.69), (3.71), and (3.75) promise that an overlayer (oxide) thickness can be estimated if its optical constants are known, and vise versa. Figure 3.37 illustrates the Ψ/Δ trajectory for native oxide on the GaAs

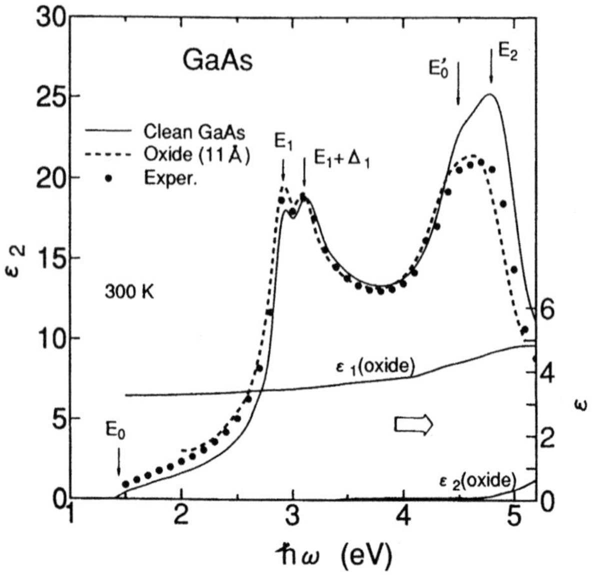

Figure 3.36 Pseudodielectric function for GaAs etch-cleaned and exposed to room air for less than 30 min (solid circles). The experimental data for clean GaAs reported by Aspnes and Studna [60] are shown by the solid line. The dashed line represents the calculated pseudodielectric-function spectrum for GaAs using the three-layer (ambient/ oxide/substrate) model. This calculation estimates the thickness of oxide overlayer on GaAs to be ~11 Å. Optical constants of the oxide used in the calculation are assumed to be almost equal to those for electrochemically-grown anodic oxide on GaAs. They are measured by SE and are shown in the figure. (From Kato and Adachi [64].)

substrate with an angle of incidence ϕ_1=70°, E=4.0 eV (λ=3100 Å), and oxide thickness increments of 50 Å. The film-free values, Ψ=18.80° and Δ=133.05°, correspond to the GaAs substrate (n_3*=3.601+i1.920, Ref. [60]). Since the optical constants at any wavelength are not available for the GaAs native oxide, we assumed that they are almost equal to those for the electrochemically-grown anodic oxide on this material [n_2*=1.95+i0.00 (E=4.0 eV), see Fig. 3.36]. An unknown native oxide thickness on GaAs can, then, be determined by inspection by comparing the measured Ψ/Δ values with this trajectory curve. It is noted that for films with k_2=0, the trajectory closes on itself at a thickness

$$d = \frac{\lambda}{2\sqrt{n_2^2 - \sin^2\phi_1}} \tag{3.80}$$

and for larger thicknesses, the trajectory simply retraces the same path [59]. In this case (Fig. 3.37), d is about 907 Å.

With given values of n_2*(E) and film thickness d, we can use Eqs. (3.53),

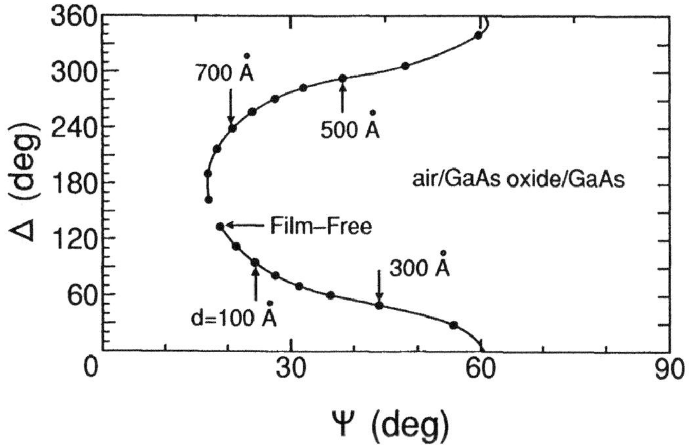

Figure 3.37 Ψ/Δ trajectory for native oxide on the GaAs substrate with an angle of incidence ϕ_1=70°, E=4.0 eV (λ=3100 Å), and oxide thickness increments of 50 Å. The film-free values (Ψ=18.80° and Δ=133.05°) correspond to the GaAs substrate (n_3*=3.601+i1.920, Ref. [60]). Since the optical constants at any wavelength are not available for the GaAs native oxide, we assumed that they are almost equal to those for the electrochemically-grown anodic oxide on this material [n_2*=1.95+i0.00 (E=4.0 eV), see Fig. 3.36].

(3.54), (3.69), (3.71), and (3.75) to calculate the pseudodielectric-function spectra for an ambient/film/substrate system. In Fig. 3.36, the dashed line represents the result of this calculation (air/GaAs oxide/GaAs substrate). This calculated result suggests that the native oxide thickness d_{ox} on GaAs will be ~11 Å. The GaAs sample exposed to room air for more than 30 min also suggested that the oxide thickness d_{ox} increases almost linearly with log t (time), indicating an oxidation rate of d_0~6.5 Å/decade. It is found that this rate is considerably larger than that reported by Kuphal and Dingers (~4.1 Å/decade) [66], but is almost the same as that reported by Adams and Pruniaux (~7.2 Å/decade) [67]. We note that the experimental ε(E) data in Figs. 3.13–17 were plotted after mathematically removing the effect of GaAs native oxide (~16.4 Å) from the substrate.

Figure 3.38 shows the calculated pseudodielectric-function spectra, ε(E)= $\varepsilon_1(E)+i\varepsilon_2(E)$, for an air/$SiO_2$/Si system with various SiO_2 film thicknesses obtained using the model shown in Fig. 3.34. The optical constants of SiO_2 and Si used in the calculation were taken from the literature [60,68]. It is evident that the pseudodielectric-function spectra ε(E) are strongly dependent on the thickness of SiO_2. The profound film effect can be found near the E_1 and E_2 CP peaks, and especially near the E_2 peak: they show a red shift together with lowering of the magnitudes. It can thus be easily understand that the E_2 peak in ε_2 provides a sensitive indication of the dielectric discontinuity between the substrate (Si) and ambient [65].

The $\varepsilon(E)$ data of clean GaAs and Si measured by Aspnes and Studna [60] were not corrected for any surface oxide layer. The SE system they used was a rotating-analyzer type. If no compensating element is placed in the optical path, this instrument measures the quantities

$$\alpha_1 = \frac{\cos 2\theta_p - N}{1 - N\cos 2\theta_p} \qquad (3.81a)$$

$$\alpha_2 = \frac{C\cos 2\theta_p}{1 - N\cos 2\theta_p} \qquad (3.81b)$$

with

$$N = \cos 2\psi \qquad (3.82a)$$

$$C = \sin 2\psi \cos\Delta \qquad (3.82b)$$

where the quantities Ψ and Δ are the conventional ellipsometry parameters. The rotating-analyzer ellipsometer thus does not measure a quantity proportional to sin Δ, and therefore will be very insensitive to Δ when Δ is close to zero or 180°.

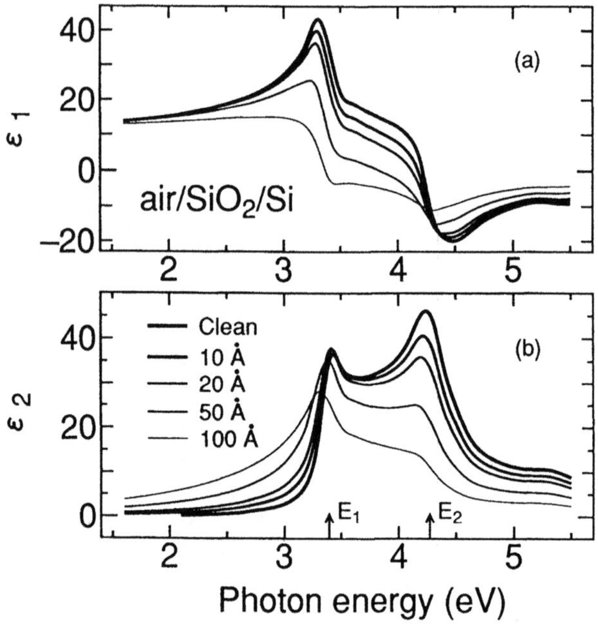

Figure 3.38 Calculated pseudodielectric-function spectra, $\varepsilon(E)=\varepsilon_1(E)+i\varepsilon_2(E)$, for an air/$SiO_2$/Si system with various SiO_2 film thicknesses (d=0–100 Å) obtained from the model depicted in Fig. 3.34. The optical constants of SiO_2 and Si used in the calculation are taken from Refs. [68] and [60], respectively.

Jellison [69], therefore, used a two-channel polarization modulation spectroscopy to measure the optical functions of GaAs, GaP, and Ge. This instrument makes an independent determination of all the associated ellipsometry parameters. This is particularly important for systems which have a thin overlayer over a material with a small value of the extinction coefficient k. He observed a noticeable difference with the data of Aspnes and Studna in the photon-energy region above E~4.0 eV. This difference may come from the fact that the correction for the oxide film was made or not. He made additional nulling ellipsometry measurements at 1.152 μm to get an independent measurement of the oxide thickness. He calculated the values of the oxide thickness d_{ox} and refractive index of the substrate from the measured Ψ and Δ, assuming that the oxide refractive index n_{ox} and substrate extinction coefficient k_{sub} were known. The oxide thickness he estimated from this analysis for GaAs was d_{ox}=16.3±1.5 Å.

Jellison [69] measured the SE data for both GaAs(100) and GaAs(111), and found that there are small but statistically significant differences between these two data, occurring mostly at higher photon energies. This difference may be due to the fact that the GaAs oxide and/or its interface are dependent upon the face on which it is grown. Another possibility stems from the fact that the absorption coefficient is extremely large in this region, and the 1–2% difference could come from the near-surface reconstruction of the surface.

Aqueous HF treatment is commonly used for the etching of Si native oxide. This agent is known to rapidly etch SiO_2, but not attack Si vigorously. The hydrogen-(H-) terminated Si surface, which can be simply obtained by immersing in aqueous HF solution, has been attractive from both technological and surface-scientific points of view. Many experimental techniques, such as XPS, LEED, HREELS, STM, and IR spectroscopy, indicated that the etching of Si in HF solution efficiently removes any oxide layer and produces stable, H-terminated surfaces. It is known that the ellipsometry is a very surface-sensitive technique that enables detection of submonolayer coverage of surface by adsorbed chemical species. SE has been used to study HF-treated Si surfaces [70,71]. The measured SE data indicated that aqueous HF etching results in the removal of the surface oxide and leaves behind Si surfaces terminated by atomic hydrogen.

A necessary first step in device technology is the preparation of smooth, clean, and undamaged surfaces on semiconductor wafers. The cleaning procedure that is most often used for Si wafers is the so-called RCA cleaning [72]. This cleaning consists of two steps, commonly called SC1 and SC2. The SC1 step involves cleaning in an $NH_4OH:H_2O_2:H_2O$=1:1:5 mixture at temperatures between 75 and 90°C, immediately followed by a deionized water rinse. The SC2 step consists of a cleaning in an $HCl:H_2O_2:H_2O$ mixture at 75 to 90°C, again followed by a deionized water rinse. The SE data clearly indicated [73] that the SC1 and SC2 solutions result in chemical oxidation of Si surface. The chemical oxidation occurred immediately upon immersing the samples in the solutions. The oxide thickness was found to show a saturated value of ~9.5–11 Å against immersion time.

3.4 Effect of Surface Roughness: Effective-Medium Approximation

3.4.1 Effective-Medium Model

A factor that can significantly influence the optical constants of *real* materials, particularly as determined by ellipsometry, is surface microscopic roughness [59]. As suggested in Fig. 3.39, surface microscopic roughness can be characterized approximately by a mean height of irregularities about an average plane and a correlation length between irregularities. Thus, the roughened surface layer on a substrate can be regarded as a physical mixture of voids and substrate material. We are concerned here with surface microscopic roughness, where the mean height and correlation length of the irregularities are both less than the wavelength of light.

Simple approaches to the description of the dielectric properties of a multiphase mixture were given by Lorentz–Lorenz (LL), Maxwell–Garnett (MG), and Bruggeman (see Refs. [58,74,75]). They all have the same generic form

$$\frac{\langle\varepsilon\rangle-\varepsilon_h}{\langle\varepsilon\rangle+2\varepsilon_h} = f_1\frac{\varepsilon_1-\varepsilon_h}{\varepsilon_1+2\varepsilon_h} + f_2\frac{\varepsilon_2-\varepsilon_h}{\varepsilon_2+2\varepsilon_h} + \ldots\ldots \tag{3.83}$$

where $\langle\varepsilon\rangle$, ε_h, ε_1, ε_2, ..., are the complex dielectric functions of the effective medium, host medium, and inclusions of types 1, 2, ..., in the host, respectively. The f_1, f_2, ..., represent the volume fractions of inclusions 1, 2, ..., in the total volume. As we will see next, the primary difference in the LL, MG, and Bruggeman EMA (effective-medium approximation) model is choice of the "host" medium.

The LL approximation was developed to describe point polarizable entities of polarizability α embedded in vacuum. In this case, $\varepsilon_h=1+i0$ and Eq. (3.83) can be

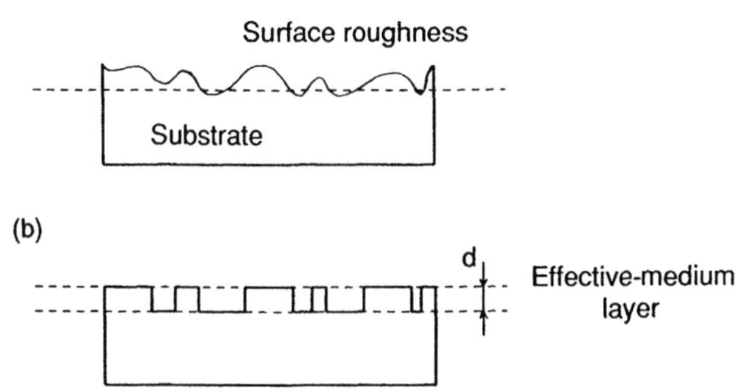

Figure 3.39 (a) A roughened surface and (b) the equivalent effective medium on a substrate of the original material.

written as

$$\frac{\langle\varepsilon\rangle-1}{\langle\varepsilon\rangle+2} = f_1\frac{\varepsilon_1-1}{\varepsilon_1+2} + f_2\frac{\varepsilon_2-1}{\varepsilon_2+2} \quad (3.84)$$

It should be noted that the LL relation is usually expressed in terms of the polarizabilities α_i, which are related by the Clausius–Mosotti relation [see Eq. (2.4)]. In the case of a roughened surface, this is a reasonable approximation up to a volume fraction of about 20% [59].

The MG approximation corresponds to macroscopic dielectric inclusions in a host dielectric background and the quantities in Eq. (3.83) have their obvious interpretations. In the case of a single inclusion in a host background, the equation becomes

$$\frac{\langle\varepsilon\rangle-\varepsilon_h}{\langle\varepsilon\rangle+2\varepsilon_h} = f_1\frac{\varepsilon_1-\varepsilon_h}{\varepsilon_1+2\varepsilon_h} \quad (3.85)$$

It is noted that the above expression degenerates to the LL form if the host happens to be vacuum or air. Then, this is not a bad approximation if the inclusions (voids) make up a small fraction of the total volume.

For a single type of inclusion in a single host, $\langle\varepsilon\rangle$ has different values in the MG approximation if the role of host and inclusion is interchanged—even if the respective volume fractions stay the same. To avoid this ambiguity in bulk applications Bruggeman suggested replacing ε_h in Eq. (3.83) with $\langle\varepsilon\rangle$, i.e., letting the effective medium itself acts as host medium. Under this assumption, Eq. (3.83) becomes

$$\sum_{i=1}^{n} f_i\frac{\varepsilon_i-\langle\varepsilon\rangle}{\varepsilon_i+2\langle\varepsilon\rangle} = 0 \quad (3.86)$$

with

$$\sum_{i=1}^{n} f_i = 0 \quad (3.87)$$

This is known as the Bruggeman EMA. The Bruggeman EMA gives a rigorous result of self-consistency that can be shown to be varied under fairly general conditions [75].

The dielectric function of an effective-medium layer or unknown parameters such as layer thickness and volume fractions of constituent phases can be numerically determined by minimizing the following mean-squares deviation using an LRA program [58]:

$$\sigma^2 = \frac{1}{N-P-1}\sum_{j=1}^{N}[(\tan\Psi_j^{\exp} - \tan\Psi_j^{calc})^2 + (\cos\Delta_j^{\exp} - \cos\Delta_j^{calc})^2] \quad (3.88)$$

where N is the number of data points and P is the number of unknown model parameters. Here, Ψ and Δ are the ellipsometric variables with "*exp*" and "*calc*" stand for measured and calculated values, respectively. The best-fit model is selected which yields a minimum value of unbiased estimator σ.

3.4.2 Roughened Surface

We use again Eqs. (3.53), (3.54), (3.69), (3.71), and (3.75) to calculate the pseudodielectric-function spectra for an air/effective layer/substrate system. In Fig. 3.40 (a), the solid lines represent the results of this calculation. The effective layer considered here consists two constituents, namely, *voids* (or *density deficit*) and bulk Si ($f_v+f_{Si}=1$). The dielectric functions of the effective layers (d=1 nm, see Fig. 3.39) are obtained from the Bruggeman EMA with several different void fractions, f_v=0, 0.3, and 0.6. Note that the f_v=0 and 1.0 correspond to the case for the clean, flat Si. The bulk Si data are taken from Ref. [60]. We can understand from Fig. 3.40 (a) that the pseudodielectric-function spectra $\varepsilon(E)$ are strong function of surface microroughness.

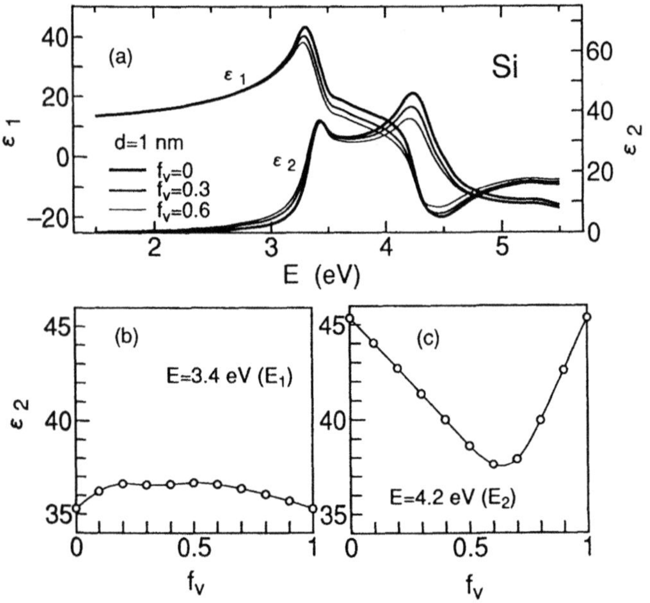

Figure 3.40 (a) Calculated pseudodielectric-function spectra, $\varepsilon(E)=\varepsilon_1(E)+i\varepsilon_2(E)$, for an air/roughened Si/Si substrate system. The effective-medium surface layers consist two constituents, namely, voids (density deficit) and bulk Si ($f_v+f_{Si}=1$). The dielectric functions of these layers (d=1 nm, see Fig. 3.39) are obtained from the Bruggeman EMA with several different void fractions, f_v=0, 0.3, and 0.6. (b) ε_2 value at the E_1 peak (3.4 eV) vs. void fraction f_v. (c) ε_2 value at the E_2 peak (4.2 eV) vs. f_v.

The corresponding ε_2 values at the E_1 and E_2 peaks vs. void fraction f_v are plotted in Figs. 3.40 (b) and (c), respectively. The ε_2 at the E_2 peak decreases linearly as the void fraction f_v increases, showing a minimum at about f_v=0.6, and increases with further increase of the void fraction. The $\varepsilon_2(E_1)$ peak shows opposite trend; i.e., it increases as the f_v increases, keeping the nearly constant value at f_v=0.2–0.6, and decreases with further increase of f_v. Note, however, that the change in $\varepsilon_2(E_1)$ is ~1.4 at the most, much smaller than the case for $\varepsilon_2(E_2)$ (~7.7).

The effects of the roughened overlayer thickness d on the pseudodielectric function of Si are shown in Fig. 3.41. We assumed here the effective roughness layer of $f_v : f_{Si}$=0.5:0.5. The open circles in Figs. 3.41 (b) and (c) are at 0.5 nm intervals from zero d thickness up to 5 nm. We again recognize that the ε_2 value at the E_2 peak provides a very sensitive and clear indication of the dielectric discontinuity between the substrate and ambient (not only surface cleanness but also surface flatness) [65].

As mentioned before, etching of Si in HF (or buffered-HF) solution produces stable, H-terminated surfaces. Recent studies [76,77] demonstrated that aqueous

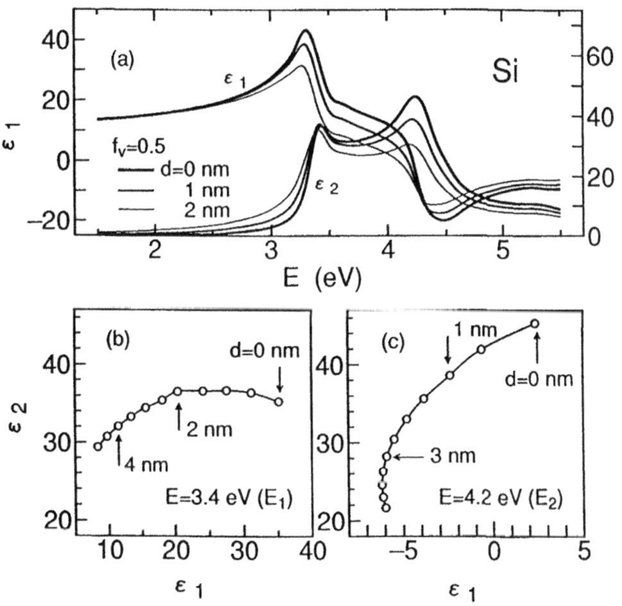

Figure 3.41 (a) Calculated pseudodielectric-function spectra, $\varepsilon(E)=\varepsilon_1(E)+i\varepsilon_2(E)$, for an air/roughened Si/Si substrate system. The effective-medium surface layers consist two constituents, namely, voids (density deficit) and bulk Si ($f_v+f_{Si}=1$). The dielectric functions of these layers (f_v=0.5) are obtained from the Bruggeman EMA with several different effective-medium layer thicknesses, d=0, 1, and 2 nm. (b) ε_2 vs. ε_1 at the E_1 peak (3.4 eV). (c) ε_2 vs. ε_1 at the E_2 peak (4.2 eV). In (b) and (c), the ε_2–ε_1 values are plotted at 0.5 nm intervals from zero d thickness up to 5 nm.

NH_4F and HBF_4 solutions can also produce H-terminated Si surfaces with no detectable impurities such as oxygen, fluorine, or hydrocarbons.

Figure 3.42 shows the $\varepsilon_2(E_2)$ value vs. immersion time t in aqueous HF (1.5%), NH_4F (40%), and HBF_4 (42%) solutions [78]. We understand from this figure that in the case of HF, the removal of the native SiO_2 film is achieved immediately upon immersing the sample in the solution. Immersions in the NH_4F and HBF_4 solutions, on the other hand, required approximately 60 s (NH_4F) and 100 s (HBF_4) to completely remove the native oxide. After longer exposures, these solutions provide a very slight degradation of the $\varepsilon_2(E_2)$ strength. This degradation may be due to an increase in the degree of surface microroughness. We also found that the saturated $\varepsilon_2(E_2)$ values for HF- and NH_4F-treated surfaces (~43–44) are considerably smaller than those reported by Yasuda and Aspnes [79]. This may be due to the fact that our chemical treatment indeed removes the oxide and leaves a bare Si surface, but as soon as the crystal is brought out into the air (i.e., set on an SE goniometer) a new oxide begins to form. Yasuda and Aspnes used a windowless cell in which a dry N_2 ambient was contained to keep the chemically treated sample surfaces unchanged. During transfer to the cell the samples were exposed to air, but for no longer than 30 s. This duration is considerably shorter than the case of ours (≤5 min). We also note in Fig. 3.42 that the saturated $\varepsilon_2(E_2)$ value for the HBF_4-treated surface is substantially lower (~42).

We plot in Fig. 3.43 the native oxide thickness on Si(111), as measured by SE, as a function of immersion time t in the HF, NH_4F, and HBF_4 solutions [78].

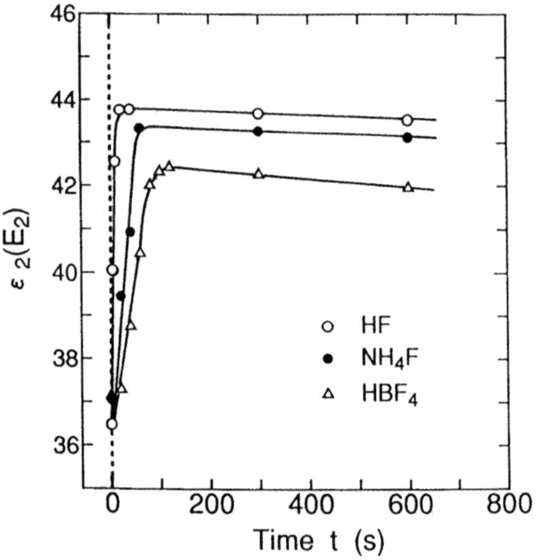

Figure 3.42 Plots of the ε_2 value at the E_2 peak maximum, $\varepsilon_2(E_2)$, vs. immersion time t in aqueous HF (1.5%), NH_4F (40%), and HBF_4 (42%) solutions for Si(111). (From Adachi *et al.* [78].)

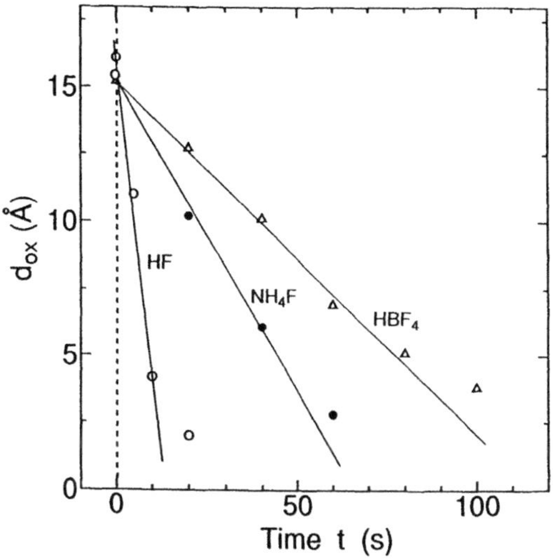

Figure 3.43 Plots of the native oxide thickness d_{ox} on Si(111), as measured by SE, vs. immersion time t in aqueous HF (1.5%), NH_4F (40%), and HBF_4 (42%) solutions. These plots give etch rates of ~72 Å/min (HF), ~13 Å/min (NH_4), and ~8 Å/min (HBF_4), respectively. (From Adachi *et al.* [78].)

These plots suggest that the oxide thickness gradually decreases with increasing t, yielding etch rates of ~72 Å/min (HF), ~13 Å/min (NH_4), and ~8 Å/min (HBF_4), respectively. We have, however, never obtained bare Si surfaces (i.e., d_{ox}~0 Å) even after sufficient etching. Note that not only the native oxide film but also roughened overlayer have effects upon the ellipsometric data. Thus, the film thickness plotted in Fig. 3.43 must be considered to be the *effective* oxide thickness but not necessarily the real one. The oxide thickness d_{ox} referred to here is of the modeled equivalent dielectric layer of pure SiO_2, including possible Si surface roughness.

In order to take into account the roughened overlayer, we used the LRA–EMA analysis with a three-phase model (ambient/roughened Si overlayer/bulk Si). Figure 3.44 shows, as an example, the simulated result for the sample etched in the HBF_4 solution for t=600 s [78]. The dashed lines show the $\varepsilon(E)$ spectra for clean, bare Si taken from Ref. [80]. The solid lines represent the result of the LRA–EMA analysis. The bulk density deficit and roughened overlayer thickness determined here are 66% and 3.7 Å, respectively. We also found that the F-based solutions (HF, NH_4F, and HBF_4) provide nearly the same overlayer parameters (i.e., roughened overlayer thicknesses of ~2.4–3.7 Å and void fractions of ~62–66%) [78].

Adachi and co-workers [78] used *ex situ* AFM to independently assess surface morphologies produced by the F-based chemical solutions. The observed AFM images revealed rms roughnesses of 2.5–3.1 Å, in reasonable agreement with those

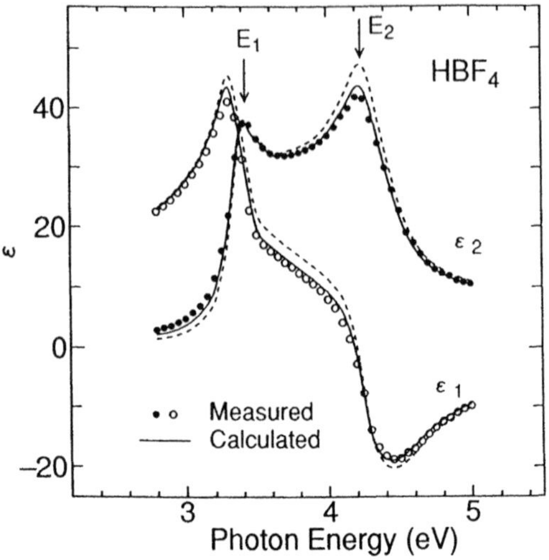

Figure 3.44 $\varepsilon(E)$ spectra for Si(111) treated with 42% HBF_4 solution for t=600 s (solid and open circles) and for clean, bare Si taken from Ref. [80] (dashed lines). The solid lines represent the result of the LRA–EMA analysis. The bulk density deficit and roughness overlayer thickness obtained from this analysis are 66% and 3.7 Å, respectively. (From Adachi *et al.* [78].)

obtained from the LRA–EMA analysis. The contact-angle measurements also suggested that the HF- and NH_4F-treated surfaces are hydrophobic, while the HBF_4-treated surface is, if anything, hydrophilic [78].

The usual way of describing the optical response in an amorphous–crystalline intermediate state is to assume that the intermediate layer consists of a physical mixture of crystalline and amorphous regions in order to use an effective-medium theory. In Sec. 4.2, we will use again the LRA–EMA to analyze such an intermediate state produced in *c*-Si by As^+-ion implantation and also in *a*-Si by thermal annealing.

REFERENCES

1. See, for instance, G. Harbake, in *Optical Properties of Solids*, edited by F. Abelès (North-Holland, Amsterdam, 1972), p. 21.
2. M. Erman, J. B. Theeten, P. Chambon, S. M. Kelso, and D. E. Aspnes, *J. Appl. Phys.* **56**, 2664 (1984).
3. H. Ehrenreich and M. L. Cohen, *Phys. Rev.* **115**, 786 (1959).

4. D. E. Aspnes, S. M. Kelso, R. A. Logan, and R. Bhat, *J. Appl. Phys.* **60**, 754 (1986).
5. F. L. Terry, Jr., *J. Appl. Phys.* **70**, 409 (1991).
6. P. Lautenschlager, M. Garriga, L. Viña, and M. Cardona, *Phys. Rev. B* **36**, 4821 (1987).
7. P. Lautenschlager, M. Garriga, S. Logothetidis, and M. Cardona, *Phys. Rev. B* **35**, 9174 (1987).
8. S. Adachi, *Physical Properties of III–V Semiconductor Compounds: InP, InAs, GaAs, GaP, InGaAs, and InGaAs* (Wiley-Interscience, New York, 1992).
9. S. Adachi, *GaAs and Related Materials: Bulk Semiconducting and Superlattice Properties* (World Scientific, SI, 1994).
10. S. Adachi, *Phys. Rev. B* **35**, 7454 (1987).
11. S. Adachi, *Phys. Rev. B* **38**, 12345 (1988).
12. S. Adachi, *J. Appl. Phys.* **66**, 6030 (1989).
13. S. Adachi, *Phys. Rev. B* **41**, 1003 (1990).
14. B. Velický and J. Sak, *Phys. Status Solidi* **16**, 147 (1966).
15. E. O. Kane, *Phys. Rev.* **180**, 852 (1969).
16. J. E. Rowe and D. E. Aspnes, *Phys. Rev. Lett.* **25**, 162 (1970).
17. R. M. Martin, J. A. Van Vechten, J. E. Rowe, and D. E. Aspnes, *Phys. Rev. B* **6**, 2500 (1972).
18. W. Hanke and L. J. Sham, *Phys. Rev. B* **21**, 4656 (1980).
19. M. del Castillo-Mussot and L. J. Sham, *Phys. Rev. B* **31**, 2092 (1985).
20. S. Adachi, *Phys. Rev. B* **38**, 12966 (1988).
21. S. Adachi, *J. Appl. Phys.* **66**, 813 (1989).
22. T. Aoki and S. Adachi, *J. Appl. Phys.* **69**, 1574 (1991).
23. S. Adachi and K. Sato, *Jpn. J. Appl. Phys.* **31**, 3907 (1992).
24. T. Kimura and S. Adachi, *Jpn. J. Appl. Phys.* **32**, 2740 (1993).
25. N. Suzuki, K. Sawai, and S. Adachi, *J. Appl. Phys.* **77**, 1249 (1995).
26. M. Miyazaki and S. Adachi, *J. Appl. Phys.* **77**, 1741 (1995).
27. S. Adachi and T. Miyazaki, *Phys. Rev. B* **51**, 14317 (1995).
28. S. Ozaki and S. Adachi, *J. Appl. Phys.* **78**, 3380 (1995).
29. S. Ozaki, S. Adachi, M. Sato, and K. Ohtsuka, *J. Appl. Phys.* **79**, 439 (1996).
30. D. W. Jenkins, *J. Appl. Phys.* **68**, 1848 (1990).
31. A. D. Rakić and M. L. Majewski, *J. Appl. Phys.* **80**, 5909 (1996).
32. C. C. Kim, J. W. Garland, H. Abad, and P. M. Raccah, *Phys. Rev. B* **45**, 11749 (1992).
33. H. R. Philipp and H. Ehrenreich, *Phys. Rev.* **129**, 1550 (1963).
34. D. L. Greenaway and G. Harbeke, *Optical Properties and Band Structure of Semiconductors* (Pergamon, Oxford, 1968).
35. Y. P. Varshni, *Physica* **34**, 149 (1967).
36. H. Yao, P. G. Snyder, and J. A. Woollam, *J. Appl. Phys.* **70**, 3261 (1991).
37. M. D. Sturge, *Phys. Rev.* **127**, 768 (1962).

38. D. E. Aspnes, W. E. Quinn, and S. Gregory, *Appl. Phys. Lett.* **56**, 2569 (1990).
39. M. Hanfland, K. Syassen, and N. E. Christensen, *J. Phys. (Paris) Colloq.* **45**, C8-57 (1984).
40. M. Alouani, L. Brey, and N. E. Christensen, *Phys. Rev. B* **37**, 1167 (1988).
41. J. Z. Hu and I. L. Spain, *Solid State Commun.* **51**, 263 (1984).
42. H. Olijnyk, S. K. Sikka, and W. B. Holzapfel, *Phys. Lett.* **103A**, 137 (1984).
43. Y. K. Vohra, K. E. Brister, S. Desgreniers, A. L. Ruoff, K. J. Chang, and M. L. Cohen, *Phys. Rev. Lett.* **56**, 1944 (1986).
44. S.-C. Yu, I. L. Spain, and E. F. Skelton, *J. Appl. Phys.* **49**, 4741 (1978).
45. See, M. I. McMahan, R. J. Nelmes, N. G. Wright, and D. R. Allan, *Phys. Rev. B* **48**, 16246 (1993), and references cited therin.
46. M. Hanfland, M. Alouani, K. Syassen, and N. E. Christensen, *Phys. Rev. B* **38**, 12864 (1988).
47. P. Etchegoin, J. Kircher, M. Cardona, C. Grein, and E. Bustarret, *Phys. Rev. B* **46**, 15139 (1992).
48. P. Etchegoin, J. Kircher, M. Cardona, and C. Grein, *Phys. Rev. B* **45**, 11721 (1992).
49. C. Pickering, R. T. Carline, M. T. Emeny, N. S. Garawal, and L. K. Howard, *Appl. Phys. Lett.* **60**, 2412 (1992).
50. See, L. Viña and M. Cardona, *Phys. Rev. B* **34**, 2586(1986), and references cited therein.
51. D. E. Aspnes, A. A. Studna, and E. Kinsbron, *Phys. Rev. B* **29**, 768 (1984).
52. L. Viña and M. Cardona, *Phys. Rev. B* **29**, 6739 (1984).
53. F. Lukeš, S. Gopalan, and M. Cardona, *Phys. Rev. B* **47**, 7071 (1993).
54. S. J. Pearton, J. W. Corbett, and M. Stavola, *Hydrogen in Crystalline Semiconductors* (Springer, New York, 1992).
55. P. de Mierry and M. Stutzmann, *Phys. Rev. B* **46**, 13142 (1992).
56. S. Zollner, M. Cardona, and S. Gopalan, *Phys. Rev. B* **45**, 3376 (1992).
57. C. Parks, A. K. Ramdas, S. Rodriguez, K. M. Itoh, and E. E. Haller, *Phys. Rev. B* **49**, 14244 (1994), and references cited therein.
58. R. M. A. Azzam and N. M. Bashara, *Ellipsometry and Polarized Light* (North-Holland, Amsterdam, 1977).
59. H. G. Tompkins, *A User's Guide to Ellipsometry* (Academic, Boston, 1993).
60. D. E. Aspnes and A. A. Studna, *Phys. Rev. B* **27**, 985 (1983).
61. S. Logothetidis, M. Cardona, P. Lautenschlager, and M. Garriga, *Phys. Rev. B* **34**, 2458 (1986).
62. S. Ninomiya and S. Adachi, *J. Appl. Phys.* **78**, 1183 (1995).
63. A. N. Saxena, *J. Opt. Soc. Am.* **55**, 1061 (1965).
64. H. Kato and S. Adachi (unpublished).
65. D. E. Aspnes and A. A. Studna, *Appl. Phys. Lett.* **39**, 316 (1981).
66. E. Kuphal and H. W. Dinges, *J. Appl. Phys.* **50**, 4196 (1979).
67. A. C. Adams and B. R. Pruniaux, *J. Electrochem. Soc.* **120**, 408 (1973).

68. H. R. Philipp, in *Handbook of Optical Constants of Solids*, edited by E. D. Palik (Academic, Orlando, 1985), p. 749.
69. G. E. Jellison, Jr., *Opt. Mat.* **1**, 151 (1992).
70. K. Utani, T. Suzuki, and S. Adachi, *J. Appl. Phys.* **73**, 3467 (1993).
71. H. Yao, J. A. Woollam, and S. A. Alterovitz, *Appl. Phys. Lett.* **62**, 3324 (1993).
72. W. Kern and D. Puotinen, *RCA Rev.* **31**, 187(1970); W. Kern, *RCA Rev.* **31**, 207 (1970).
73. S. Adachi and K. Utani, *Jpn. J. Appl. Phys.* **32**, L1189 (1993); K. Kobayashi, T. Suzuki, and S. Adachi, *Jpn. J. Appl. Phys.* **33**, L15 (1994).
74. C. G. Granqvist and O. Hunderi, *Phys. Rev. B* **16**, 3513 (1977).
75. D. E. Aspnes, J. B. Theeten, and F. Hottier, *Phys. Rev. B* **20**, 3292 (1979).
76. G. S. Higashi, R. S. Becker, Y. J. Chabal, and A. J. Becker, *Appl. Phys. Lett.* **58**, 1656 (1991); P. Dumas and Y. J. Chabal, *Chem. Phys. Lett.* **181**, 537 (1991).
77. V. V. Levenets, V. I. Beklemishev, E. P. Kirilenko, I. I. Makhonin, A. Yu. Trifonov, B. A. Loginov, and V. V. Protasenko, *Jpn. J. Appl. Phys.* **34**, 1723 (1995).
78. S. Adachi, T. Arai, and K. Kobayashi, *J. Appl. Phys.* **80**, 5422 (1996).
79. T. Yasuda and D. E. Aspnes, *Appl. Opt.* **33**, 7435 (1994).
80. V. Nayar, W. Y. Leong, C. Pickering, A. J. Pidduck, R. T. Carline, and D. J. Robbins, *Appl. Phys. Lett.* **61**, 1304 (1992).

Chapter 4

THE INTERBAND TRANSITION REGION: AMORPHOUS AND MICROCRYSTALLINE MATERIALS

4.1 AMORPHOUS MATERIALS

4.1.1 Optical Band Gap

(a) Definition—The analysis of optical spectra is one of the most useful tools for understanding electronic structure of amorphous semiconductors. Pierce and Spicer [1] performed photoemission measurements on *a*-Si to study the electronic structure and optical properties of the material. They used a conventional retarding-field energy analyzer and a high-resolution screened-emitter analyzer to measure the energy distribution curves from this material. When the energy distribution curves from *c*-Si are examined, one finds variations in the position and strength of structure as a result of the conservation of wave vector $\boldsymbol{k}$ in the crystal. However, in amorphous materials no such variations had been found in the energy distribution curves. This is not unexpected because the absence of long-range order in amorphous materials renders the Bloch theorem inapplicable and leaves the crystalline momentum $\hbar\boldsymbol{k}$ undefined. They suggested optical transitions in amorphous semiconductors to be described, to a first approximation, by the nondirect transition model in which conservation of the energy but not wave vector is significant [1].

The absence of ***k*** space in amorphous semiconductors does not imply that the density of electronic states is not a meaningful concept. In fact, techniques such as the photoemission showed [1,2] that, just as for crystals, the electronic states fall into bands separated by energy gaps. A schematic illustration of the DOS $N(E)$ for an amorphous semiconductor compared to that for a crystal, is shown in Fig. 4.1.

The most noteworthy features found in the electronic states of amorphous semiconductors are [3]: (1) the absence of sharp band edges and tailing of the DOS into the gap; (2) localization of states in energy regions where the DOS is low; (3) the existence of broad bands of defect levels in the gap (in contrast to sharp discrete levels in crystals) through which hopping conduction may occur.

The form of the corresponding optical absorption edge is illustrated schematically in Fig. 4.2 [3,4]. In the strong absorption edge (region A in Fig. 4.2), the absorption coefficient α can be described by the relation

$$E\alpha \sim (E-E_o)^n \qquad (4.1)$$

where the constant n is usually equal to 2 or 3, and E_o is the optical band-gap energy.

At lower energy (region B), α shows an exponential tail written in the form:

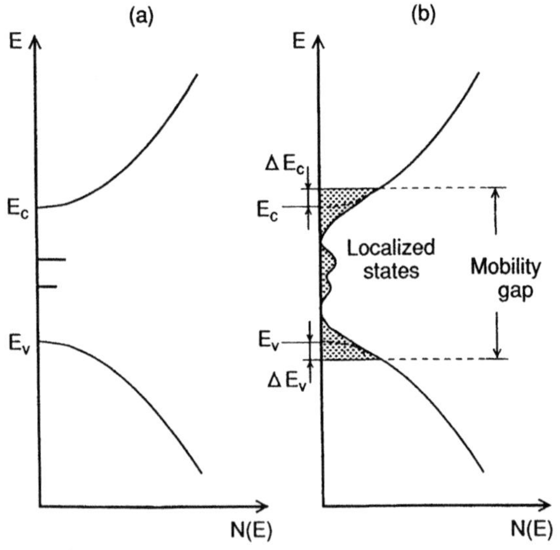

Figure 4.1 DOS $N(E)$ as a function of energy E for (a) crystalline and (b) amorphous semiconductors. Dotted area indicates localized states. ΔE_c and ΔE_v are band tails due to disorder.

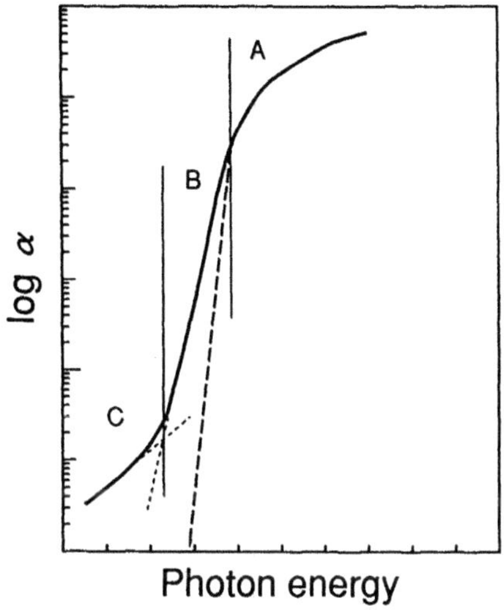

Figure 4.2 Three principal regions, A, B, and C, of optical absorption in an amorphous semiconductor. The dashed line represents extrapolation of part A, using Eq. (4.1).

$$\alpha \sim e^{E/E_u} \qquad (4.2)$$

This region is called *Urbach* region (Urbach tail, see Sec. 5.1.3) and corresponds to absorption involving one or both of the bad tails, with additional broadening expected from any potential fluctuations arising from charged defect states and lattice vibrations (phonons). The exponential factor E_u indicates the width of the band tails of the localized states.

At very low absorption part (region C), one observes another exponential part of the absorption edge where

$$\alpha \sim e^{E/E_t} \qquad (4.3)$$

and the energy E_t is always larger than E_u. This region arises from transitions involving defect states within the gap. In contrast to region B (which is similar in different materials and easily reproducible), region C may show a very structure-sensitive property.

The optical band-gap energy, E_o, is one of the most important physical parameters in amorphous semiconductors. Expression (4.1) is a somewhat semiempirical equation, because it is not based on a rigid theoretical model; consequently, E_o deduced from experimental data is a somewhat vague parameter that has no clear physical meaning, but its variation with preparation conditions, temperature, pressure, etc., conveys a good deal of information about the material of interest [5].

The real part of the frequency-dependent electrical conductivity $\sigma(E)$ can be given by [6]

$$\mathrm{Re}\,\sigma(E) = \frac{2\pi e^2\hbar^3\Omega}{m^2} \int N_v(E')N_c(E'+E)\,|P_{cv}|^2\, dE'/E \qquad (4.4)$$

where $N_v(E')$ and $N_c(E'+E)$ are the densities of the initial and final states, respectively, $|P_{cv}|^2$ is the squared momentum-matrix element, and Ω is the volume of the specimen. The absorption coefficient $\alpha(E)$ can also be defined, from electromagnetic theory, in terms of Re $\sigma(E)$ as (see more details in Sec. 5.3.1)

$$\alpha(E) = \frac{4\pi}{n(E)c}\,\mathrm{Re}\,\sigma(E) \qquad (4.5)$$

where $n(E)$ is the refractive index of the material and c is the light velocity in vacuum. Substituting Eq. (4.4) into Eq. (4.5), we obtain

$$\alpha(E) = \frac{8\pi^2 e^2\hbar^3\Omega}{n(E)cm^2} \int N_v(E')N_c(E'+E)\,|P_{cv}|^2\, dE'/E \qquad (4.6)$$

If we assume that $N(E)$ behaves like some power of E at the extremity of the band, we obtain [5]

$$N_c(E) = N_c^0\,[(E-E_c)/\Delta E\,]^{r_1} \qquad (4.7a)$$

$$N_v(E) = N_v^0\,[(E_v-E)/\Delta E\,]^{r_2} \qquad (4.7b)$$

where ΔE_c and ΔE_v are the band-tail widths due to disorder. If the bands are identical so that $r_1{=}r_2{\equiv}r$, $N_c^0{=}N_v^0$, and $\Delta E_c{=}\Delta E_v{\equiv}\Delta E$ (see Fig. 4.1), and $|P_{cv}|^2$ is assumed to be energy-independent, one obtains [5]

$$\alpha(E) = \frac{8\pi^2 e^2\hbar^3\Omega\,|P_{cv}|^2}{n(E)cm^2}\, 2\int_{E_c-E}^{E_v} (N_c^0)^2[(E_v-E)(E'+E-E_c)/(\Delta E)^2]^r\, dE'/E \qquad (4.8)$$

Equation (4.8) can, then, be modified by

$$\alpha(E) = \frac{16\pi^2 e^2\hbar^3\Omega\,|P_{cv}|^2}{n(E)cm^2}\,(N_c^0)^2\,\frac{[E-(E_c-E_v)]^{2r+1}}{E(\Delta E)^{2r}}\int_0^1 [y(1-y)]^r\, dy \qquad (4.9)$$

with

$$y = \frac{E_c-E-E'}{E_c-E-E_v} \qquad (4.10)$$

The optical absorption $\alpha(E)$ is finally given by [5]

$$\alpha(E) = \frac{16\pi e^2\hbar^3\Omega\,|P_{cv}|^2}{n(E)cm^2}\,\frac{[E-(E_c-E_v)]^{2r+1}}{E(\Delta E)^{2r}}\,\frac{[\Gamma(r+1)]^2}{\Gamma(2r+2)} \qquad (4.11)$$

where $[\Gamma(r+1)]^2/\Gamma(2^r+2)$ is a solution of the integral in Eq. (4.9).

Assuming that the DOS distribution is parabolic (i.e., r=1/2) as in crystal, we obtain (Tauc expression)

$$E\,\alpha(E)\,n(E) = \frac{2\pi^2 e^2\hbar^3\Omega\,|P_{cv}|^2}{cm^2\Delta E}(E-E_o)^2 \qquad (4.12)$$

with $E_o \equiv E_c - E_v$.

Davis and Mott [7] argued that the parabolic-band (r=1/2) assumption is improbable, although for states beyond ($E_c+\Delta E_c$) and ($E_v+\Delta E_v$) it may be valid. They assumed that $N(E) \propto E$ at band edges. If we put r=1 in Eq. (4.7), we obtain the Davis–Mott expression [5]:

$$E\,\alpha(E)\,n(E) = \frac{8\pi e^2\hbar^3\Omega\,|P_{cv}|^2}{3cm^2(\Delta E)^2}(E-E_o)^3 \qquad (4.13)$$

In Fig. 4.3, $(E\alpha n)^{1/3}$ and $(E\alpha n)^{1/2}$ are plotted vs. E for a-Si. The experimental data are taken from Aspnes *et al.* [8]. These data were obtained from SE on a low-pressure chemical-vapor-deposited sample on a thermally oxidized c-Si wafer. It is understood from this figure that the cubic law [Eq. (4.13)] offers better fit than the quadratic law [Eq. (4.12)] for a wide photon-energy range beyond the absorption edge. These plots give the optical energy gaps of E_o=1.35 eV [$(E\alpha n)^{1/3}$] and 1.8 eV [$(E\alpha n)^{1/2}$].

Klazes *et al.* [9] found that the cubic law provides better fit to their experimental a-Si data than the quadratic law, from which they deduced optical band gap E_c (1.1–1.8 eV) using a linear extrapolation of $(E\alpha n)^{1/3}$ as a function of photon energy E. Khawaja and Hogarth [10] suggested that the cubic law shows relatively good agreement with the experimental data for binary vanadate glasses, a-Si, a-Ge, a-Ta_2O_5, and a-As_2S_3, whereas the square law holds for chalcogenide glasses.

It is noted that the extracted E_o value depends strongly on the theoretical model used (Ref. [5], also see Fig. 4.3). Unfortunately, there is no well-established physical model owing to the complexity of the electronic states in disordered system. However, the fact does not change the conclusion that the position and shape of the optical absorption edge are sensitive and quantitative measures of the structural and electrical disorder of amorphous materials [11]. Ibrahim and Al-Ani

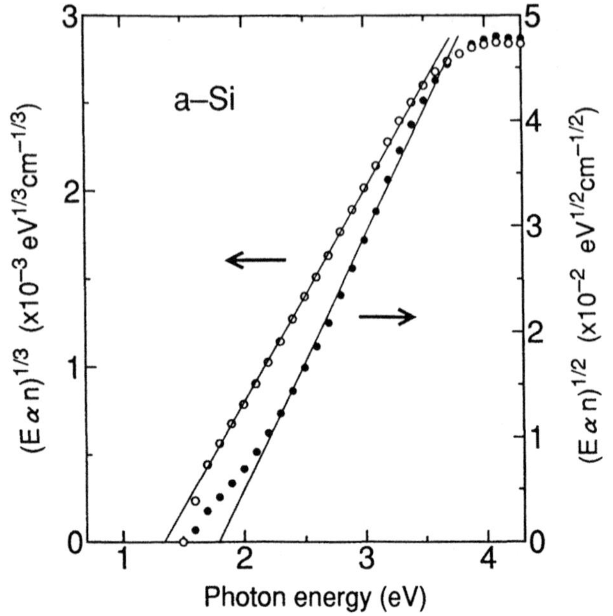

Figure 4.3 Plots of $(E\alpha n)^{1/3}$ and $(E\alpha n)^{1/2}$ vs. E for *a*-Si. The experimental data are taken from Aspnes *et al.* [8].

[5] reviewed various models with regard to their theoretical foundations, assumptions, mathematical derivations, and results.

(b) Temperature and Pressure Effects

1. Temperature effect—Cody *et al.* [12] reported experimental data on the temperature variation of E_0 for *a*-Si:H. It is well known that the introduction of atomic hydrogen into tetrahedrally-bonded amorphous semiconductors is effective in reducing the density of midgap states by several orders of magnitude [13]. They obtained E_0 from a plot of $(\alpha/E)^{1/2}$ against E for $\alpha \gg 10^4$ cm^{-1} [i.e., $(\alpha/E)^{1/2} \propto (E-E_0)$]. Note that this expression is not commonly used. They found that the distinction of between $(\alpha/E)^{1/2}$ and $(\alpha E)^{1/2}$ as a definition of E_0 is important only when films of widely differing thicknesses are compared and that these two formulas give identical results if one considers only changes in E_0 for the same film [11].

Figure 4.4 reproduces the experimental results of Cody *et al.* [12]. In the figure, the red shift of E_0 [$\Delta E_0(T)=E_0(0)-E_0(T)$] is plotted as a function of T. It is noted that the variation of $E_0(T)$ for *a*-Si:H is very similar to those for *c*-Si [14] (see also Fig. 3.12 for *c*-GaAs).

As mentioned in Sec. 3.2.1, the temperature dependence of the band-gap energy E_g in crystalline semiconductors can be given by Eq. (3.40). The first and second

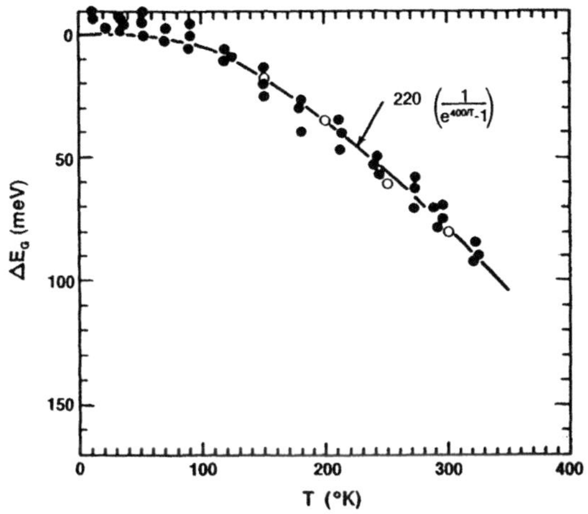

Figure 4.4 Red shift of E_o [$\Delta E_o(T)=E_o(0)-E_o(T)$] as a function of T for *a*-Si:H. (From Cody *et al.* [12].)

terms on the right-hand side of this expression are the explicit and implicit contributions, respectively. The temperature variation of E_g in crystalline materials can also be explained by the empirical Varshni equation [Eq. (3.42)]. The effect of thermal expansion is a trivial increase of the lattice constant with increasing T, thereby, shrinking the band-gap energies. Beside the changes of band structure induced by thermal expansion of the lattice, the temperature dependence of E_g is mainly due to the electron–phonon interactions [Debye–Waller term (two-phonon process) and Fan (self-energy) term (one-phonon process)] [15,16]. We can, thus, propose a temperature dependence of E_g with an equation containing the Bose–Einstein occupation factor for phonons:

$$E_g(T) = E_B - a_B\left(1 + \frac{2}{e^{\theta/T} - 1}\right) \qquad (4.14)$$

where the parameter θ describes the mean frequency of the phonons involved and a_B is the strength of the electron–phonon interaction. From this equation, we obtain

$$\Delta E_g = E_g(0) - E_g(T) = \frac{2a_B}{e^{\theta/T} - 1} \qquad (4.15)$$

The solid line in Fig. 4.4 shows the calculated result of this equation with $2a_B$=2.20×10^{-1} eV and θ=400 K. Cody *et al.* [12] suggested that the disorder, rather than hydrogen content, is the fundamental determining factor in the optical band gap of *a*-Si:H. Lautenschlager *et al.* [14] also obtained good agreement with Eq.

(4.14) for *c*-Si.

We can understand from numerical calculation that the Varshni equation and Eq. (4.14) give essentially identical results. The $E_0(T)$ data taken for *e*-beam-evaporated *a*-Si can thus be explained by Eq. (4.14) and also by the Varshni equation [17].

Differentiating Eq. (3.42) with respect to T, we obtain

$$\frac{dE_g(T)}{dT} = -\alpha\left[\frac{2T}{T+\beta} - \frac{T^2}{(T+\beta)^2}\right] \tag{4.16}$$

Thus, if $T \gg \beta$, this equation becomes

$$\frac{dE_g(T)}{dT} \sim -2\alpha \tag{4.17}$$

Equation (4.17) gives the linear temperature coefficient of the band-gap energy E_g. While we are aware that results for *a*-Si are in general very sample dependent (i.e., depending on the preparation technique, the sample thickness, etc.), we expect that the qualitative temperature behavior of E_0 for amorphous semiconductors should have some general validity. We list in Table 4.1 the linear temperature coefficient dE_0/dT obtained for some tetrahedrally-coordinated amorphous semiconductors.

2. *Pressure effect*—High-pressure studies have significantly contributed to our understanding of the properties of solids. The primary effect of pressure is to reduce the interatomic distance. As a consequence, the compression gives rise to a change in electronic properties and a phase transition from semiconductor to metal.

TABLE 4.1 *Linear temperature and pressure coefficients of the optical band-gap energy E_0 for some tetrahedrally-coordinated amorphous semiconductors. The lowest-gap crystalline values are also given in parentheses.*

	a-Si	*a*-Ge	*a*-GaP	*a*-GaAs
dE_0/dT (10^{-4} eV/K)	-4.8[a]	-4.6[c]	-3.1[e]	-3.8[e]
	(-2.8)[f]	(-3.7)[g]	(-6.2)[f]	(-3.95)[h]
dE_0/dp (10^{-3} eV/GPa)	-10[b]	8–15[d]		
	(-15)[f]	(50)[g]	(-11)[f]	(115)[h]

[a]G. D. Cody *et al.*, *Phys. Rev. Lett.* **47**, 1480 (1981).
[b]S. Minomura, *J. Phys. (Paris) Colloq.* **42**, C4-181 (1981).
[c]P. Persans *et al.*, *Bull. Am. Phys. Soc.* **28**, 532 (1983).
[d]K. Tanaka, *Phys. Rev. B* **43**, 4302 (1991).
[e]N. Connell, *Phys. Status Solidi B* **53**, 213 (1972).
[f]E_g^X ($\Gamma^v \to X^c$)
[g]E_g^L ($\Gamma^v \to L^c$)
[h]E_0 ($\Gamma^v \to \Gamma^c$)

Note that this type of phase transition occurs both in crystalline and amorphous semiconductors [18].

In crystalline semiconductors, the pressure coefficients of CP energies are strongly dependent upon the identity of the symmetry points, such as, Γ, X, and L, in the BZ (see, *e.g.*, Ref. [19]). By comparing the pressure coefficients of the optical band gap E_0 for amorphous materials with those for crystalline ones, thus, it is possible to obtain information about the effect of structural disorder on the electronic properties of the amorphous materials, as we will discuss below.

There have been reported several experimental data on the hydrostatic-pressure dependence of the optical band gap E_0 in the tetrahedrally-bonded amorphous semiconductors (*a*-Si [20,21], *a*-Ge [21,22], and *a*-SiGe [23]). The Γ point of the conduction-band minima in the tetrahedral crystalline semiconductors occurs at the center of the BZ and has pressure coefficients (with respect to the Γ-valence maximum) which are always positive and of a magnitude $dE_0/dp \sim 10^{-2}$–10^{-1} eV/GPa. On the other hand, X points have negative pressure coefficients about an order of magnitude smaller than the Γ point [20]. Finally, L points have positive pressure coefficients about half as small as those at Γ. This simple rule holds without serious exception (InSb) in all tetrahedrally-bonded crystalline semiconductors.

The fundamental absorption edge of *c*-Si at atmospheric pressure corresponds to indirect transitions from the highest valence band at the Γ point to the lowest conduction band near X (E_g^X, $\Gamma^v \rightarrow X^c$). The pressure coefficient of the lowest indirect gap, dE_g^X/dp, is thus expected to be negative (Table 4.1). *c*-Ge is also an indirect-gap semiconductor, but its lowest indirect gap E_g^L ($\Gamma^v \rightarrow L^c$) has positive pressure coefficient.

In amorphous semiconductors, a short-range order is maintained to considerably extent, but a long-range order vanished completely. Hydrostatic compression induces only the decrease in the covalent bond length of crystalline materials which increases the band-gap energy. By contrast, amorphous tetrahedral network may suffer not only the bond compression but also other structural modifications, such as the bond-angle fluctuation. This is because the environment around a tetrahedral unit is not symmetric in the amorphous network [22]. However, it is expected that the short-range order in amorphous materials does not differ so largely from that in crystalline materials. Then, pressure variation of the short-range order in an amorphous material may be nearly the same as in the crystalline state. If the pressure dependence of the band-gap energy is dominated mainly by the effect of short-rang order, the amorphous and crystalline states will show similar pressure dependence. In fact, the pressure coefficient of the optical band gap dE_0/dp for *a*-Si has negative value while that for *a*-Ge has positive one, as in the case of the crystalline states (see Table 4.1). An estimation of the sign of dE_0/dp is possible by using this simple rule. We can expect, for example, negative sign for *a*-GaP and positive one for *a*-GaAs. To our knowledge, however, no experimental data has been reported so far for these amorphous semiconductors.

4.1.2 Optical Dispersion Model

(a) Density-Of-States Model—The valence and conduction bands are known to retain their meaning even in amorphous state. Assuming that the basic volume B_0 of the specimen contains the same number of atoms in the amorphous as in the crystalline state, we can express the imaginary part of the complex dielectric function as [24]:

$$\varepsilon_2(E) = \left(\frac{2\pi e\hbar}{mE}\right)\frac{2}{(2\pi)^3}\frac{1}{B_0}\,|P_{cv}(E)|^2\int dE'\,N_v(E')\,N_c(E'+E) \qquad (4.18)$$

Equation (4.18) ensures that, if detailed information on the $|P_{cv}(E)|^2$, $N_v(E)$, and $N_c(E)$ are known, $\varepsilon_2(E)$ can be calculated numerically or analytically. The valence-band states in unhydrogenated a-Si have been studied extensively with photoemission techniques [2]. The results show a peak about 2 eV below the top of valence band with a full width at half maximum of about 3 eV. Based on the photoemission results, Feng *et al.* [25] constructed the valence-band DOS distribution of a-Si as an asymmetric Gaussian:

$$N_v(E) = \begin{cases} A_v \exp[-(E-E_{vm})^2/2\sigma_1^2] & \text{if } E<E_{vm} \\ A_v \exp[-(E-E_{vm})^2/2\sigma_2^2] & \text{if } E>E_{vm} \end{cases} \qquad (4.19)$$

Yield spectroscopy measurements [2] revealed that the conduction-band DOS of a-Si remains flat beyond a sharply rising onset above E_c, the conduction-band edge. To simulate such a distribution, Feng *et al.* [25] used a steplike function for a-Si conduction band, formulated as

$$N_c = A_c(E - E_c)^{0.2} \qquad (4.20)$$

Feng *et al.* [25] also analyzed $\varepsilon_2(E)$ spectrum of a-Si:H with and without the contribution from the Si–H antibonding states. The valence-band DOS distribution of a-Si:H is still assumed to be determined from fits to the $\varepsilon_2(E)$ spectra of a-Si:H samples. The Si–Si bonds have the same conduction-band distribution as in a-Si. The distribution of the antibonding states is assumed to be given by a Gaussian:

$$A_{ab} \exp[-(E-E_{ab})^2/2\sigma_{ab}^2] \qquad (4.21)$$

The last parameter we need in order to calculate $\varepsilon_2(E)$ is the squared-matrix element $|P_{cv}(E)|^2$. A detailed knowledge of this quantity for a-Si and a-Si:H is not possible in the literature. Because of this, a smooth extrapolation of c-Si matrix element calculated by Kramer *et al.* [26] is used [25].

Having established the formulas for the DOS distribution and matrix element, we derive the DOS for *a*-Si and *a*-Si:H. We calculate the $\varepsilon_2(E)$ spectrum with an initial set of DOS parameters, and then perform a least-squares fit to experimental spectrum. The final DOS parameters can be determined from the best fit. Figure 4.5 shows, as an example, the fitted result of $\varepsilon_2(E)$ spectrum for *a*-Si:H [25]. The experimental $\varepsilon_2(E)$ spectrum was taken for *a*-Si:H (H content=22 at.%) grown by dc magnetron sputtering deposition, using *in situ* SE. The DOS parameters determined from the fit were as follows: E_{vm}=-2.06 eV, σ_1=0.64 eV, σ_2=0.51 eV, A_{ab}=1.28, E_{ab}=3.2 eV, σ_{ab}=0.79 eV, E_c=1.5 eV, and A_v/A_c=3.4.

As seen in Fig. 4.5, the DOS model shows excellent agreement with the experimental ε_2 data over the entire range of photon energies. It should be noted, however, that an expression for $\varepsilon_1(E)$ cannot be obtained from this model in analytic form. Numerical calculation of the KK transformation is seemed to be the only way to evaluate $\varepsilon_1(E)$ spectrum.

(b) Plasma-Frequency (Campi–Coriasso) Model—The model proposed by Campi and Coriasso [27] provides an analytic expression for the imaginary part of the dielectric function $\varepsilon_2(E)$ of tetrahedrally-bonded amorphous semiconductors. The model obtained by them is as follows:

$$\varepsilon_2(E) = \frac{E_p^2}{E} \frac{\Gamma(E-E_0)^2}{[E_{0c}^2-(E-E_0)^2]^2 + \Gamma^2(E-E_0)^2} H(E-E_0) \qquad (4.22)$$

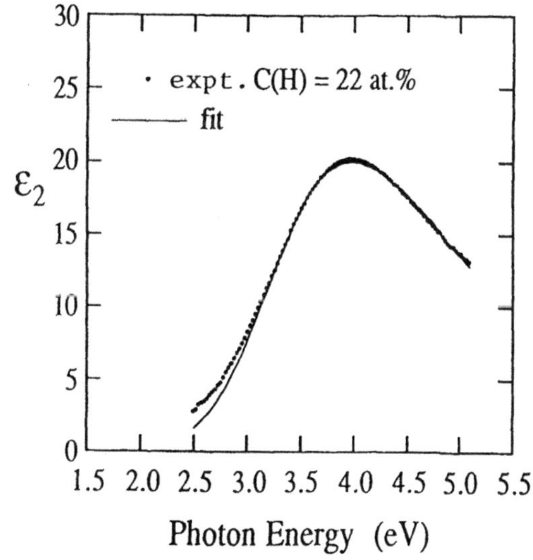

Figure 4.5 $\varepsilon_2(E)$ spectrum for *a*-Si. The experimental $\varepsilon_2(E)$ spectrum was taken for *a*-Si:H (H content=22 at.%) grown by dc magnetron sputtering deposition, using ***in situ*** SE at room temperature. (From Feng *et al.* [25].)

where the parameters E_p and E_o stand for the plasma energy (frequency) and optical band-gap energy of the material, respectively, and $H(z)$ denotes the Heaviside step function [see Eq. (3.15)]. The other two parameters Γ and E_{oc} are directly related to the electronic DOS and especially to the energy distance between the bonding and antibonding states and the valence-band width. Moreover, Γ determines the broadening of $\varepsilon_2(E)$ peak, as well as the energy separation between the maximum and minimum of the $\varepsilon_1(E)$ spectrum, whereas the $\varepsilon_2(E)$ peak is located at around $E\sim(E_{oc}+E_o)$ (see Fig. 4.6).

Equation (4.22) is a phenomenological expression substantiated by a proper definition of the parameters Γ and E_{oc}. As suggested in Ref. [27], in the limiting case $E_o=0$, this reduces to the familiar expression for the imaginary part of the dielectric function in the context of the Lorentz model. Furthermore, the Drude model can be recovered simply by setting $E_{oc}=E_o=0$.

Equation (4.22) represents the simplest available $\varepsilon_2(E)$ model that has all the physical elements of a theory of dielectric response for materials exhibiting a band gap. The exact analytical evaluation of the real part of the dielectric function $\varepsilon_1(E)$ was obtained by Campi and Coriasso [27] using the KK transformation [see Eq. (1.16)]. Nevertheless, the resulting expression was so huge and cumbersome as to render it computationally impractical. From this fact, they concluded that numerical evaluation is a more appropriate way to proceed.

We reproduce in Fig. 4.6 the calculated $\varepsilon(E)$ spectra of *a*-Si compared to literature data [27]. The $\varepsilon_1(E)$ spectrum was obtained by the KK transformation of the model [Eq. (4.22)]. The dashed lines represent the experimental data [28]. As can be seen in the figure, the agreement between the model and experiment is impressive, especially at photon energies higher than ~5 eV.

The inset of Fig. 4.6 shows the valence- and conduction-band DOS of *a*-Si. The splitting S was evaluated as the energy separation between the center of mass of the conduction band and that of the valence band. The width W was also evaluated as the total valence-band width. In the model, these quantities were used to predict values for E_{oc} and Γ in the following manner:

$$E_{oc}+E_o \sim 0.60S \qquad (4.23a)$$

$$\Gamma \sim 0.52W \qquad (4.23b)$$

(c) Penn-Gap (Forouhi–Bloomer) Model—Forouhi and Bloomer [29] obtained the dispersion relations for $n(E)$ and $k(E)$ which depend on only five parameters. These relations are consistent with KK analysis. Assuming that the valence and conduction bands are parabolic, and using the KK relationships [Eq. (1.20)], these authors obtained expressions for $n(E)$ and $k(E)$:

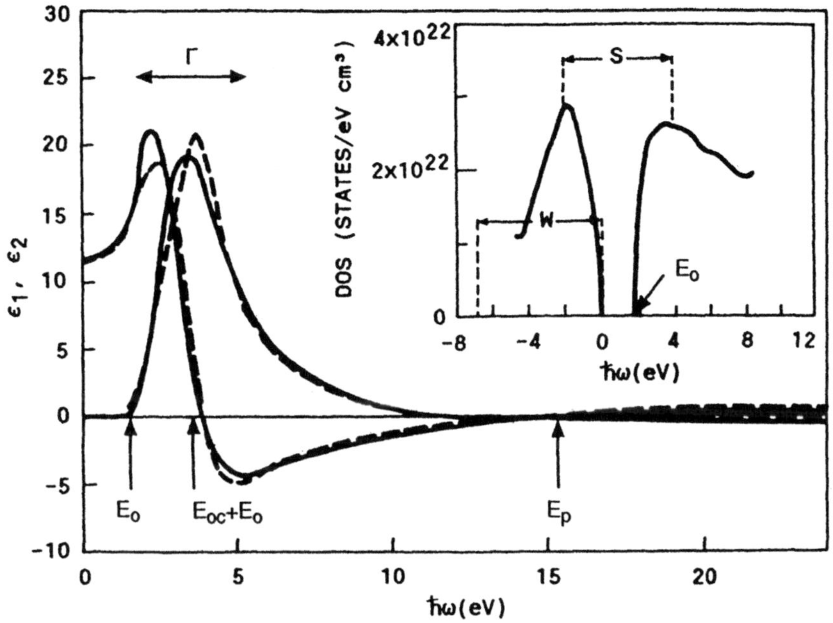

Figure 4.6 $\varepsilon_1(E)$ and $\varepsilon_2(E)$ spectra for *a*-Si calculated using the plasma-frequency model. The $\varepsilon_1(E)$ spectrum was obtained by the KK transformation of the model [Eq. (4.22)]. The dashed lines represent the experimental data. The positions of E_o, E_p, and $E_{oc}+E_o$ are indicated by the vertical arrows. An energy interval whose extension is Γ is also indicated. The inset shows the valence- and conduction-band DOS of *a*-Si. (From Campi and Coriasso [27].)

$$n(E) = n(\infty) + \frac{B_0 E + C_0}{E^2 - BE + C} \tag{4.24a}$$

$$k(E) = \frac{A(E-E_o)^2}{E^2 - BE + C} \tag{4.24b}$$

where E_o is the optical band-gap energy and $n(\infty)$ is a constant usually greater than unity. $n(E)$ and $k(E)$ given by Eq. (4.24) can be fully and simultaneously determined if only four of the six parameters E_o, A, B, C, B_0, C_0, together with $n(\infty)$ are specified. It is noted here that B_0 and C_0 are not independent parameters according to the equations

$$B_0 = \frac{A}{Q}\left[-\frac{B^2}{2} + E_o B - E_o^2 + C\right] \tag{4.25a}$$

$$C_0 = \frac{A}{Q}\left[(E_0^2+C)\frac{B}{2}-2E_0C\right] \tag{4.25b}$$

with Q=1/2$(4C-B^2)^{1/2}$.

We can rewrite Eq. (4.24) in the following forms [30,31]:

$$n(E) = n(\infty) + \frac{B_0E+C_0}{(E-E_{pg})^2+\Gamma_0^2} \tag{4.26a}$$

$$k(E) = \frac{A(E-E_0)^2}{(E-E_{pg})^2+\Gamma_0^2} \tag{4.26b}$$

This expression includes the energy E_{pg}, where the absorption coefficient becomes maximum, which is proportional to the mean distance between the valence and conduction bands and is often called *"Penn gap."* The parameter Γ_0 then represents the broadening of the Penn-gap absorption peak.

Figure 4.7 shows the numerically calculated $n(E)$ and $k(E)$ spectra for *a*-Si, compared with the experimental data. The theoretical curves are obtained from Eq. (4.26). The numerical parameters used in this calculation are as follows: E_0=1.0 eV, E_{pg}=3.5 eV, Γ=1.5 eV, A=1.03, B_0=-3.34, C_0=19.97, and $n(\infty)$=1.20. The experimental data are taken from Ref. [8]. We can see in the figure that the Penn-gap model shows excellent agreement with the experimental data, except in the lower photon-energy region of $n(E)$.

The real [$\varepsilon_1(E)$] and imaginary [$\varepsilon_2(E)$] parts of the complex dielectric function can be easily calculated from $n(E)$ and $k(E)$ using the relations of Eq. (1.18). We show in Fig. 4.8 the results of this calculation. As in the case for $n(E)$ and $k(E)$ (Fig. 4.7), the model can successfully explain the peculiar experimental $\varepsilon_1(E)$ and $\varepsilon_2(E)$ spectra over a wide range of the photon energy.

(d) Model Dielectric Function—The MDF to be described here is an extension of Tauc, Grigorovici, and Vancu [24,32]. The $\varepsilon_2(E)$ spectrum is assumed to yield a continuous absorption obeying the power law of $1/E^2(E-E_0)^2$ and have a steep high-energy end at the high-energy cutoff E_{hc}. The corresponding $\varepsilon_1(E)$ spectrum shows clear structures both at the E_0 and at the E_{hc} edges. Introducing the damping (broadening) effect into the model, the optical spectra become structureless which are typically observed in amorphous semiconductors.

The optical transitions in amorphous semiconductors are described, to a first approximation, by the nondirect transition model in which conservation of energy but not wave vector is significant. Assuming that the conduction and valence bands are parabolic, Tauc *et al.* [24,32] obtained a linear variation of $E\varepsilon_2(E)$ with E:

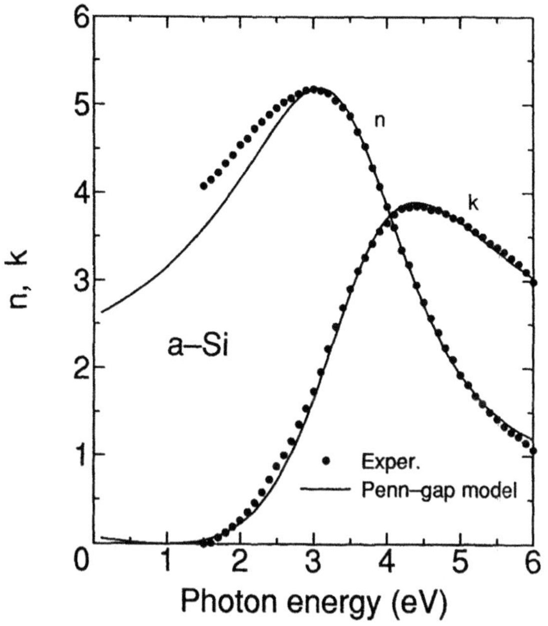

Figure 4.7 Numerically calculated $n(E)$ and $k(E)$ spectra for a-Si, compared with the experimental data. The theoretical curves are obtained from Eq. (4.26). The numerical parameters used in this calculation are as follows: E_0=1.0 eV, E_{pg}=3.5 eV, Γ=1.5 eV, A=1.03, B_0=-3.34, C_0=19.97, and $n(\infty)$=1.20. The experimental data are taken from Ref. [8].

$$E\,\varepsilon_2(E)^{1/2} = D_a^{1/2}(E-E_0)\,H(E-E_0) \qquad (4.27)$$

where D_a is a nondimensional strength parameter and $H(z)$ is the Heaviside function defined by Eq. (3.15).

It should be noted that the expression (4.27) is almost identical to that of the indirect band-gap transitions in crystalline semiconductors (see Sec. 3.1.1). Therefore, we get

$$\varepsilon_2(E) = \frac{D_a}{E^2}(E-E_0)^2\,H(E-E_0)\,H(E_{hc}-E) \qquad (4.28)$$

Assuming that the strength parameter D_a is independent of the photon energy, the KK transformation of Eq. (4.28) gives [33,34]

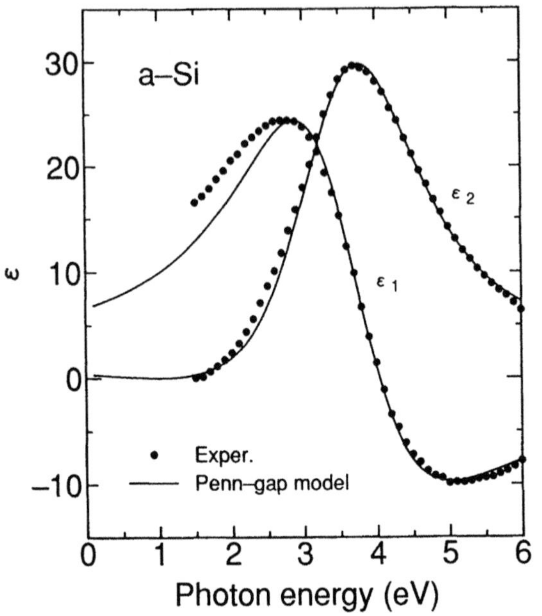

Figure 4.8 Numerically calculated $\varepsilon_1(E)$ and $\varepsilon_2(E)$ spectra for *a*-Si, compared with the experimental data. The theoretical curves are obtained from Eqs. (4.26) and (1.18). The numerical parameters are the same as those used in Fig. 4.7. The experimental data are taken from Ref. [8].

$$\varepsilon(E) = 1 + \frac{2D_a}{\pi}\Big[-\frac{E_o^2}{(E+i\Gamma)^2}\ln\Big(\frac{E_{hc}}{E_o}\Big) + \frac{1}{2}\Big(1+\frac{E_o}{E+i\Gamma}\Big)^2\ln\frac{E+i\Gamma+E_{hc}}{E+i\Gamma+E_o}$$

$$+ \frac{1}{2}\Big(1-\frac{E_o}{E+i\Gamma}\Big)^2\ln\frac{E+i\Gamma-E_{hc}}{E+i\Gamma-E_o}\Big] \qquad (4.29)$$

In Eq. (4.29), the damping effect (Γ) is taken into consideration in a phenomenological manner ($E \rightarrow E+i\Gamma$). Figure 4.9 shows the numerically calculated $\varepsilon(E)$ spectra [Eq. (4.29)] with three different damping values Γ=0, 0.4, and 0.8 eV (E_o=1.5 eV, E_{hc}=4.0 eV, and D_a=50). The $\varepsilon_1(E)$ [Re $\varepsilon(E)$] spectrum with Γ=0 eV exhibits a divergence at the high-energy cutoff E_{hc}. Such a divergence can be successfully taken off by introducing the damping effect in the model. The damping effect smears out the E_o and E_{hc} structures and, as a result, the optical spectra become structureless which are typically observed in amorphous semiconductors.

Figure 4.10 shows the effect of E_{hc} on the spectral feature of $\varepsilon(E)$. As evident from this figure, increasing E_{hc} increases value for $\varepsilon_1(0)$ [$=\varepsilon_1(E\rightarrow 0$ eV)], the zero-frequency (long-wavelength) dielectric constant.

If we consider the Davis–Mott cubic law [Eq. (4.13)] in the model, we obtain

$$\varepsilon_2(E) = \frac{D_a}{E^2}(E-E_0)^3 H(E-E_o) H(E_{hc}-E) \qquad (4.30)$$

The KK transformation of Eq. (4.30) finally gives an expression for both $\varepsilon_1(E)$ [Re $\varepsilon(E)$] and $\varepsilon_2(E)$ [Im $\varepsilon(E)$]:

$$\varepsilon(E) = 1 + \frac{2D_a}{\pi}\Big[(E_{hc}-E_o) + \frac{E_o^3}{(E+i\Gamma)^2}\ln(\frac{E_{hc}}{E_o}) - \frac{E+i\Gamma}{2}(1+\frac{E_o}{E+i\Gamma})^3 \ln\frac{E+i\Gamma+E_{hc}}{E+i\Gamma+E_o}$$
$$+ \frac{E+i\Gamma}{2}(1-\frac{E_o}{E+i\Gamma})^3 \ln\frac{E+i\Gamma-E_{hc}}{E+i\Gamma-E_o}\Big] \qquad (4.31)$$

Figure 4.11 shows the MDF $\varepsilon(E)$ curves fitted to the experimental data of *a*-Si. The solid and dashed lines are obtained from Eqs. (4.29) and (4.31), respectively, with Γ=0 eV. The band-gap and strength parameters used in these calculations are

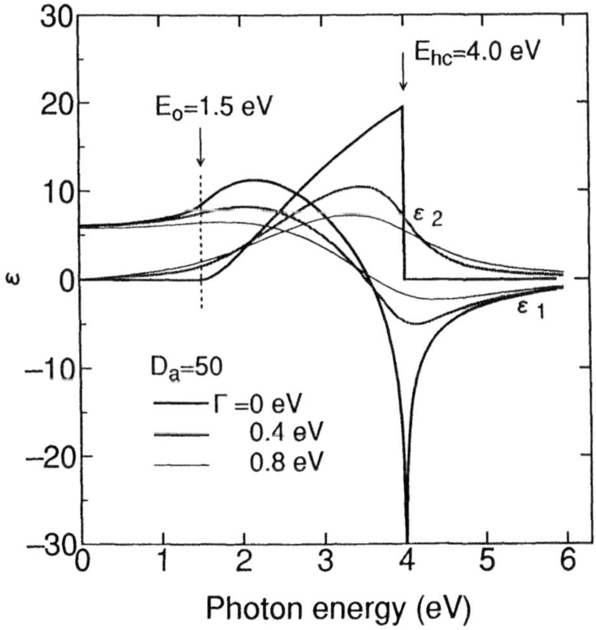

Figure 4.9 Numerically calculated $\varepsilon(E)$ spectra [Eq. (4.29)] with three different damping values Γ=0, 0.4, and 0.8 eV (E_o=1.5 eV, E_{hc}=4.0 eV, and D_a=50).

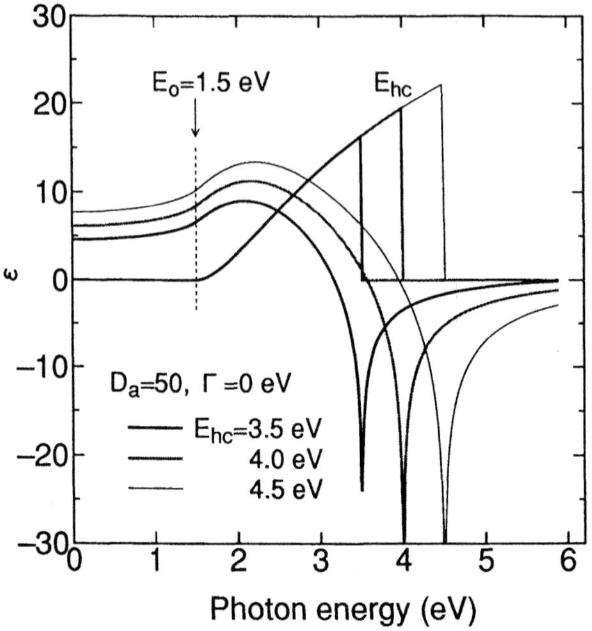

Figure 4.10 Numerically calculated $\varepsilon(E)$ spectra [Eq. (4.29)] with three different high-energy cutoff values E_{hc}=3.5, 4.0, and 4.5 eV (E_0=1.5 eV, Γ=0 eV, and D_a=50).

as follows: E_0=1.8 eV, E_{hc}=4.8 eV, and D_a=117 [Eq. (4.29)]; E_0=1.35 eV, E_{hc}=4.5 eV, and D_a=37 [Eq. (4.31)]. The experimental data are taken from Ref. [8].

The MDF curves shown in Fig. 4.11 provide a hard cutoff in the $\varepsilon_2(E)$ spectrum at the E_{hc} edge. The curves also exhibit a divergence in the $\varepsilon_1(E)$ spectrum at this edge. However, they show good agreement with the experimental $\varepsilon_1(E)$ and $\varepsilon_2(E)$ data at low photon energies. The resultant zero-frequency dielectric constant $\varepsilon_1(0)$ is estimated to be ~12. Note that this value is in good agreement with that of *c*-Si [$\varepsilon_1(0)$=11.6, see Fig. 4.11 (solid circle)]. It is also noted that in the limit $\Gamma \to 0$ eV the calculated $\varepsilon_2(E)$ spectrum of Eq. (4.29) [Eq. (4.31)] exactly agrees with that of Eq. (4.28) [Eq. (4.30)].

As mentioned before, we can remove the disagreeable features at the E_{hc} edge region seen in Fig. 4.11 by taking account of the damping effect in the model. Comparison of this calculated result [Eq. (4.29)] to the experimental data [8] is shown in Fig. 4.12 by the solid lines [33]. The experimental data give a maximum value of ε_2~30 near 3.7 eV. The dielectric function exhibiting the biggest ε_2 value is most representative of a bulk sample of completely coordinated amorphous material. The value of ε_2~30 has been shown to be an apparently limiting peak value for *a*-Si, i.e., its spectrum is thought to be the "*intrinsic*" dielectric properties of dense *a*-Si.

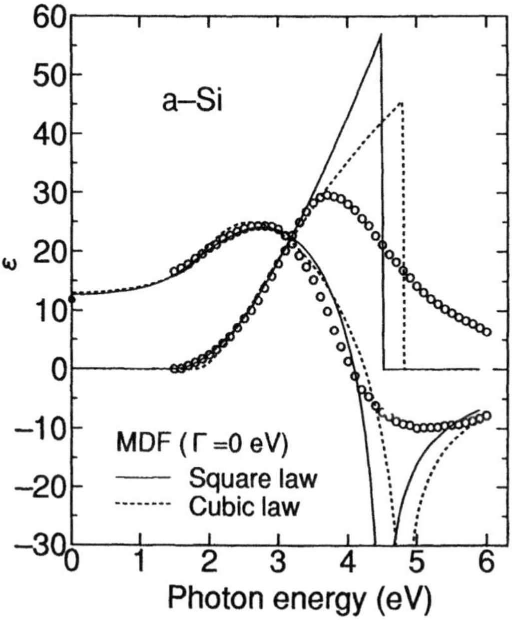

Figure 4.11 MDF $\varepsilon(E)$ spectrum fitted to the experimental data of *a*-Si. The solid and dashed lines are obtained from Eqs. (4.29) and (4.31), respectively, with Γ=0 eV. The band-gap and strength parameters used here are as follows: E_0=1.8 eV, E_{hc}=4.6 eV, D_a=117 [Eq. (4.29)]; E_0=1.35 eV, E_{hc}=4.5 eV, D_a=37 [Eq. (4.31)]. The experimental data are taken from Ref. [8]. The solid circle corresponds to the value $\varepsilon_1(0)$=11.6 for *c*-Si.

As shown in Fig. 4.9, the damping effect can greatly change the spectral features of $\varepsilon(E)$. To achieve a better fit, therefore, we must use different D_a from that used for the dashed line (Γ=0 eV). The solid lines in Fig. 4.12 represent the calculated results with Γ=0.6 eV and D_a=172.

The fits with our MDF model to the experimental $\varepsilon(E)$ of *a*-Ge are shown in Fig. 4.13 [33]. The solid lines are calculated from Eq. (4.29) with Γ=0.9 eV (E_0=0.85 eV, E_{hc}=4.1 eV, and D_a=81). The experimental data of Connell *et al.* [35] on sputter-deposited *a*-Ge for nearly ideal, void-free material are quoted here (solid circles). These data yield a peak value of ε_2~23 at ~2.9 eV.

The plots of $E\varepsilon_2(E)^{1/2}$ vs. E give the optical gap energy E_0=0.85 eV and strength parameter D_a=50 for this material (*a*-Ge). The zero-frequency $\varepsilon_1(0)$ for *c*-Ge is ~16.0. This value is shown in Fig. 4.13 by the solid circle. The cutoff energy E_{hc}=4.1 eV provides the value ε_1~16 when $E \rightarrow 0$ eV. The dashed lines in Fig. 4.13 are calculated with these numerical values.

Figure 4.14 shows a comparison of the MDF model [Eq. (4.29)] to the experimental $\varepsilon_1(E)$ and $\varepsilon_2(E)$ spectra for *a*-GaAs [33]. The dashed and solid lines are calculated with Γ=0 eV (E_0=0.8 eV, E_{hc}=5.1 eV, and D_a=29) and 0.9 eV (E_0=0.8

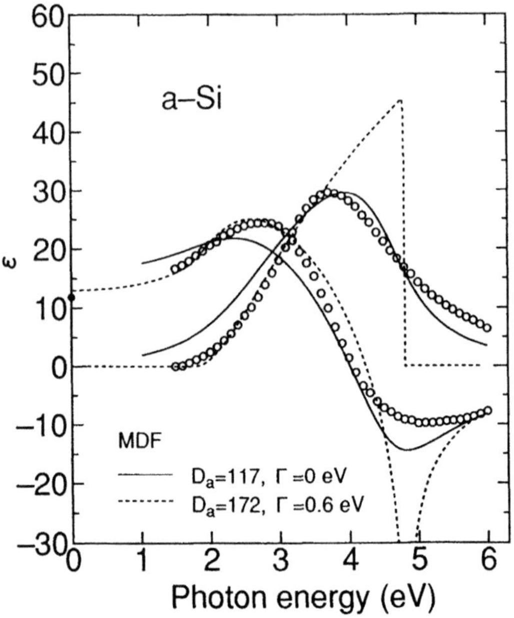

Figure 4.12 $\varepsilon(E)$ spectrum of *a*-Si. The experimental data are taken from Ref. [8]. The solid lines are obtained from Eq. (4.29) with the following numerical values: E_o=1.8 eV, E_{hc}=4.6 eV, Γ=0.6 eV, and D_a=172. The solid circle corresponds to the value $\varepsilon_1(0)$=11.6 for *c*-Si. The dashed lines are obtained from Eq. (4.29) with E_o=1.8 eV, E_{hc}=4.6 eV, Γ=0 eV, and D_a=117. (From Adachi [33].)

eV, E_{hc}=5.1 eV, and D_a=44), respectively. The experimental data are taken from Ref. [36]. These data correspond to implantation-amorphized GaAs produced by high-frequency (1×10^{14} cm^{-2}) high-energy (270 keV) As$^+$ bombardment.

As in the case of *a*-Si (Fig. 4.12) and *a*-Ge (Fig. 4.13), the MDF model with Γ=0 eV can explain the spectral dependence of ε in the low-energy region of *a*-GaAs. The model with Γ≠0 eV improves the fit at the E_{hc} region, resulting in reasonable agreement with the experimental data over the entire range of photon energies.

The fits with our MDF model to the experimental $n^*(E)=n(E)+ik(E)$ and $\alpha(E)$ for *a*-GaAs are shown in Figs. 4.15 and 4.16, respectively [34]. The experimental data are taken from Ref. [36]. Like $\varepsilon(E)$, these data show no clear structural feature. Reasonable agreement can be achieved between the MDF calculation and experimental data.

The model described here requires four parameters, namely, E_o, E_{hc}, Γ, and D_a. These parameters are commonly used as adjustable constants for the calculations of both $\varepsilon_1(E)$ and $\varepsilon_2(E)$, simultaneously. The optical gap E_o and strength parameter

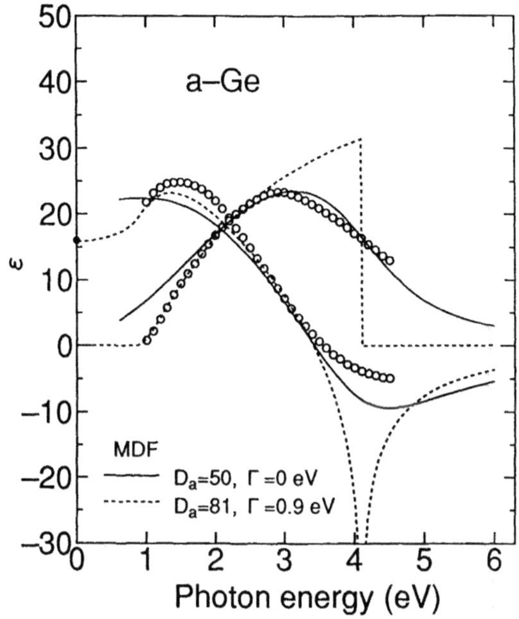

Figure 4.13 $\varepsilon(E)$ spectrum of *a*-Ge. The experimental data are taken from Ref. [8]. The solid lines are obtained from Eq. (4.29) with the following numerical values: E_o=0.85 eV, E_{hc}=4.1 eV, Γ=0.9 eV, and D_a=81. The dashed lines are obtained from Eq. (4.29) with E_o=0.85 eV, E_{hc}=4.1 eV, Γ=0 eV, and D_a=50. (From Adachi [33].)

D_a are determined from the Tauc plot of $E\varepsilon^2(E)^{1/2}$ [or $E^{2/3}\varepsilon^2(E)^{1/3}$] vs. E; this plot gives a straight line of slope $D_a^{1/2}$ [$D_a^{1/3}$] and intercept E_0 on the energy axis [see Eqs. (4.27) and (4.30)]. In principle, the parameter values D_a and E_{hc} can be determined from the sum rule of Eq. (1.30). The $\varepsilon(0)_{eff}$ with $E_m \rightarrow \infty$ [Eq. (1.30)] corresponds to $\varepsilon_1(E\rightarrow 0$ eV). We obtained that the amorphous materials reported here have nearly the same $\varepsilon(0)_{eff}$ values as those for the crystalline materials. This means that the optical properties discussed here indeed correspond to the "*intrinsic*" properties of *dense* materials. If an amorphous material contains considerable amount of void networks in the bulk medium (i.e., not *dense* sample), such sample may usually show smaller $\varepsilon_1(E\rightarrow 0$ eV) value [and, via Eq. (1.29), smaller n_{eff}] than the crystalline one.

The high-energy cutoff E_{hc} obtained here are 4.6 eV (*a*-Si), 4.1 eV (*a*-Ge), and 5.1 eV (*a*-GaAs). These energies are nearly the same as the E_2 CP energies in the crystalline substances [E_2=4.27 eV (*c*-Si), 4.35 eV (*c*-Ge), and 4.7 eV (*a*-GaAs)]. As mentioned in Chapter 3 (see, *e.g.*, Fig. 3.7), the E_2 peaks in the crystalline substances usually dominate in the optical spectra of most semiconductors. The $\varepsilon_2(E)$ spectra of As^+-implanted GaAs showed [36] that one of the main features

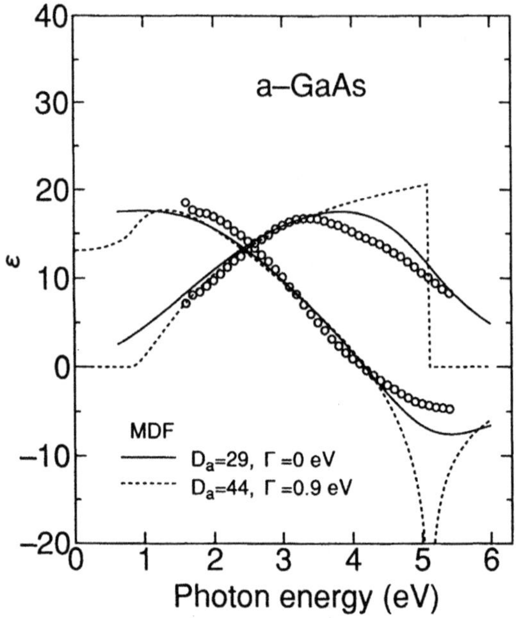

Figure 4.14 $\varepsilon(E)$ spectrum of *a*-GaAs. The experimental data are taken from Ref. [36]. The solid lines are obtained from Eq. (4.29) with the following numerical values: E_o=0.8 eV, E_{hc}=5.1 eV, Γ=0.9 eV, and D_a=44. The solid circle corresponds to the value $\varepsilon_1(0)$=16 for *c*-Ge. The dashed lines are obtained from Eq. (4.29) with E_o=0.8 eV, E_{hc}=5.1 eV, Γ=0 eV, and D_a=29. (From Adachi [33].)

induced by ion implantation is a gradual smearing out of the E_1 and E_2 CP structures by damage clusters. The ε_2 peak value fell drastically with increasing damage. The E_2-related peak finally disappeared for high ion doses.

In conclusion, we have presented and discussed various calculation models for the optical constants of amorphous semiconductors. It is found that the Penn-gap (Forouhi–Bloomer) model could account for the peculiar optical-constant spectra of amorphous materials, except in the lower photon-energy region. The MDF with Γ=0 eV, on the other hand, can successfully explain the experimental spectra at low photon energies. This advantageous feature should be used to estimate the zero-frequency $\varepsilon_1(E\rightarrow 0 \text{ eV})$ value $[=n(E\rightarrow 0 \text{ eV})^2]$ [27]. The damping (broadening) energies obtained both from the Penn-gap and MDF models seem to be considerably larger than those for the interband transitions in crystalline materials. We note that the damping effect considered here reflects the distribution of electronic states rather than being the consequence of its general concept (i.e., disorder broadening).

Since the MDF expression is purely analytic function of the electronic energy-

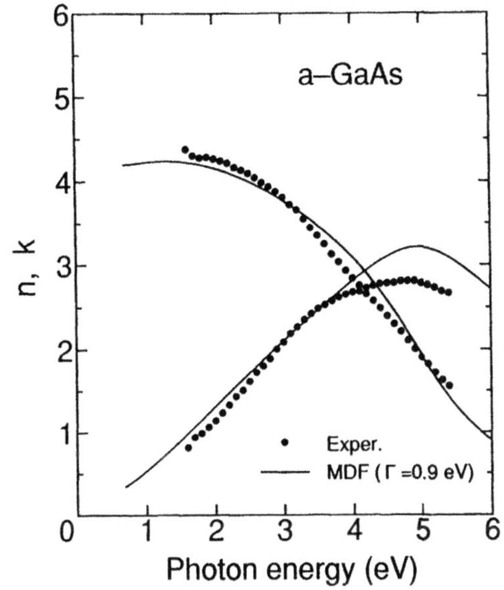

Figure 4.15 Numerically calculated $n^*(E)=n(E)+ik(E)$ for *a*–GaAs. The solid circles represent the experimental data taken from Ref. [36]. (From Adachi [34].)

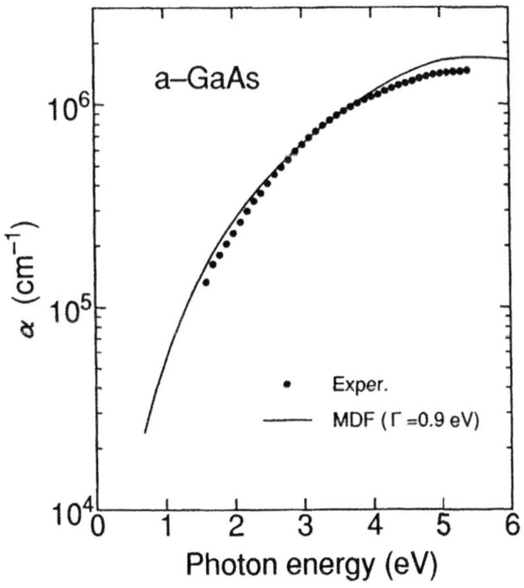

Figure 4.16 Numerically calculated $\alpha(E)$ for *a*–GaAs. The solid circles represent the experimental data taken from Ref. [36]. (From Adachi [34].)

band parameters, the model can be applicable to the analysis of some perturbation-induced effects of the optical constants [37]. In the MDF model, the strength parameter D_a is taken to be constant over energies of pair states between the conduction and valence bands. This assumption is usually made in the theory of optical properties of crystals. If we assume that D_a is a proper function of energy, the fit would be greatly improved. However, to check the validity of its energy dependence, we will require an exact electronic energy-band structure of the amorphous materials, which is an open question even now [28,38].

More recently, Jellison and Modine [39] presented a parameterization of the optical constants of amorphous tetrahedral semiconductors. In their model, the imaginary part of the dielectric function $\varepsilon_2(E)$ was determined by multiplying Tauc joint-DOS by the $\varepsilon_2(E)$ obtained from the Lorentz oscillator model. The real part of the dielectric function $\varepsilon_1(E)$ was calculated from $\varepsilon_2(E)$ using the KK relation. Analyses were presented for *a*-Si, *a*-SiO, *a*-As_2S_3, and *a*-Si_3N_4, and results were excellent agreement with the experimental data over the entire range of photon energies (0.8–6.0 eV).

4.1.3 Temperature, Pressure, and Doping Effects on Optical Constants

The temperature dependence of optical properties for crystalline semiconductors has been studied extensively. On the contrary, amorphous substances are much less investigated in this aspect, although there is an extensive literature on the absorption coefficient near or at the optical band-gap region [see Sec. 4.1.1(b)]. Previous studies have reported that there is almost no temperature variation of the absorption coefficient for a thin *a*-Si samples at an Nd:YAG laser wavelength (1.06 μm). Then, it has been sometimes assumed that unlike *c*-Si, the optical constants of *a*-Si are indeed taken as independent of temperature in such spectral region (see Ref. [17]).

The damping energies derived in Sec. 4.1.2 are found to be considerably larger than those for the interband transitions in crystalline substances. The damping energy in crystalline semiconductors can be successfully explained by the sum of two different contributions: $\Gamma(T)=\Gamma_0+\Gamma_\theta(T)$, where Γ_0 is independent of the temperature T, arising mainly from crystalline imperfections, and $\Gamma_\theta(T)$ is a contribution through emission and absorption of lattice vibrations (phonons) of average frequency θ, proportional to $[\exp(\theta/T)-1]$ [see Eq. (3.43)]. This expression ensures that the higher the temperature, the larger the Γ value. Amorphous materials preserve the short-range order (in the present case, tetrahedral coordination), but does not preserve the long-range order. Therefore, it is natural to consider that the larger Γ required in the amorphous analysis is due mainly to the effects of long-range disorder in the amorphous materials. If the main damping mechanism is due to such structural defects, its energy value may be not dependent of the temperature and, therefore, one would predict no significant temperature dependence of the optical constants in the interband transition region ($E>E_o$) of amorphous materials.

To our knowledge, no experimental or theoretical data has been reported on the pressure dependence of the optical constants in the interband transition region of amorphous semiconductors (i.e., except near or at the E_0 region). We can, however, expect no dramatic change in the optical constants with pressure until the substances show structural phase transition. This is because of the absence of long-range order in the amorphous states. Moreover, it has been reported [8] that the optical constants of amorphous semiconductors are not strongly influenced by the doping of impurities.

4.2 MICROCRYSTALLINE MATERIALS

4.2.1 General Remarks

As mentioned in Sec. 1.1.2, most models of amorphous solids can be classified into two categories: microcrystalline and random networks. In the microcrystalline model, atoms are considered to exist in small regions of crystalline order with disordered boundaries (see Fig. 1.4). On the other hand, the random network model contains no regions of crystalline order. In the previous section, we discussed the optical properties of amorphous materials in the random network category (i.e., purely amorphous). In this section, we will discuss those in the microcrystalline category. It is noted that polycrystalline semiconductors with crystallite sizes in the range below 500 Å (say 50–500 Å, or 5–50 nm) are usually termed *microcrystalline*, although *nanocrystalline* is also sometimes used [40]. Throughout this section we use the term "microcrystalline" (symbolized here by, *e.g.*, μc-Si). Although we present here only on the results for μc-Si, it is of course possible to extend the ideas to other classes of microcrystalline semiconductors.

μc-Si has attracted much attention in the past few years due to its unique properties intermediate between those of *a*-Si and *c*-Si. In particular, as compared to *a*-Si, μc-Si exhibits higher doping efficiency and electrical conductivity together with a lower optical absorption [41]. Thus, μc-Si is potentially of great importance for many device applications. μc-Si is also interesting from a fundamental standpoint, because it can serve as a model system for studying the effects of crystalline particle or grain size on the optical and electronic properties of semiconductors.

Various techniques have been used to fabricate μc-Si (μc-Si:H), which include chemical-vapor transport, photochemical vapor deposition, rf and microwave discharge [42]. When an energetic ion penetrates a crystalline target, sufficient kinetic energy may be imparted to lattice atoms during nuclear collisions to cause atomic displacements. Such violent displacement processes cause the accumulation of radiation damage within the lattice and, as a result, amorphous damaged layers may be built up on the surfaces. By performing ion implantation at various fluences or acceleration energies, it is possible to obtain materials with properties continuously variable from crystalline to amorphous.

In Sec. 4.2.2, we will show that the optical constants of μc-Si measured by SE in the visible–near-UV region can be accurately simulated with the Bruggeman EMA using a model of the material as a three component, microscopic composite consisting of *a*-Si, *c*-Si, and void [43]. The samples to be reported here were produced by As^+-ion implantation into Si(111). Results on the isothermal annealing will also be presented.

Although the three-composite Bruggeman EMA yields excellent fits to the measured SE, attempts to correlate the resulting bulk structural information with that deduced from Raman analysis have revealed an ambiguity in the SE interpretation [44]. Specifically, some films that exhibit a considerable volume fraction of *a*-Si (*c*-Si) from SE have shown no such detectable phase from Raman spectroscopy. This result suggests that the *c*-Si component in the SE analysis may serve the additional role of simulating the broadening of the optical transitions that arises from a reduction in the excited-state lifetime of electrons due to scattering from grain boundaries and/or defects within the grains [44].

In Sec. 4.2.3, therefore, we will try to analyze the optical properties of *a*-Si partially crystallized by thermal annealing with the assumption that the material consists of a physical mixture of microcrystalline and amorphous regions [45]. We will find that the complex dielectric function of μc-Si differs appreciably from that of *c*-Si, especially in the vicinity of the sharp CP structures. Analyses will also be presented on the dielectric-function spectra of μc-Si based on the MDF.

4.2.2 Amorphization of Crystalline Si by Ion Implantation

(a) As-Implanted Samples—A number of reports have been reported describing the optical properties of ion-implanted Si measured by means of optical absorption, reflectance, differential reflectance, and SE (see Ref. [43]). Although a better understanding has been gained in the area of Si ion-implantation technology, little is known about the damage by As^+-ion implantation. It is well known that VLSI devices require both vertical and lateral down scaling of dimensions; in order to achieve this purpose, it is necessary to obtain shallow junctions. The As^+-ion species is the *n*-type dopant easily enabling the formation of shallow n^+-*p* junctions in Si.

The usual way of describing the optical response in the amorphous–crystalline intermediate states is to assume that the film layer consists of a physical mixture of crystalline and amorphous regions in order to use the EMA. We use the Bruggeman EMA as described in Sec. 3.4 to analyze the optical properties of Si implanted by As^+ ions. The Si(111) wafers were implanted with singly charged As^+ ions at 80 keV at several doses from 3×10^{13} to 3×10^{15} cm^{-2} at room temperature. SE measurements were carried out after cleaning the surface with organic solvents and then rinsing in a dilute HF (1.5%) solution to remove the native oxide layer.

Figures 4.17 and 4.18 show, respectively, the imaginary and real parts of the dielectric functions, $\varepsilon(E)=\varepsilon_1(E)+i\varepsilon_2(E)$, for As^+-ion implanted Si(111) with various ion fluences measured by SE [43]. For the purpose of comparison, we show in the figures the dielectric behaviors of perfectly crystalline and fully amorphous

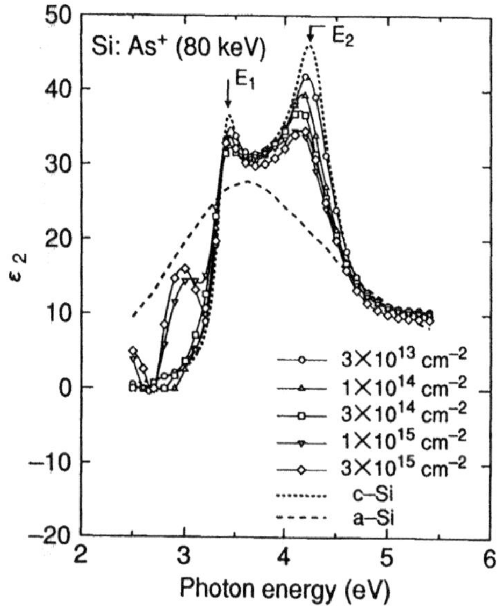

Figure 4.17 Imaginary parts of the dielectric functions for As^+-ion implanted Si with doses from 3×10^{13} to 3×10^{15} cm^{-2} as measured by SE at room temperature, together with those for *c*- and *a*-Si. The arrows indicate the positions of E_1 and E_2 CPs. (From Adachi *et al.* [43].)

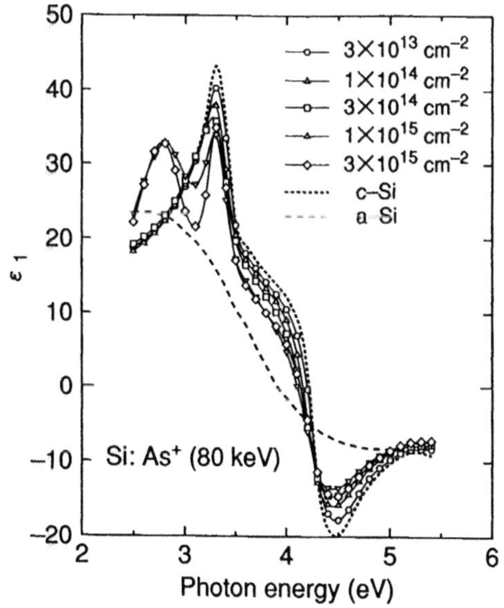

Figure 4.18 Real parts of the dielectric functions for As^+-ion implanted Si with doses from 3×10^{13} to 3×10^{15} cm^{-2} as measured by SE at room temperature, together with those for *c*- and *a*-Si. (From Adachi *et al.* [43].)

materials. They were taken from Aspnes and Studna (*c*-Si) [46] and from Bagley *et al.* (*a*-Si) [47].

As seen in Figs. 4.17 and 4.18, the dielectric behaviors of the implanted samples are substantially different from those of the perfectly crystalline material, especially in the vicinity of the sharp CP structures (E_1 and E_2). This difference becomes more noticeable if samples are implanted with higher fluences. We can also recognize interference bands on the low energy side of the spectra (<3.5 eV) taken for the two highest dose samples (1×10^{15} and 3×10^{15} cm^{-2}).

In order to present more qualitative and quantitative information on these spectral differences, we performed the LRA–EMA analysis using the four-phase model as shown in Fig. 4.19. In this model, damaged layer is considered to be two discrete layers (Layers 1 and 2) and each is inhomogeneous with the first a mixture of *c*-Si, *a*-Si, and void, and the second a mixture of *c*-Si and *a*-Si. The reference data for *c*-Si and *a*-Si used in the analysis are the same as those shown in Figs. 4.17 and 4.18. Figure 4.20 shows the results of this analysis for the sample implanted with 3×10^{13}-cm^{-2} As^+ ions [43]. The measured SE data are plotted by solid (ε_1) and open circles (ε_2). The LRA–EMA calculation results are shown by the solid lines. The structural parameters obtained in this fit are listed in Table 4.2. The corresponding values of unbiased estimator of a least-squares function σ [see Eq. (3.88)] are shown in the last column of this table.

The LRA–EMA analysis shown in Fig. 4.20 gives an unbiased estimator of σ=0.0101. The analysis also gives f_{a1}=9.0% (d_1=234 Å) and f_{a2}=24.1% (d_2=419 Å) in the first- (Layer 1) and second-phase (Layer 2) layer, respectively. The mean projected range of the implanted As^+ ions R_p is about 500 Å, while the damaged depth d_1+d_2 (~650 Å) obtained from this analysis is slightly larger than this value. It is obvious that the implanted sample cannot be considered as optically homogeneous in the 2.5–5.4-eV region. We also find that the two-phase [ambient/(*a*-Si+*c*-Si)] model provides no good agreement with the experimental data. Optical spectra

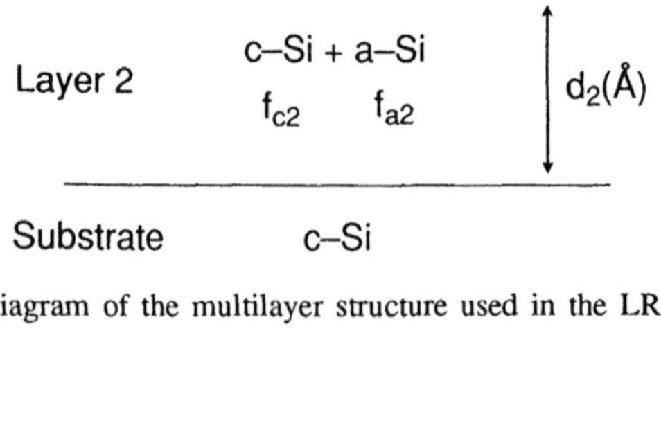

Figure 4.19 Schematic diagram of the multilayer structure used in the LRA–EMA analysis (As^+-ion implanted Si).

of Si partially amorphized by As^+ ions at 150 keV, on the other hand, had been successfully interpreted by such simple two-phase model [48]. This may be due to the fact that the higher energy ions can produce damaged layers deeper into the substrate.

The best-fit results of the LRA–EMA analysis for a sample implanted with 3×10^{15}-cm^{-2} As^+ ions are shown in Fig. 4.21 [43]. It is easily expected that higher dose implantation produces larger *a*-Si composition in the layer. The *a*-Si percentages obtained here are f_{a1}=29.5% and f_{a2}=100% that are obviously larger than the 3×10^{13}-cm^{-2} dose case (f_{a1}=9.0% and f_{a2}=24.1%, Fig. 4.20). The 3×10^{15}-cm^{-2} sample contains about 66% *c*-Si in the first layer and 0% in the second layer (i.e., fully amorphous). A buried amorphous layer can best be understood as forming as a result of an implantation range profile where the region of the implant profile that produces enough damage to exceed the amorphous threshold of the substrate material is below the surface [49]. It was also reported that the thickness of the buried amorphous layer formed by implantation can be identified by interference bands seen at lower photon energies (<3.5 eV) [49,50]. We can see from Fig. 4.21 that the four-phase model successfully explains the experimentally observed interference bands.

In Fig. 4.22, we plot the amorphous fraction f_{a2} in the second layer (Layer 2) as a function of As^+-ion fluence [43]. From this plot, we can obtain the following relation between f_{a2} (%) and the As^+-ion fluence [As^+] (in cm^{-2}):

$$f_{a2} = 21.5 \ln([As^+]) - 640 \qquad (4.32)$$

The straight line in Fig. 4.22 represents the calculated result of Eq. (4.32). An amorphous-threshold fluence can thus be expected to occur at around ~9×10^{14} cm^{-2}.

(b) Implanted and Annealed Samples—After As^+-ion implantation at 80 keV, the highest-dose (3×10^{15} cm^{-2}) wafer was cut into a number of smaller specimens for subjecting them to isothermal annealing experiments. The isothermal annealing was carried out at 1100°C in a dry N_2 atmosphere for durations between 50 and 400 min. Figures 4.23 and 4.24 show the annealing-induced change in the dielectric-function spectra of As^+-ion implanted Si (3×10^{15} cm^{-2}) [43]. For comparison, we show in the figures the $\varepsilon(E)$ spectrum of unimplanted Si (*c*-Si) [46] by the dotted lines.

As mentioned before, the as-implanted sample reveals the interference bands on the low energy side of the spectra (<3.5 eV). These interference bands are found to disappear very quickly after annealing. It is also found that the ε_2 value at the E_2 peak increases with increasing annealing time. The ε_2 value for the best annealed sample is, however, still smaller than that for the unimplanted Si.

Figures 4.25 and 4.26 show the four-phase LRA–EMA analysis results for the samples annealed for 100 and 400 min, respectively [43]. It is seen that our LRA–EMA analysis shows excellent agreement with the experimental data. The

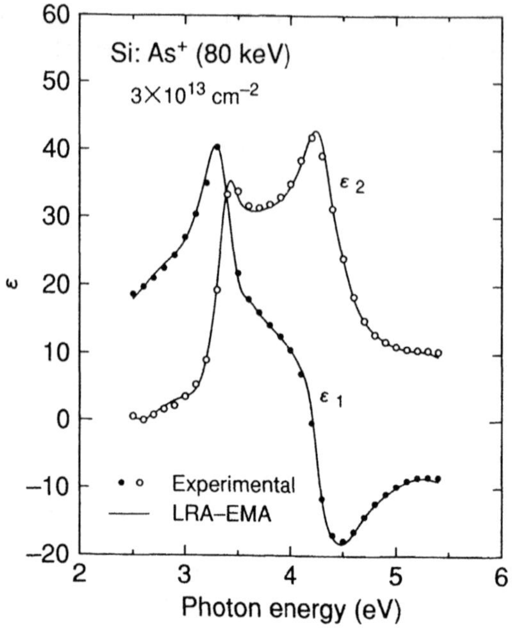

Figure 4.20 Results of the LRA–EMA analysis for As^+-ion implanted Si (3×10^{13} cm^{-2}). The SE data measured at 300 K are plotted by the solid (ε_1) and open circles (ε_2). The LRA–EMA results are shown by the solid lines. The structural parameters obtained from this fit are listed in Table 4.2. (From Adachi *et al.* [43].)

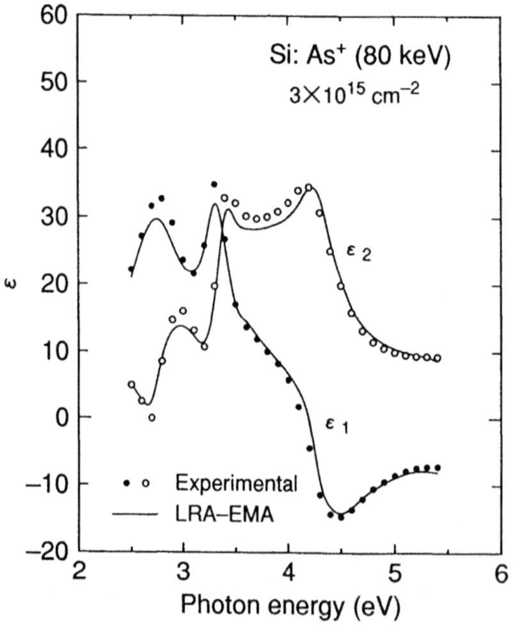

Figure 4.21 Results of the LRA–EMA analysis for As^+-ion implanted Si (3×10^{15} cm^{-2}). The SE data measured at 300 K are plotted by the solid (ε_1) and open circles (ε_2). The LRA–EMA results are shown by the solid lines. The structural parameters obtained from this fit are listed in Table 4.2. (From Adachi *et al.* [43].)

TABLE 4.2 *LRA–EMA parameters of the best-fit four-phase model (Fig. 4.19) for Si:As$^+$ with various ion fluences. The values of the unbiased estimator of the least-squares function (σ) are also shown in the last column of this table.*

Dose (cm^{-2})	Layer 1	Layer 2	σ
3×10^{13}	d_1=234 Å	d_2=419 Å	0.0101
	f_{c1}=82.2%	f_{c2}=75.9%	
	f_{a1}=9.0%	f_{a2}=24.1%	
	f_{v1}=8.8%		
1×10^{14}	d_1=242 Å	d_2=437 Å	0.0218
	f_{c1}=79.2%	f_{c2}=42.9%	
	f_{a1}=18.1%	f_{a2}=57.1%	
	f_{v1}=2.7%		
3×10^{14}	d_1=246 Å	d_2=422 Å	0.0261
	f_{c1}=70.2%	f_{c2}=21.2%	
	f_{a1}=27.6%	f_{a2}=78.8%	
	f_{v1}=2.2%		
1×10^{15}	d_1=350 Å	d_2=311 Å	0.0345
	f_{c1}=62.0%	f_{c2}=1.8%	
	f_{a1}=35.9%	f_{a2}=98.2%	
	f_{v1}=2.1%		
3×10^{15}	d_1=440 Å	d_2=162 Å	0.0271
	f_{c1}=65.8%	f_{c2}=0%	
	f_{a1}=29.5%	f_{a2}=100%	
	f_{v1}=4.7%		

structural parameters obtained from these fits are summarized in Table 4.3. As seen in Fig. 4.21, however, the LRA–EMA analysis for the as-implanted sample (3×10^{15} cm^{-2}) provided relatively poor agreement with the experimental data. This may be due to the fact that the complex dielectric function of the crystalline component considered here differs appreciably from that of the perfectly crystalline Si, especially in the vicinity of the sharp CP structures. The change in the optical properties of the crystalline component (μc-Si) may arise from the structural defects caused by the ion implantation (i.e., an increase in the broadening parameters), as

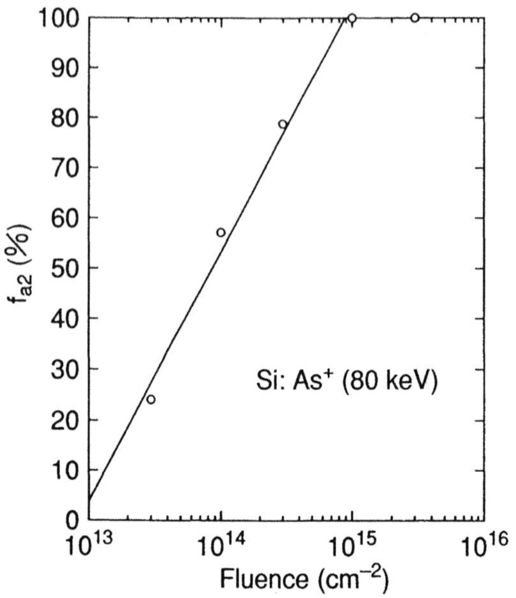

Figure 4.22 Amorphous fraction f_{a2} in the second layer (Layer 2) as a function of As$^+$-ion fluence. The straight line represents the calculated result of Eq. (4.32). (From Adachi *et al.* [43].)

we will see in the next section.

Figure 4.27 plots the percent composition of *c*-Si (f_{c2}) in the second layer as a function of annealing time for As$^+$-ion implanted Si (3×10^{15} cm^{-2}) [43]. We can understand from this figure that the recrystallization occurs very fast ($\leq$30 min) and then gradually with increasing annealing time. The perfect recrystallization (f_{c2}=100%) can be expected to occur for ~550 min. We also find essentially the same recrystallization behavior in the partially amorphized overlayer (f_{c1}, see Table 4.3).

4.2.3 Recrystallization of Amorphous Si by Thermal Annealing

(a) LRA–EMA Analysis—Knowledge of the crystallization process of amorphous materials is of considerable interest not only from a fundamental point of view but also for improving device performance using these materials. Investigations of the amorphous–to–microcrystalline transition of Si films during growth have been performed using SE [42,51]. The optical functions of several forms of thin-film Si (*a*-Si, fine-grain, and large-grain polycrystalline Si) grown on oxidized Si have been determined by means of SE [52].

We now present here the optical properties of *a*-Si partially crystallized by

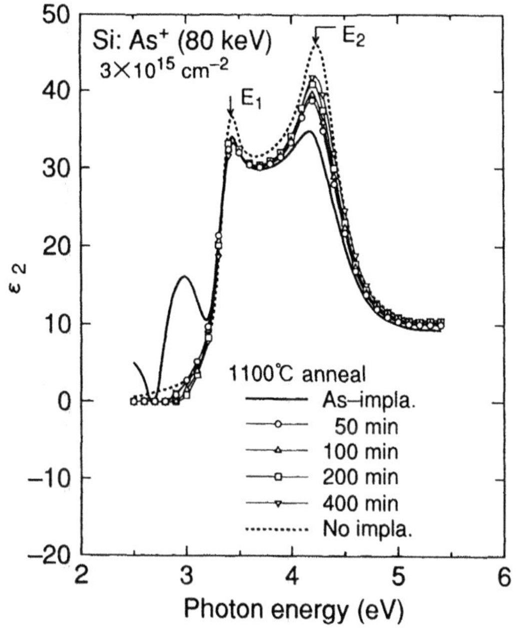

Figure 4.23 Imaginary parts of the dielectric functions for As^+-ion (3×10^{15} cm^{-2}) implanted Si after annealing at 1100°C for various times as measured by SE at 300 K. The arrows indicate the positions of E_1 and E_2 CPs. (From Adachi *et al.* [43].)

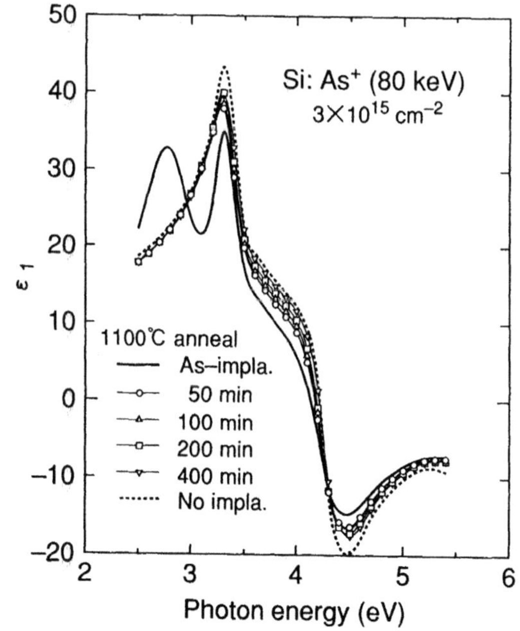

Figure 4.24 Real parts of the dielectric functions for As^+-ion (3×10^{15} cm^{-2}) implanted Si after annealing at 1100°C for various times as measured by SE at 300 K. (From Adachi *et al.* [43].)

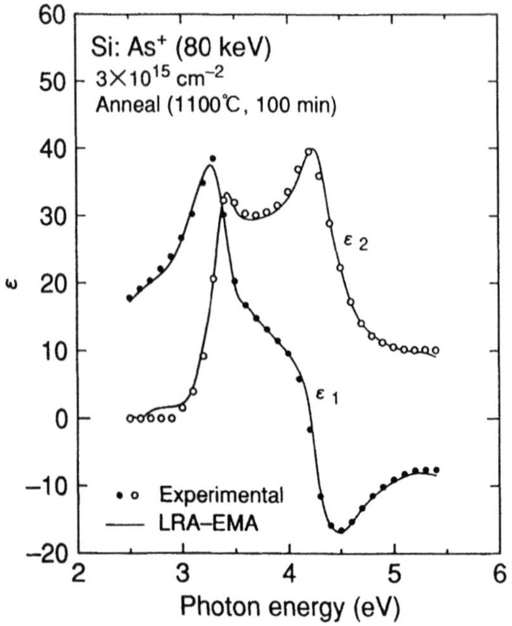

Figure 4.25 Results of the LRA–EMA analysis for As^{+}-ion (3×10^{15} cm^{-2}) implanted and annealed (1100°C, 100 min) Si. The SE data measured at 300 K are plotted by the solid (ε_1) and open circles (ε_2). The LRA–EMA calculation results are shown by the solid lines. The structural parameters obtained from this fit are listed in Table 4.3. (From Adachi *et al.* [43].)

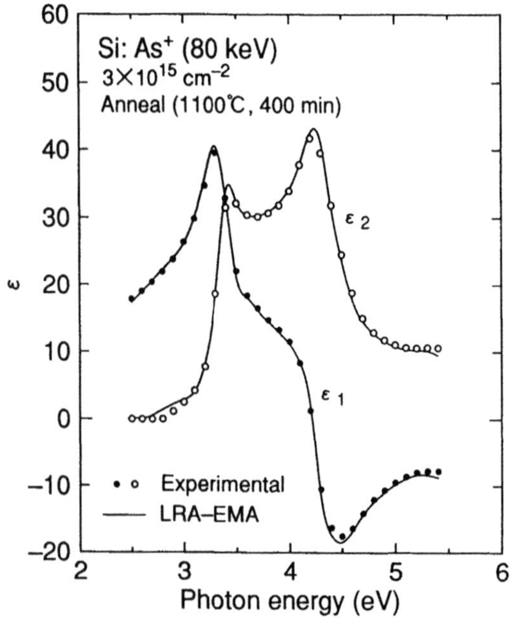

Figure 4.26 Results of the LRA–EMA analysis for As^{+}-ion (3×10^{15} cm^{-2}) implanted and annealed (1100°C, 400 min) Si. The SE data measured at 300 K are plotted by the solid (ε_1) and open circles (ε_2). The LRA–EMA calculation results are shown by the solid lines. The structural parameters obtained from this fit are listed in Table 4.3. (From Adachi *et al.* [43].)

TABLE 4.3 *LRA–EMA parameters of the best-fit four-phase model (Fig. 4.19) for $Si:As^+$ (3×10^{15} cm^{-2}) annealed at 1100°C for various annealing times. The values of the unbiased estimator of the least-squares function (σ) are also shown in the last column of this table.*

Annealing Time (min)	Layer 1	Layer 2	σ
50	d_1=267 Å	d_2=444 Å	0.0164
	f_{c1}=82.7%	f_{c2}=52.4%	
	f_{a1}=14.9%	f_{a2}=47.6%	
	f_{v1}=2.4%		
100	d_1=265 Å	d_2=447 Å	0.0150
	f_{c1}=84.3%	f_{c2}=60.1%	
	f_{a1}=11.9%	f_{a2}=39.9%	
	f_{v1}=3.8%		
200	d_1=248 Å	d_2=443 Å	0.0137
	f_{c1}=87.7%	f_{c2}=66.3%	
	f_{a1}=9.4%	f_{a2}=33.7%	
	f_{v1}=2.8%		
400	d_1=268 Å	d_2=394 Å	0.0141
	f_{c1}=93.6%	f_{c2}=86.7%	
	f_{a1}=4.1%	f_{a2}=13.3%	
	f_{v1}=2.3%		

furnace annealing studied using SE. To obtain both qualitative and quantitative information on the effects of thermal annealing, we use the Bruggeman EMA in which the partially crystallized layer is assumed as a physical mixture of fully amorphous Si and μc-Si. The best values of the structural parameters are obtained using the LRA technique.

The a-Si samples used here were deposited by rf sputtering on Si(100) substrates. The sputtering system was evacuated to a base pressure of ~5×10^{-7} Torr before the introduction of Ar sputtering gas. The sputter deposition was carried out at 50°C at a working pressure of 5×10^{-3} Torr and a power of 200 W. The thickness of the deposited a-Si films was ~3600 Å. After deposition, the Si(100) wafers were cut into a number of smaller specimens to be subjected to isothermal annealing experiments. The isothermal annealing was carried out at 1100°C in a dry N_2 atmosphere for time periods between 30 and 400 min.

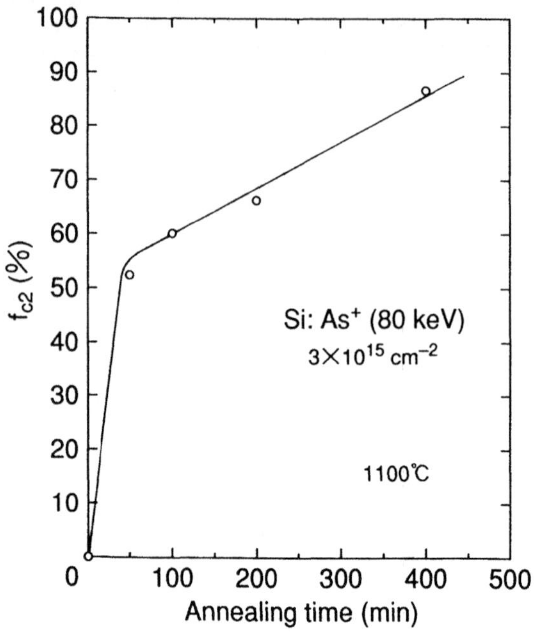

Figure 4.27 Crystalline fraction f_{c2} in the second layer (Layer 2) as a function of annealing time for As^+-ion implanted Si (3×10^{15} cm^{-2}). (From Adachi *et al.* [43].)

Figures 4.28 and 4.29 show, respectively, the imaginary and real parts of the dielectric functions for *a*-Si films annealed for four different times (30, 60, 200, and 400 min) measured by means of SE [45]. A single broad peak observed in the $\varepsilon_2(E)$ spectrum of the as-deposited sample (*a*-Si) is typical behavior of amorphous tetrahedrally-bonded semiconductors, as have already been mentioned in Sec. 4.1.2. For purpose of comparison, we show in these figures the $\varepsilon(E)$ spectrum of *c*-Si taken from Ref. [46]. It is understood that the dielectric behavior of the annealed sample is substantially different from that of the unannealed, fully amorphous material (*a*-Si). This difference becomes more noticeable when samples are annealed for longer times.

We perform the LRA–EMA analysis using the two-phase model as shown in Fig. 4.30. The *a*-Si $\varepsilon(E)$ data used here are the same as those shown in Figs. 4.28 and 4.29. From this analysis, we deduce the $\varepsilon(E)$ spectra of μc-Si which will be discussed in detail in the next subsection. As mentioned before (Sec. 4.1.2), the optical properties of amorphous materials are strongly influenced by the bulk density deficit. The "*intrinsic*" dielectric properties of dense *a*-Si show the largest ε_2 peak value (ε_2~30, see Fig. 4.12). We can therefore consider that our sputter-deposited *a*-Si film (ε_2~20) may contain a large number of voids in the film medium.

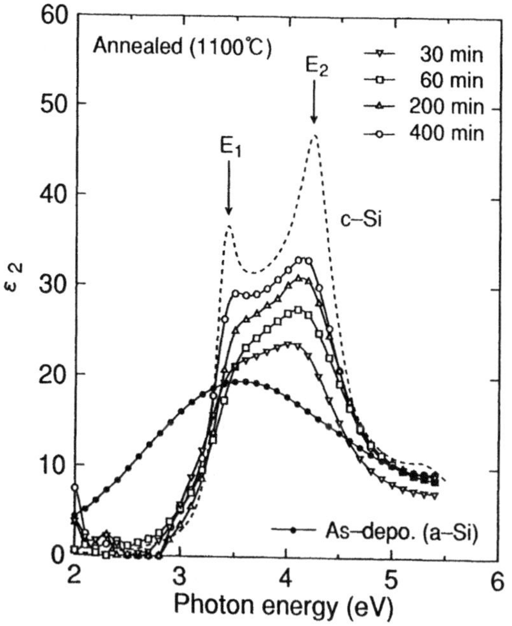

Figure 4.28 Imaginary parts of the dielectric functions (ε_2) for *a*-Si partially crystallized by thermal annealing at 1100°C for 30, 60, 200, and 400 min as measured by SE at 300 K, together with those for *c*-Si and as-deposited *a*-Si. The arrows indicate the positions of E_1 and E_2 CPs. (From Suzuki and Adachi [45].)

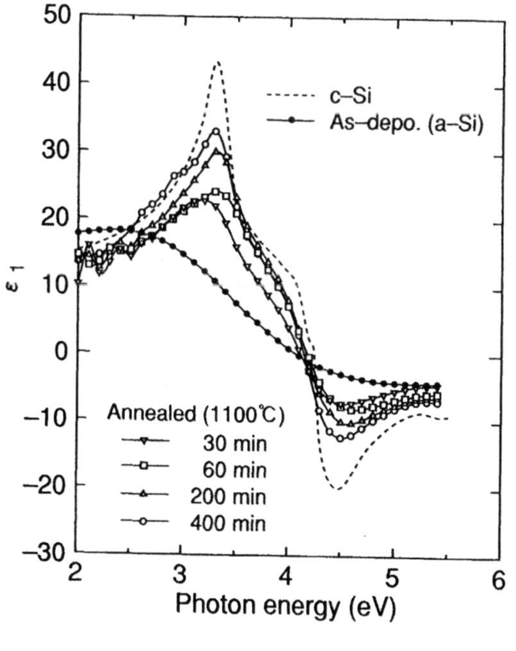

Figure 4.29 Real parts of the dielectric functions (ε_1) for *a*-Si partially crystallized by thermal annealing at 1100°C for 30, 60, 200, and 400 min as measured by SE at 300 K, together with those for *c*-Si and as-deposited *a*-Si. (From Suzuki and Adachi [45].)

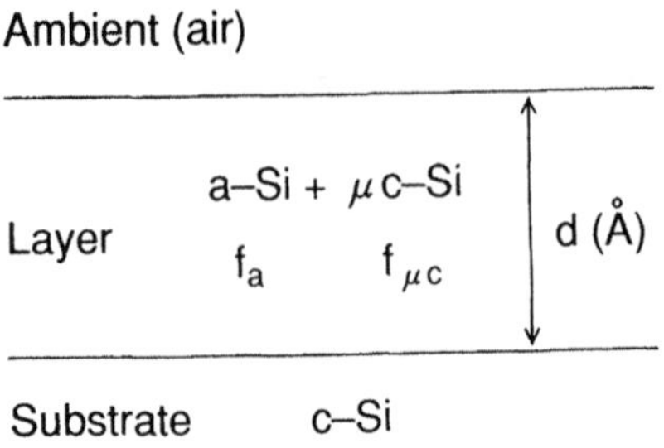

Figure 4.30 Schematic diagram of the multilayer structure used in the LRA–EMA analysis [*a*-Si film (*d*~3600 Å) partially crystallized by thermal annealing].

Figure 4.31 shows the results of the LRA–EMA analysis for a sample annealed for 60 min [45]. The measured SE data are plotted by the solid (ε_1) and open circles (ε_2). The calculated LRA–EMA results are shown by the solid lines. The obtained structural parameters and σ are listed in Table 4.4.

The best-fit results of the LRA–EMA analysis for a sample annealed for 400 min are shown in Fig. 4.32 [45]. It is easily expected that annealing of longer time leads to a smaller value of *a*-Si component in the layer. The *a*-Si percentage obtained here is f_a=18.4% (see Table 4.4), which is obviously smaller than the value obtained from the 60-min-annealed sample (f_a=40.3%, Fig. 4.31).

We plot in Fig. 4.33 the microcrystalline fraction $f_{\mu c}$ as a function of annealing time for samples annealed at 1100°C [45]. We can understand from this plot that the crystallization occurs quickly (≤30 min) and slowly with further increase in annealing time.

(b) MDF Analysis—It is of both technological importance and scientific interest to model the $\varepsilon(E)$ spectra of *μc*-Si. In the following, we analyze in the following the $\varepsilon(E)$ spectra of *μc*-Si using the MDF [see details in Sec. 3.1.1 (c)].

The $\varepsilon(E)$ spectra of *μc*-Si deduced from the LRA–EMA analysis are shown in Figs. 4.34 [$\varepsilon_2(E)$] and 4.35 [$\varepsilon_1(E)$] [45]. For purpose of comparison, the $\varepsilon(E)$ spectra of *c*-Si [46] are shown in these figures. The arrows in Fig. 4.34 indicate the positions of E_1 and E_2 CPs. As seen in Figs. 4.34 and 4.35, the $\varepsilon(E)$ spectra of *μc*-Si differ clearly from that of *c*-Si. The $\varepsilon(E)$ spectra of *μc*-Si are also found to depend largely on annealing time. In particular, the longer the annealing time, the stronger E_1 is. These *μc*-Si spectra resemble those reported in the literature [52–4].

Figure 4.36 shows fits of our MDF model to the experimental $\varepsilon(E)$ of *c*-Si (Ref. [46]) and *μc*-Si (annealed for 400 min) [45]. The solid and dashed lines are taken from the sum of Eqs. (3.20), (3.23), (3.26), and (3.30). The CP parameters of the fits are listed in Table 4.5. It is seen that the MDF calculation shows satisfactory agreement with the experimental data over the entire range of photon energies.

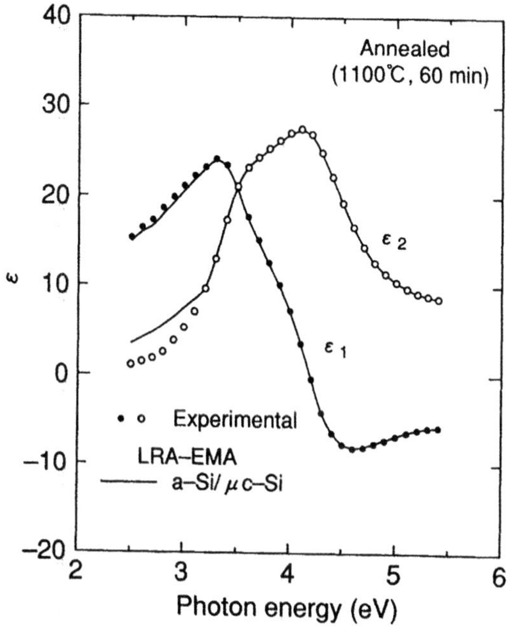

Figure 4.31 Results of the LRA–EMA analysis for a sample annealed at 1100°C for 60 min. The SE data measured at 300 K are plotted by the solid (ε_1) and open circles (ε_2). The calculated LRA–EMA results (*a*-Si/*μc*-Si) are shown by the solid lines. The structural parameters and unbiased estimator of a least-squares function σ obtained from the fit are listed in Table 4.4. (From Suzuki and Adachi [45].)

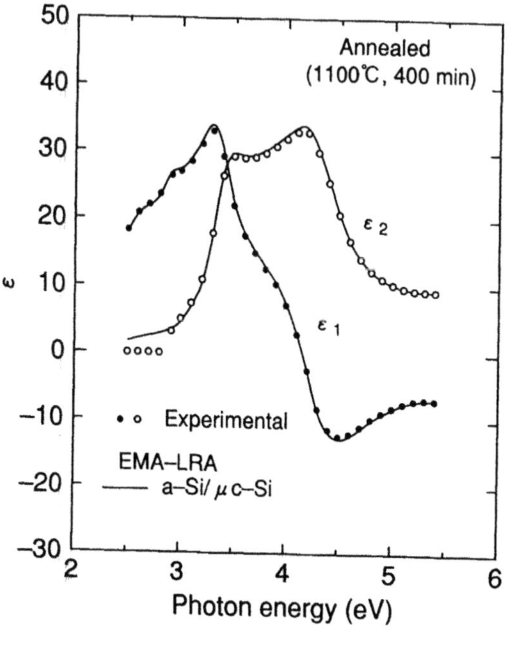

Figure 4.32 Results of the LRA–EMA analysis for a sample annealed at 1100°C for 400 min. The SE data measured at 300 K are plotted by the solid (ε_1) and open circles (ε_2). The calculated LRA–EMA results (*a*-Si/*μc*-Si) are shown by the solid lines. The structural parameters and unbiased estimator of a least squares function σ obtained from the fit are listed in Table 4.4. (From Suzuki and Adachi [45].)

TABLE 4.4 *LRA–EMA parameters of the best-fit two-phase model (Fig. 4.30) for a-Si partially crystallized by thermal annealing. The values of the unbiased estimator σ (E=3.0–5.4 eV) are also shown in the last column of this table.*

Annealing Time (min)	f_a (%)	$f_{\mu c}$ (%)	σ
30	46.6	53.4	0.0213
60	40.3	59.7	0.0163
100	25.5	74.5	0.0078
200	25.3	74.7	0.0060
400	18.4	81.6	0.0080

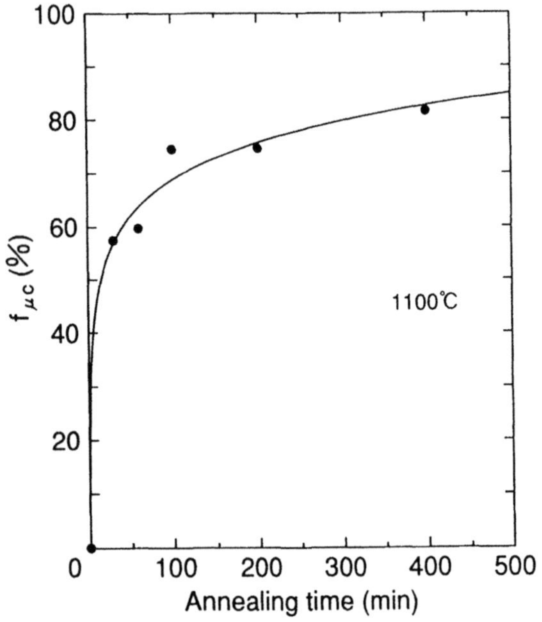

Figure 4.33 Microcrystalline fraction $f_{\mu c}$ as a function of annealing time for samples annealed at 1100°C. (From Suzuki and Adachi [45].)

In Fig. 4.37, the solid and dashed lines represent individual contributions to $\varepsilon_2(E)$ of the E_1, E_2, E_0', and E_1' for *c*-Si and *μc*-Si, respectively [45]. They are obtained from Eqs. (3.20) and (3.23) for the E_1-gap contribution (2D-M_0 CP), from Eqs. (3.26) (DHO) and (3.30) (2D-M_2 CP) for the E_2-gap one, and from Eq. (3.26) for the E_0'- and E_1'-gap contributions.

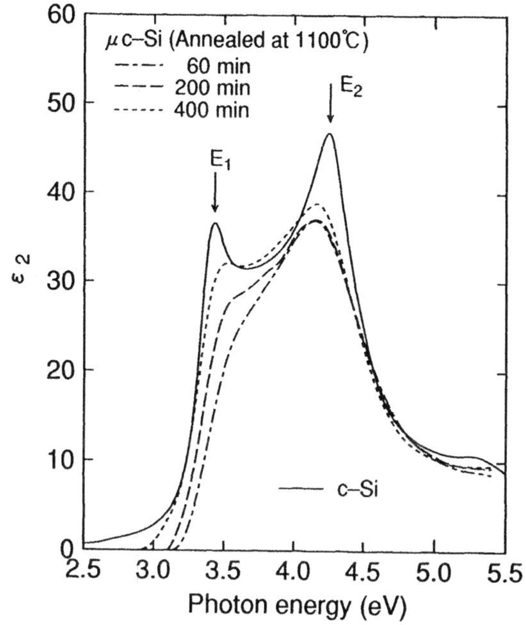

Figure 4.34 Imaginary parts of the dielectric functions (ε_2) for μc-Si deduced from the LRA–EMA analysis, together with that for c-Si. The arrows indicate the positions of E_1 and E_2 CPs. (From Suzuki and Adachi [45].)

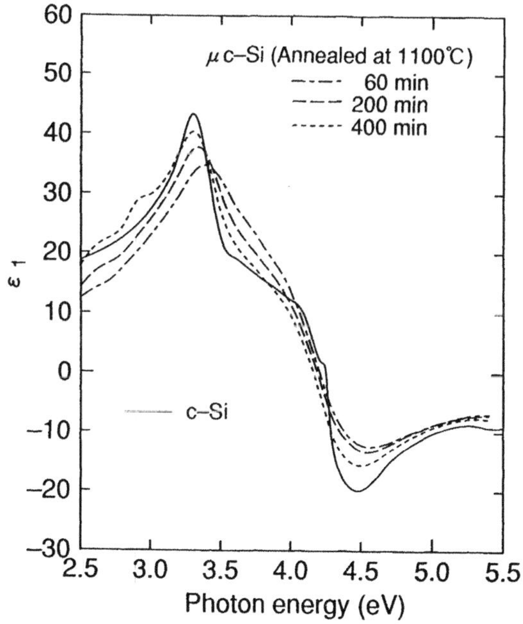

Figure 4.35 Real parts of the dielectric functions (ε_1) for μc-Si deduced from the LRA–EMA analysis, together with that for c-Si. (From Suzuki and Adachi [45].)

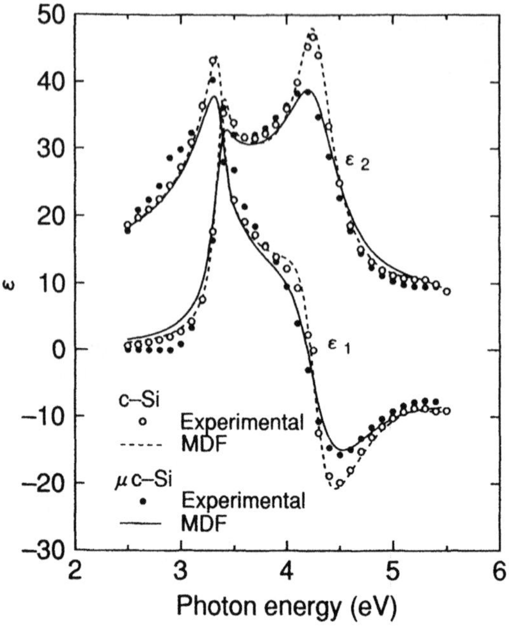

Figure 4.36 Fits of our MDF model to the experimental ε(*E*) of *c*-Si [46] and *μc*-Si (annealed at 1100°C for 400 min). The solid and dashed lines are taken from the sum of Eqs. (3.20), (3.23), (3.26), and (3.30). The CP parameters of the fits are listed in Table 4.5. (From Suzuki and Adachi [45].)

The E_1 gap is of the 3D-M_1 (2D-M_0) type. Hence, the line shape of the corresponding $\varepsilon_2(E)$ spectrum should be characterized by a steep low-energy side and a broader high-energy side. It is evident from Fig. 4.37 that the sharp peak appearing at the E_1 edge of *c*-Si could not be explained only by the one-electron approximation [Eq. (3.20)]. Additional contribution of the E_1-edge exciton [Eq. (3.23)] can provide a sharp peak in the $\varepsilon_2(E)$ spectrum, in good agreement with the experimental data. The E_2 structure in Si can be successfully fitted by a mixture of the DHO and 2D-M_2 CP. The steep low-energy end of the 2D-M_2 CP contribution at the E_1 edge is a result of the E_{cl} cutoff-energy modification. The strongest component in the $\varepsilon_2(E)$ spectrum is found to be the E_2-gap contribution (DHO). It is also recognized that the strengths of the E_0'- and E_1'-gap contributions are very weak.

As seen in Table 4.5, each CP in *μc*-Si has the same strength parameter value as that of *c*-Si. This result is consistent with the requirement of an oscillator-strength sum rule. On the contrary, the broadening parameter values of *μc*-Si are found to be not the same as those of *c*-Si. Figure 4.38 plots the broadening parameters of some prominent interband transitions, $\Gamma(E_1)$ and $\gamma(E_2)$, as a function of annealing

TABLE 4.5 *Parameters used in the MDF analysis of ε(E) for c-Si and a-Si partially crystallized by thermal annealing.*

		Annealing Time (min)				
Parameter	c-Si	30	60	100	200	400
E_1 (eV)	3.39	3.39	3.39	3.39	3.39	3.39
B_1	5.22	5.22	5.22	5.22	5.22	5.22
B_{1x} (eV)	1.44	1.44	1.44	1.44	1.44	1.44
Γ (eV)	0.08	1.30	0.50	0.30	0.26	0.11
E_2 (eV)	4.28	4.28	4.28	4.28	4.28	4.28
C	3.08	3.08	3.08	3.08	3.08	3.08
γ	0.100	0.190	0.150	0.157	0.155	0.145
F	3.82	3.82	3.82	3.82	3.82	3.82
Γ (eV)	0.100	0.190	0.150	0.157	0.155	0.145
E_0' (eV)	3.35	3.35	3.35	3.35	3.35	3.35
C	0.07	0.07	0.07	0.07	0.07	0.07
γ	0.09	0.17	0.17	0.17	0.17	0.09
E_1' (eV)	5.33	5.33	5.33	5.33	5.33	5.33
C	0.30	0.30	0.30	0.30	0.30	0.30
γ	0.12	0.30	0.30	0.20	0.20	0.17

time [45]. It is evident that the broadening parameters decrease with increasing annealing time. This result is not surprising because long-time annealing may sufficiently improve their crystalline qualities. Our previous analysis suggested that the Γ values in amorphous semiconductors are considerably larger than those of the interband transitions in crystalline materials (Sec. 4.1.2). It is important to note that the amorphous materials preserve the short-range order, but do not preserve the long-range order. Therefore, it is possible to consider that the large broadening values required in amorphous materials are due to the effects of long-range disorder (and short-range defects) in these materials. If the main broadening mechanism in μc-Si is due to such structural defects, its energy value will be strongly dependent on annealing temperature and also on time.

Aspnes *et al.* [55] explained the decrease in the strengths of CP structures in ion-implanted crystals by means of the projected areas of several valence and core excitons in GaAs. They found that the projected areas for the E_1 and E_2 exciton resonances in GaAs are $(10\ \text{nm})^2$ and $(6.3\ \text{nm})^2$, respectively. The structures at the

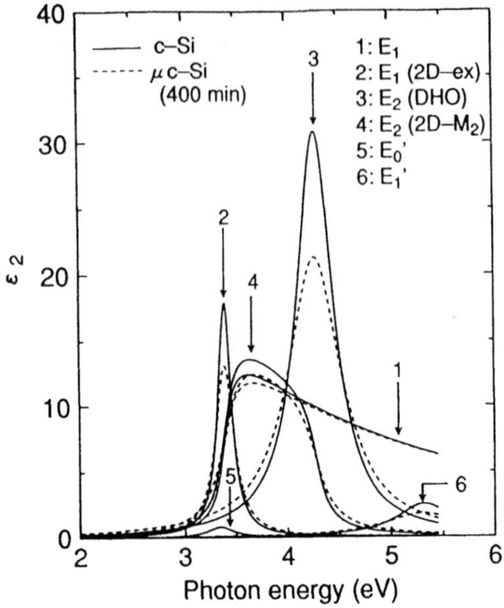

Figure 4.37 Individual contributions to $\varepsilon_2(E)$ of the E_1, E_2, E_0', and E_1' for *c*-Si (solid lines) and *μc*-Si (dashed lines). They are obtained from Eqs. (3.20) and (3.23) for the E_1-gap contribution (2D-M_0 CP), from Eqs. (3.26) (DHO) and (3.30) (2D-M_2 CP) for the E_2-gap one, and from Eq. (3.26) for the E_0'- and E_1'-gap contributions. (From Suzuki and Adachi [45].)

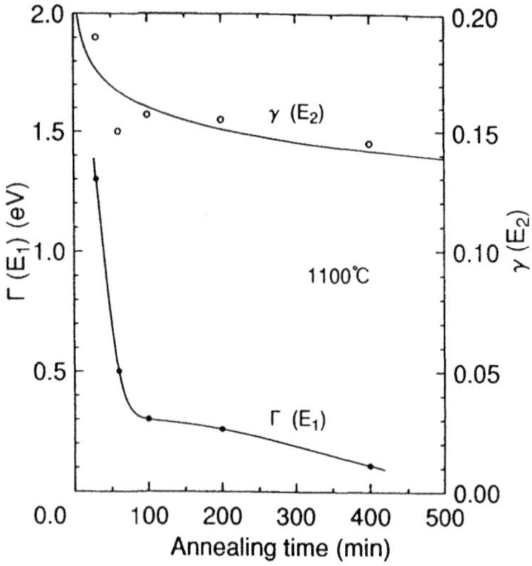

Figure 4.38 Broadening parameters of some prominent interband transitions, $\Gamma(E_1)$ and $\gamma(E_2)$, as a function of annealing time for *μc*-Si. (From Suzuki and Adachi [45].)

E_1-exciton resonance hence vanish via lifetime broadening for shorter annealing times than do those at the E_2 resonance, in qualitative agreement with the present results (see Figs. 4.34 and 4.38). Nguyen and Collins [44] also reported such finite-size effects on the dielectric functions of Si microcrystallites.

In conclusion of this section, we reported the optical properties of micro-crystalline semiconductors (μc-Si). One would expect the optical constants of microcrystalline semiconductors to be near those of single crystals. It is shown that the LRA–EMA analysis yields good fits to the experimental SE data. The MDF analysis also shows that the optical constants of μc-Si differ appreciably from those of c-Si, especially in the vicinity of the sharp CP structures. This is thought to be due to the lifetime broadening of the excited states at the grain boundaries and/or defects within the grains.

REFERENCES

1. D. T. Pierce and W. E. Spicer, *Phys. Rev. B* **5**, 3017 (1972).
2. L. Ley, in *Semiconductors and Semimetals*, edited by R. K. Willardson and A. C. Beer (Academic, Orland, 1984), Vol. 21 (Volume Editor, J. I. Pankove), Part B, p. 385.
3. D. A. Davis, *Jpn. J. Appl. Phys.* **32**, 178 (1993).
4. D. L. Wood and J. Tauc, *Phys. Rev. B* **5**, 3144 (1972).
5. A. Ibrahim and S. K. J. Al-Ani, *Int. J. Optoelectron.* **9**, 25 (1994).
6. N. F. Mott and E. A. Davis, *Electronic Processes in Non-Crystalline Materials* (Clarendon, Oxford, 1979).
7. E. A. Davis and N. F. Mott, *Phil. Mag.* **22**, 903 (1970).
8. D. E. Aspnes, A. A. Studna, and E. Kinsbron, *Phys. Rev. B* **29**, 768 (1984).
9. R. H. Klazes, M. H. L. M. van den Broek, J. Bezemer, and S. Radelaar, *Phil. Mag. B* **45**, 377 (1982).
10. E. E. Khawaja and C. A. Hogarth, *J. Phys. C: Solid State Phys.* **21**, 607 (1988).
11. G. D. Cody, in *Semiconductors and Semimetals*, edited by R. K. Willardson and A. C. Beer (Academic, Orland, 1984), Vol. 21 (Volume Editor, J. I. Pankove), Part B, p. 11.
12. G. D. Cody, T. Tiedje, B. Abeles, B. Brooks, and Y. Goldstein, *Phys. Rev. Lett.* **47**, 1480 (1981).
13. See, for instance, D. Adler, in *Semiconductors and Semimetals*, edited by R. K. Willardson and A. C. Beer (Academic, Orland, 1984), Vol. 21 (Volume Editor, J. I. Pankove), Part A, p. 291.
14. P. Lautenschlager, M. Garriga, L. Vina, and M. Cardona, *Phys. Rev. B* **36**, 4821 (1987).

15. C. K. Kim, P. Lautenschlager, and M. Cardona, *Solid State Commun.* **59**, 797 (1986).
16. S. Zollner, S. Gopalan, and M. Cardona, *Solid State Commun.* **77**, 485 (1991).
17. N. Do, L. Klees, P. T. Leung, F. Tong, W. P. Leung, and A. C. Tam, *Appl. Phys. Lett.* **60**, 2186 (1992).
18. V. A. Sidorov, V. V. Brazhkin, L. G. Khvostantsev, A. G. Lyapin, A. V. Sapelkin, and O. B. Tsiok, *Phys. Rev. Lett.* **73**, 3262 (1994), and references cited therein.
19. S. Adachi, *GaAs and Related Materials: Bulk Semiconducting and Superlattice Properties* (World Scientific, SI, 1994).
20. B. Welber and M. H. Brodsky, *Phys. Rev. B* **16**, 3660 (1977).
21. S. Minomura, *J. Phys. (Paris) Colloq.* **42**, C4-181 (1981).
22. K. Tanaka, *Phys. Rev. B* **43**, 4302 (1991).
23. M. Imai, K. Tsuji, and T. Yagi, *Jpn. J. Appl. Phys.* **32**, 191 (1993), and references cited therein.
24. J. Tauc, in *Optical Properties of Solids*, edited by F. Abelès (North-Holland, Amsterdam, 1972), p. 277.
25. G. F. Feng, M. Katiyar, J. R. Abelson, and N. Maley, *Phys. Rev. B* **45**, 9103 (1992).
26. B. Kramer, K. Maschke, and P. Thomas, *Phys. Status Solidi B* **49**, 525 (1972).
27. D. Campi and C. Coriasso, *J. Appl. Phys.* **64**, 4128 (1988).
28. W. B. Jackson, S. M. Kelso, C. C. Tsai, J. W. Allen, and S.-J. Oh, *Phys. Rev. B* **31**, 5187 (1985).
29. A. R. Forouhi and I. Bloomer, *Phys. Rev. B* **34**, 7018 (1986).
30. S. Logothetidis, J. Petalas, A. Markwitz, and R. L. Johnson, *J. Appl. Phys.* **73**, 8514 (1993).
31. J. Petalas and S. Logothetidis, *Phys. Rev. B* **50**, 11801 (1994).
32. J. Tauc, R. Grigorovici, and A. Vancu, *Phys. Status Solidi* **15**, 627 (1966).
33. S. Adachi, *Phys. Rev. B* **43**, 12316 (1991).
34. S. Adachi, *J. Appl. Phys.* **70**, 2304 (1991).
35. G. A. N. Connell, R. J. Temkin, and W. Paul, *Adv. Phys.* **22**, 643 (1973).
36. M. Erman, J. B. Theeten, P. Chambon, S. M. Kelso, and D. E. Aspnes, *J. Appl. Phys.* **56**, 2664 (1984).
37. C. A. Alejo-Armenta, C. Vázquez-López, and A. Pérez-González, *Solid State Commun.* **86**, 417 (1993).
38. K. Maschke and P. Thomas, *Phys. Status Solidi* **41**, 743 (1970).
39. G. E. Jellison, Jr. and F. A. Modine, *Appl. Phys. Lett.* **69**, 371 (1996); ibid. **69**, 2137 (1996).
40. See, G. F. Feng and R. Zallen, *Phys. Rev. B* **40**, 1064 (1989).
41. See, M. Fang and D. Drévillon, *J. Appl. Phys.* **70**, 4894 (1991).
42. T. V. Herak, J. J. Schellenberg, P. K. Shufflebotham, and K. C. Kao, *J. Appl. Phys.* **64**, 688 (1988), and references cited therein.

43. S. Adachi, T. Matsumura, and T. Suzuki, *Jpn. J. Appl. Phys.* **33**, 1931 (1994).
44. H. V. Nguyen and R. W. Collins, *Phys. Rev. B* **47**, 1911 (1993).
45. T. Suzuki and S. Adachi, *Jpn. J. Appl. Phys.* **32**, 4900 (1993).
46. D. E. Aspnes and A. A. Studna, *Phys. Rev. B* **27**, 985 (1983).
47. B. G. Bagley, D. E. Aspnes, and A. C. Adams, *Bull. Am. Phys. Soc.* **25**, 12 (1980).
48. S. Adachi and T. Aoki, *J. Appl. Phys.* **71**, 3313 (1992).
49. T. M. Burns, S. Chongsawangvirod, J. W. Andrews, E. A. Irene, G. McGuire, and S. Chevacharoeukul, *J. Vac. Sci. Technol. B* **9**, 41 (1991).
50. R. E. Hummel, W. Xi, P. H. Holloway, and K. A. Jones, *J. Appl. Phys.* **63**, 2591 (1988).
51. M. Fang and B. Drevillon, *J. Appl. Phys.* **71**, 5445 (1992).
52. G. E. Jellison, Jr., M. F. Chisholm, and S. M. Gorbatkin, *Appl. Phys. Lett.* **62**, 3348 (1993).
53. S. Logothetidis, G. Kiriakidis, and E. C. Paloura, *J. Appl. Phys.* **70**, 2791 (1991).
54. B. Boultadakis, S. Logothetidis, and S. Ves, *J. Appl. Phys.* **72**, 3648 (1992).
55. D. E. Aspnes, S. M. Kelso, C. G. Olson, and D. W. Lynch, *Phys. Rev. Lett.* **48**, 1863 (1982).

Chapter 5

AT OR BELOW THE FUNDAMENTAL ABSORPTION EDGE

5.1 OPTICAL ABSORPTION

5.1.1 One-Electron Approximation

(a) Direct-Gap Edge—In obtaining Eq. (3.5), we assumed that the matrix element $P_{cv}(\boldsymbol{k})$ is constant in its value so that we can obtain the well-known line shape of ε_2 at the 3D M_0 CP [Eq. (3.14)]. In cases where transitions are forbidden in the electric-dipole approximation, one can make use of the $\boldsymbol{k}$ dependence of the matrix element and expand around the CP at $\boldsymbol{k}_c$

$$P_{cv}(\boldsymbol{k}) = P_{cv}(\boldsymbol{k}_0) + \sum_{j=1}^{3} \left(\frac{\partial P_{cv}(\boldsymbol{k})}{\partial k_j} \right)_{\boldsymbol{k}=\boldsymbol{k}_c} (k_j - k_{cj}) + \dots \qquad (5.1)$$

and keep only the linear term. We can, then, obtain for *"forbidden"* direct transitions at the 3D M_0 CP [1]

$$\varepsilon_2(E) \propto \left| \frac{\partial P_{cv}(\boldsymbol{k})}{\partial |\boldsymbol{k}|} \right|^2 (E - E_0)^{3/2} \qquad (5.2)$$

This result is to be contrasted with the square-root energy dependence found in Eq. (3.14) for the *"allowed"* direct transitions. It is noted, however, that the strength of ε_2 is generally much lower for the *"forbidden"* transitions since we have only the contribution from the matrix element in first order in $\boldsymbol{k}$.

It is known that cuprous oxide (Cu_2O) is a material showing dipole forbidden transitions. At low temperatures, the optical absorption spectrum of Cu_2O contains several series of exciton lines of which are the *yellow* and *green* series. The lowest energy peak of the yellow series is weak and was shown by Gross and Kaplyanskii [2] to have characteristics indicating that the transition was electric quadrupole (not dipole) in nature.

We schematically show in Fig. 5.1 the line shapes of the optical absorption $\alpha(E)$ for the dipole-allowed [(a)] and forbidden [(b)] transitions at the direct gap [3D M_0 CP (E_0)], together with those for indirect allowed [(c)] and forbidden [(d)] transitions at the indirect gap [E_g^{ID}, see Sec. 5.1.1 (b) below], and impurity-induced absorption band [(e)].

The dipole-transition matrix element $P_{cv}(\boldsymbol{k})=\langle c|\boldsymbol{p}|v\rangle$ can be exactly calculated if the wave functions of initial ($|i\rangle$) and final ($|f\rangle$) electronic states are explicitly known; however, this will rarely be the case. As have already shown in Sec. 1.2.2, we can easily determine, from the symmetry of the wave functions, whether the

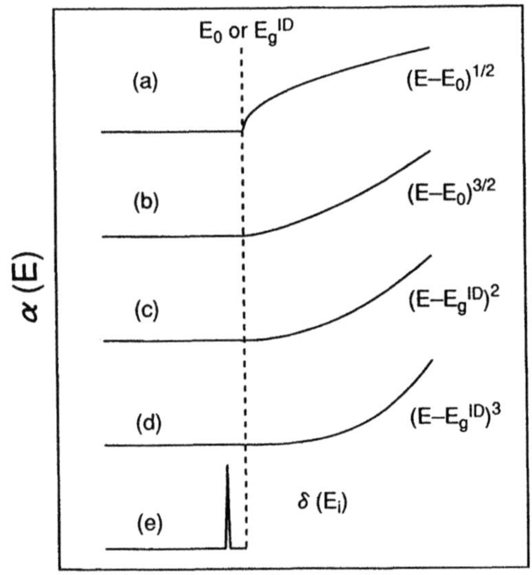

Figure 5.1 Line shapes of the optical absorption $\alpha(E)$ for the dipole-allowed [(a)] and forbidden [(b)] transitions at the direct gap [3D M_0 CP (E_0)], together with those for the indirect allowed [(c)] and forbidden [(d)] transitions at the indirect gap (E_g^{ID}), and impurity-induced absorption band (free-band–to–impurity) [(e)].

matrix element is different from zero or not, i.e., whether the transitions are allowed or forbidden, respectively [see, *e.g.*, Eq. (1.3)].

(b) Indirect-Gap Edge—Transitions between states which are not vertical in an energy-band diagram are called indirect transitions. We show in Fig. 5.2 an example of the energy-band diagram for an indirect-band-gap semiconductor. The top of the valence band VB and the direct DCB (Γ point) and indirect ICB minima (X or L point) of the conduction band are shown in the figure.

The theory of indirect optical transitions can be developed by considering the perturbation Hamiltonian [3]:

$$\mathcal{H} = \mathcal{H}_{\mathrm{eR}} + \mathcal{H}_{\mathrm{eL}} \qquad (5.3)$$

where $\mathcal{H}_{\mathrm{eR}}$ is the electron–radiation perturbation which is linear in both electron and photon creation–annihilation operators and $\mathcal{H}_{\mathrm{eL}}$ is the electron–lattice perturbation which is linear in phonon creation–annihilation operator but bilinear in electron one. The indirect optical-absorption mechanism can thus be expressed as a second-order process in the perturbation. Using the results of second-order time-dependent perturbation theory, we obtain the transition probability per unit time for a process in which the valence electron is scattered to the conduction band (ICB) and a photon of energy E and a phonon of momentum $\hbar \boldsymbol{q}=\hbar(\boldsymbol{k}_2-\boldsymbol{k}_1)$ (and energy $\hbar\omega_q$) are both absorbed as follows (except some proportionality constants):

$$W \propto D n_q \cdot \delta[E_c(\boldsymbol{k}_2) - E_v(\boldsymbol{k}_1) - E + \hbar\omega_q] \qquad (5.4)$$

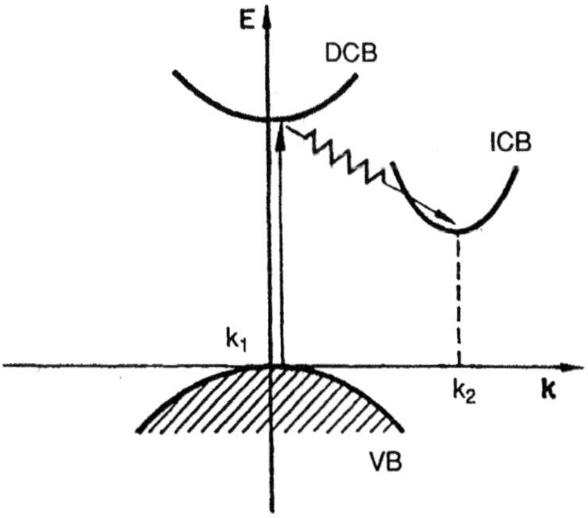

Figure 5.2 Schematic diagram showing the indirect optical-absorption process. VB is the valence band, DCB and IDC are the direct and indirect conduction bands, respectively.

with

$$D \equiv \sum_{\beta} \left| \sum_{\alpha} \frac{<\beta|\mathcal{H}_{eL}|\alpha><\alpha|\mathcal{H}_{eR}|0>}{E_{\alpha}(\boldsymbol{k}_1)-E_{v}(\boldsymbol{k}_1)-E} \right|^2 \qquad (5.5)$$

where n_q is defined by Eq. (3.33a), E_c and E_v the conduction- and valence-band energies, respectively, $|0>$ the electronic ground (initial) state, $|\alpha>$ the intermediate state in the band DCB with energy E_α, $|\beta>$ the final state in the band ICB, $<\alpha|\mathcal{H}_{eR}|0>$ the momentum-matrix element between the VB and DCB extrema located near the Γ band and separated by $E_0(\boldsymbol{k}_1)$, and $<\beta|\mathcal{H}_{eL}|\alpha>$ is the matrix element of the phonon-assisted transition from the DCB to ICB, where for GaP, *e.g.*, they involve the TA, LA, TO, and TA+LO$^\Gamma$ phonons (Ref. [4], see also Fig. 5.8). The energy denominator in Eq. (5.5) can be usually assumed to be $E_\alpha(\boldsymbol{k}_1)-E_v(\boldsymbol{k}_1)-E \sim E_0-E \sim E_0$. (The quantity D is also assumed to be independent of $\boldsymbol{k}$ in the vicinity of the band extrema.)

Considering the case of indirect-optical transitions between spherical bands and summing over $\boldsymbol{k}_1$ and $\boldsymbol{k}_2$ of Eq. (5.4) in the BZ, we obtain the following expression for the imaginary part of the complex dielectric function $\varepsilon_2(E)$ [3]:

$$\varepsilon_2(E) \left\{ \begin{array}{ll} \propto D n_q (E-E_g^{ID}+\hbar\omega_q)^2 & \text{for } E>E_g^{ID}-\hbar\omega_q \\ = 0 & \text{for } E<E_g^{ID}-\hbar\omega_q \end{array} \right. \qquad (5.6)$$

The absorption of light thus begins at $E=E_g^{ID}-\hbar\omega_q$. Another contribution to the indirect optical transitions is due to the emission of a phonon and can be obtained using the same procedure, where the only differences from the previous case are the sign of the phonon energy $\hbar\omega_q$ and $n_q \rightarrow (n_q+1)$.

We see that the *"allowed"* indirect transitions have the square energy dependence given by Eq. (5.6). In case where the indirect transitions are forbidden we proceed as for direct transitions mentioned before. If the parameter D in Eq. (5.6) is replaced by its first derivative, again a factor to the first power in E comes in and we obtain a cubic dependence on energy E:

$$\varepsilon_2(E) \propto (n_q+\tfrac{1}{2}\mp\tfrac{1}{2})(E-E_g^{ID}\pm\hbar\omega_q)^3 \qquad (5.7)$$

Summarizing the four possible cases of transitions we have the energy exponents 1/2 for *direct "allowed"* [Eq. (3.14)], 3/2 for *direct "forbidden"* [Eq. (5.2)], 2 for *indirect "allowed"* [Eq. (5.6)], and 3 for *indirect "forbidden"* transitions [Eq. (5.7), see also see Fig. 5.1]. Such behaviors are often used to distinguish experimentally between the different types. As an example, we show in Fig. 5.3 the dependence

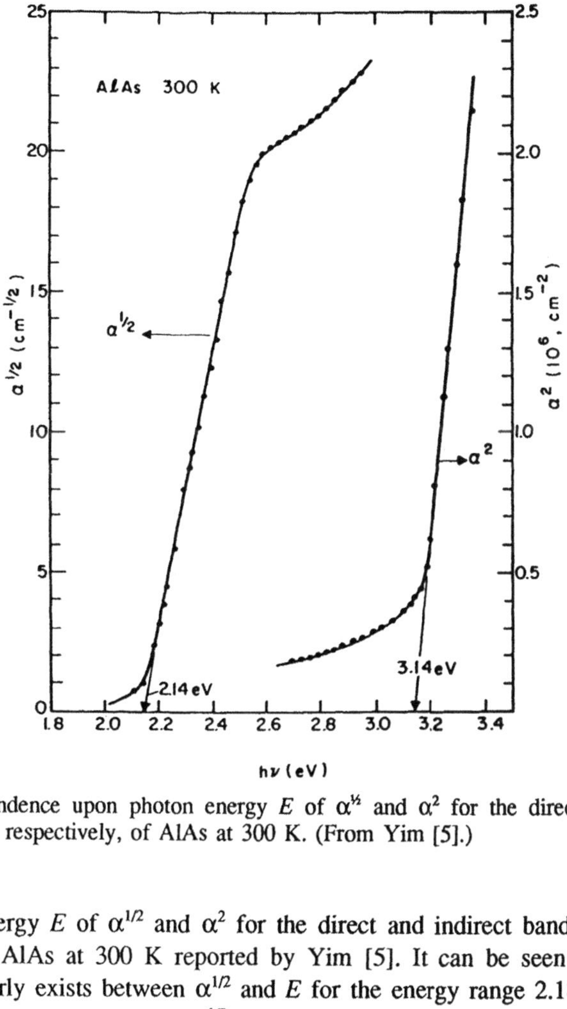

Figure 5.3 Dependence upon photon energy E of $\alpha^{1/2}$ and α^2 for the direct and indirect band-gap regions, respectively, of AlAs at 300 K. (From Yim [5].)

upon photon energy E of $\alpha^{1/2}$ and α^2 for the direct and indirect band-gap regions, respectively, of AlAs at 300 K reported by Yim [5]. It can be seen that a linear relationship clearly exists between $\alpha^{1/2}$ and E for the energy range 2.18 to 2.51 eV. By extrapolating the linear plot to $\alpha^{1/2}$=0, an indirect energy-gap value of 2.14±0.01 eV is obtained. Similarly, a linear relationship exists between α^2 and E for energies above 3.18 eV. The extrapolation of the linear portion to α^2=0 gives a direct energy-gap value of 3.14±0.01 eV.

5.1.2 Exciton Effect

(a) Direct Exciton—Let us consider the contribution of exciton transitions to $\varepsilon(E)$ at the fundamental absorption edge. The phenomenon of the excitons in crystals has been a subject of considerable interest for many years and has been summarized by Elliott [6], Knox [7], and Reynolds and Collins [8].

Because the Coulomb interaction is always present between the electrons and holes, the exciton transitions play an important role in the fundamental optical process. The exciton state can be expanded in terms of the Bloch functions for the perfect periodic lattice. The wave function of the exciton will be a linear combination of the conduction-band (Ψ_c) and valence-band wave function (Ψ_v). Let us now consider the case where the electron–hole interaction is relatively weak, i.e., the Wannier–Mott excitons. In this case, the lowest conduction band and highest valence band need to be included in the expansion of the eigenstate Ψ, so that

$$\Psi(\boldsymbol{r}_e, \boldsymbol{r}_h) = \sum_{\boldsymbol{k}_e, \boldsymbol{k}_h} C(\boldsymbol{k}_e, \boldsymbol{k}_h) \Psi^*_{c,\boldsymbol{k}_e}(\boldsymbol{r}_e) \Psi_{v,\boldsymbol{k}_h}(\boldsymbol{r}_h) \tag{5.8}$$

Only small ranges of $\boldsymbol{k}$ about the two extrema contribute appreciably to the sum of Eq. (5.8). Periodicity requires that the exciton wave vector $\boldsymbol{K}$ is a good quantum number and that

$$\boldsymbol{K} = \boldsymbol{k}_e + \boldsymbol{k}_h \tag{5.9}$$

The Fourier transformation of $C(\boldsymbol{k}_e, \boldsymbol{k}_h)$ may be written as

$$\Phi_{\boldsymbol{K},n} = \frac{1}{\sqrt{V_0 N}} e^{i\boldsymbol{KR}} \phi_n(\boldsymbol{r}) \tag{5.10}$$

where V_0 is the volume of a unit cell, N the number of such cells in the crystal, and $\boldsymbol{R}$ and $\boldsymbol{r}$ are the center-of-mass coordinate and relative position, respectively, being given by

$$\boldsymbol{R} = \frac{m_e \boldsymbol{r}_e + m_h \boldsymbol{r}_h}{m_e + m_h} \tag{5.11}$$

$$\boldsymbol{r} = \boldsymbol{r}_e - \boldsymbol{r}_h \tag{5.12}$$

Equation (5.10) represents the wave function of the exciton over-all motion in the crystal. The EMA (effective-mass approximation) gives Eq. (5.10) as a solution of the Schrödinger-like equation

$$\left[\mathcal{H}_e\left(\boldsymbol{p} + \frac{1}{2}\boldsymbol{P} - \boldsymbol{p}_e\right) + \mathcal{H}_h\left(-\boldsymbol{p} + \frac{1}{2}\boldsymbol{P} - \boldsymbol{p}_h\right) - \frac{e^2}{\varepsilon_0 \varepsilon_s |\boldsymbol{r}|} \right] \Phi = E\Phi \tag{5.13}$$

where $\boldsymbol{p}$, $\boldsymbol{P}$ are the momenta conjugate to $\boldsymbol{r}$, $\boldsymbol{R}$; $\boldsymbol{p}_e$, $\boldsymbol{p}_h$ are $\hbar$ times $\boldsymbol{k}_e$, $\boldsymbol{k}_h$; and ε_s is the static dielectric constant. E and Φ can be found by solving Eq. (5.13). The exciton energy E_{ex} of interest here is

$$E_{ex} = E_0 + E \qquad (5.14)$$

where $E_0=E_c(\boldsymbol{k}=0)-E_v(\boldsymbol{k}=0)$ is the band-gap energy at $\boldsymbol{k}=0$.

The two-particle wave equation, Eq. (5.13), can be expressed as a sum of two separate terms if it is written in a coordinate system of the center-of-mass coordinate $\boldsymbol{R}$ (*translational part*) and the electron–hole separation part $\boldsymbol{r}$ (*rotational part*). The translational part can be written as

$$-\frac{\hbar^2}{2M}\nabla_R^2\Phi_R(\boldsymbol{R}) = E_k\Phi_R(\boldsymbol{R}) \qquad (5.15)$$

which describes just the motion of a free particle with a mass

$$M = m_e + m_h \qquad (5.16)$$

Equation (5.15) gives that

$$E_K = \frac{\hbar^2|\boldsymbol{K}|^2}{2M} \qquad (5.17)$$

and

$$\Phi_R(\boldsymbol{R}) = e^{i\boldsymbol{KR}} \qquad (5.18)$$

The rotational part, on the other hand, can be written as

$$\left[-\frac{\hbar^2}{2\mu}\nabla_r^2 - \frac{e^2}{\varepsilon_0\varepsilon_s|\boldsymbol{r}|}\right]\phi_n(\boldsymbol{r}) = E_n\phi_n(\boldsymbol{r}) \qquad (5.19)$$

This equation is similar to that for the hydrogenic problem, but with the electron charge replaced by $e/\sqrt{\varepsilon_s}$ and the free electron mass m_0 replaced by the exciton reduced mass μ:

$$\frac{1}{\mu} = \frac{1}{m_{o,e}} + \frac{1}{m_{o,h}} \qquad (5.20)$$

where $m_{o,e}$ and $m_{o,h}$ are the optical masses of the electron and hole, respectively. The optical masses can now be defined in terms of the longitudinal mass, m_l, and the transverse mass, m_t, by

$$\frac{1}{m_o} = \frac{1}{3}\left(\frac{1}{m_l} + \frac{2}{m_t}\right) \qquad (5.21)$$

For holes at k=0 (Γ), one has to take the HH and LH masses [9]:

$$\frac{1}{m_{o,h}} = \frac{1}{2}\left(\frac{1}{m^s_{HH}} + \frac{1}{m^s_{LH}}\right) \quad (5.22)$$

with

$$\frac{m^s_{HH}}{m_0} = \frac{1}{\gamma_1(1-\mu_o)} \quad (5.23a)$$

$$\frac{m^s_{LH}}{m_0} = \frac{1}{\gamma_1(1+\mu_o)} \quad (5.23b)$$

where

$$\mu_o = \frac{6\gamma_3+4\gamma_2}{5\gamma_1} \quad (5.24)$$

and γ_i is the Luttinger parameter.

The eigen value of Eq. (5.19) can be given, as an analogy of the hydrogenic problem, as

$$E_n = -\frac{G^{3D}}{n^2}, \qquad n=1,2,3,\ldots\ldots \quad (5.25)$$

where G^{3D} is the exciton binding (Rydberg) energy given by

$$G^{3D} = \frac{\mu e^4}{2\hbar^2(\varepsilon_0\varepsilon_s)^2} = 13.6\frac{\mu/m_0}{\varepsilon_s^2} \quad \text{(in eV)} \quad (5.26)$$

The corresponding exciton Bohr radius $a_B{}^{3D}$ is written as

$$a_B^{3D} = \frac{\hbar^2\varepsilon_0\varepsilon_s}{\mu e^2} = 0.53\frac{\varepsilon_s}{\mu/m_0} \quad \text{(in Å)} \quad (5.27)$$

In obtaining Eqs. (5.26) and (5.27) [see also Eqs. (5.13) and (5.19)], we assumed that the relative dielectric constant is equal to the static dielectric constant ε_s. This is because the exciton binding energy is usually smaller than the lattice vibrational energy (LO phonons) in many semiconductors. If not so, we must use ε_∞ instead of ε_s. In Table 5.1, we summarize the exciton binding energy G^{3D}, 1st orbital (n=1) Bohr radius a_B^{3D}, and exciton reduced mass μ for various semiconductors.

It is clear from Eq. (5.19) that $\phi_n(r)$ is a hydrogenic wave function with substituted for $m_0 \to \mu$ and $\varepsilon_0 \to \varepsilon_0\varepsilon_s$. Since $\phi_n(r)$ is a smooth function extending over a large region of the crystal, it is usually called "*envelope function.*" Because $\phi_n(r)$ describes the rotational motion of the exciton, it is characterized by the three quantum numbers n (*principal*), l (*azimuthal*), and m (*magnetic*). However, it is known that only the s-rotational state (l=0) is important for the optically created excitons. It is, therefore, sufficient for our purpose to label the envelope function only with one quantum number n.

For the simple case of two spherical bands of masses m_e and m_h, the envelope function of the nth-exciton state $\phi_n(0)$ can be written as

$$|\phi_n(0)|^2 = \frac{V_0}{\pi (a_B^{3D})^3 n^3} \equiv |\phi_1(0)|^2 \frac{1}{n^3} \qquad (5.28)$$

A series of exciton lines can, then, be predicted at energies

$$E_{ex}^n = E_0 + \frac{\hbar^2 K^2}{2M} - \frac{G^{3D}}{n^2} \qquad (5.29)$$

with intensity falling like n^{-3} through Eq. (5.28). The contribution of the discrete excitons to ε_2 can also be given by [6]

$$\varepsilon_2^D(E) = \frac{f^D}{E^2} \sum_{n=1}^{\infty} \frac{1}{n^3} |\phi_1(0)|^2 \delta(E_{ex}^n - E) \qquad (5.30)$$

where f^D is the strength parameter proportional to the squared momentum-matrix element.

The KK transformation of Eq. (5.30) gives

$$\varepsilon_1^D(E) = \sum_{n=1}^{\infty} \frac{F^D}{n^3[(E_{ex}^n)^2 - E^2]} \qquad (5.31)$$

where, for simplicity, we replaced $2E_{ex}^n f^D |\phi_1(0)|^2/\pi$ by F^D. The contribution of the discrete exciton series to $\varepsilon(E)$ [$\varepsilon_1^D(E)$=Re $\varepsilon(E)$, $\varepsilon_2^D(E)$=Im $\varepsilon(E)$] can be finally written as

$$\varepsilon^D(E) = \sum_{n=1}^{\infty} \frac{1}{n^3} \frac{F^D}{[(E_0 - (G^{3D}/n^2)]^2 - (E + i\Gamma)^2} \qquad (5.32)$$

where Γ is the exciton lifetime-broadening energy.

As the absorption edge approached, the infinite number of lines will overlap so that it may be considered as a continuum. In the true continuum where $E - E_0 > 0$,

TABLE 5.1 *Exciton binding energy G (meV), 1st orbital Bohr radius a_B^{3D} (Å), and exciton reduced mass μ (in units of electron rest mass) for various semiconductors. W=Wurtzite, ZB=Zincblende, D=Direct exciton, and ID=Indirect exciton.*

Material	G^{3D} (G^{ID})	a_B^{3D}	μ	Remark
C (Diamond)	80[a]	16	0.191	ID
Si	14.7[b]	42	0.145	ID
Ge	4.15 (±3/2)[c]	108	0.079	ID
	3.14 (±1/2)[c]	143	0.059	ID
3C-SiC	26.7[d]	28	0.178	ID
AlAs	18[e]	40	0.134	ID
	13[e]	55	0.097	D
α-GaN (W)*	20(A)[f]	39	0.128	D
	26.7(A)[g]	29	0.171	D
	20(B)[f]	39	0.128	D
	18(C)[f]	43	0.115	D
GaP	22[h]	29	0.201	ID
	5[i]	129	0.046	D
GaAs	3.6[e]	155	0.044	D
GaSb	1.5[j]	306	0.027	D
InP	4.8[k]	120	0.055	D
InAs	1.0[l]	494	0.016	D
InSb	0.4[m]	1017	0.009	D
ZnO (W)	63.1(A)[n]	13	0.338	D
	50.4(B)[n]	17	0.270	D
	48.9(C)[n]	17	0.262	D
α-ZnS (W)	40(A)[o]	19	0.271	D
	41(B)[o]	18	0.278	D
β-ZnS (ZB)	34.2[p]	24	0.199	D
ZnSe	18.7[p]	41	0.119	D
ZnTe	12.8[q]	62	0.078	D
w-CdS (W)	30.2(A)[r]	28	0.166	D
	30.8(B)[r]	27	0.169	D
w-CdSe (W)	15(A)[r]	49	0.104	D
	16(B)[r]	46	0.111	D
CdTe	10.5[s]	65	0.085	D

*In wurtzite crystals with anisotropy of the dielectric constant and effective band masses, the exciton effect is discussed in detail by Flohrer *et al.* [*Phys. Status Solidi* **91**, 467 (1979)].

[a]P. J. Dean *et al.*, *Phys. Rev.* **140**, A352 (1965).

[b]See N. O. Lipari and M. Altarelli, *Phys. Rev. B* **15**, 4883 (1977).

[c]A. Frova *et al.*, *Phys. Rev. Lett.* **34**, 1572 (1975).

[d]Calculated value [D. Bimberg *et al.*, *Solid State Commun.* **40**, 437 (1981)]. As the exciton binding energy exceeds the optical phonon energy in this material, the high-frequency dielectric constant (ε_∞=6.52) rather than the static dielectric constant (ε_s=9.72) was used in the exciton calculation.

[e]S. Adachi, *GaAs and Related Materials: Bulk Semiconducting and Superlattice Properties* (World Scientific, Singapore, 1994).

[f]G. D. Chen *et al.*, *Appl. Phys. Lett.* **68**, 2784 (1996).

[g]D. Volm *et al.*, *Phys. Rev. B* **53**, 16543 (1996).

[h]M. D. Sturge *et al.*, *Appl. Phys. Lett.* **32**, 49 (1978).

[i]P. J. Dean *et al.*, *J. Appl. Phys.* **38**, 3551 (1967).

[j]C. Ghezzi *et al.*, *Phys. Rev. B* **52**, 1463 (1995).
[k]A. M. White *et al.*, *J. Phys. C* **5**, 1727 (1972); S. B. Nam *et al.*, *Phys. Rev. B* **13**, 1643 (1976).
[l]P. J. P. Tang *et al.*, *Phys. Rev. B* **55**, 4376 (1997).
[m]See A. Baldereschi and N. O. Lipari, *Phys. Rev. B* **3**, 439 (1971).
[n]A. Mang *et al.*, *Solid State Commun.* **94**, 251 (1995).
[o]Estimated from J. C. Miklosz and R. G. Wheeler, *Phys. Rev. B* **3**, 439 (1971) $[G^{3D}=4(E_{ex}^{2s}-E_{ex}^{1s})]$.
[p]A. Mang *et al.*, *Phys. Rev. B* **53**, 16283 (1996).
[q]H. Leiderer *et al.*, *J. Phys.: Condens. Mater.* **6**, 6057 (1994).
[r]J. Voigt *et al.*, *Phys. Status Solidi B* **91**, 189 (1979).
[s]M. Nawrocki and A. Twardowski, *Phys. Status Solidi B* **97**, K61 (1980).

one obtains for the envelope function of the continuum state

$$|\phi_k(0)|^2 = \frac{\pi\alpha_k e^{\pi\alpha_k}}{N\sinh(\pi\alpha_k)} = \frac{2\pi\alpha_k}{N(1-e^{-2\pi\alpha_k})} \qquad (5.33)$$

with

$$\alpha_k = \left|\frac{G^{3D}}{E-E_0}\right|^{1/2} = \left|\frac{G^{3D}}{(\hbar^2k^2/2\mu)}\right|^{1/2} \qquad (5.34)$$

The continuum exciton state corresponds to positive energy solutions of the hydrogen-like equation (5.19). The energy of this type of excitons may be written as

$$E_{ex}^k = E_0 + \frac{\hbar^2K^2}{2M} + \frac{\hbar^2k^2}{2\mu} \qquad (5.35)$$

The contribution of the continuum excitons to ε_2 is then written as [6]

$$\varepsilon_2^C(E) = \frac{f^C}{E^2}\sum_k |\phi_k(0)|^2\,\delta(E_{ex}^k-E) \qquad (5.36)$$

where f^C is the continuum-exciton strength parameter.

Equation (5.36) is known to give continuous exciton absorption at $E-E_0 \geq 0$ when $\alpha_k \to \infty$ [6]. For $E-E_0 \gg G^{3D}$ where $\alpha_k \to 0$, $\varepsilon_2^C(E)$ will become proportional to $(E-E_0)^{1/2}$, in agreement with that neglected the electron-hole interaction [i.e., Eq. (3.14)]. If we neglect the exponential in the denominator of Eq. (5.33), i.e.,

$$\frac{\pi\alpha_k e^{\pi\alpha_k}}{\sinh(\pi\alpha_k)} \sim 2\pi\alpha_k \qquad (5.37)$$

the KK transformation of Eq. (5.36) gives

$$\varepsilon_1^{C}(E) = \frac{F^{C}E_{ex}^{C1}}{G^{3D}E^2}\ln\frac{E_0^2}{E_0^2-E^2} \tag{5.38}$$

where E_{ex}^{C1} is the ground-state exciton energy ($\sim E_0$) and F^{C} is the strength constant having a similar physical meaning as F_n^{D} in Eq. (5.31). Finally, the contribution of the continuum excitons to $\varepsilon(E)$ [$\varepsilon_1^{D}(E)$=Re $\varepsilon(E)$, $\varepsilon_2^{D}(E)$=Im $\varepsilon(E)$] can be given by

$$\varepsilon^{C}(E) = \frac{F^{C}E_{ex}^{C1}}{4G^{3D}(E+i\Gamma)^2}\ln\frac{E_0^2}{E_0^2-(E+i\Gamma)^2} \tag{5.39}$$

In the true continuum, the rotational energy is much larger than the Coulomb interaction energy, and therefore the excitons behave like free particles with an effective mass μ. Consequently, the 3D-exciton spectrum consists of a series of discrete parabolic bands below E_0 at $\boldsymbol{K}$=0, which merges into a continuum at higher energies, as depicted in Fig. 5.4.

We show in Figs. 5.5 (b) and (c) the calculated line shapes of $\varepsilon(E)$ for the discrete- and continuum-exciton transitions, respectively. They are obtained from Eqs. (5.32) and (5.39), respectively. The contribution to $\varepsilon(E)$ of the free electron–hole pair transitions [Eq. (3.10)] is also shown in Fig. 5.5 (a). The numerical parameters used here are: E_0=1.5 eV, G^{3D}=10 meV, and Γ=2 meV. The strength parameters, $AE_0^{-3/2}$, F^{D}, and F^{C}, are properly chosen such that they give a peak or maximum value of $\varepsilon_2 \sim 5$ in their respective spectrum.

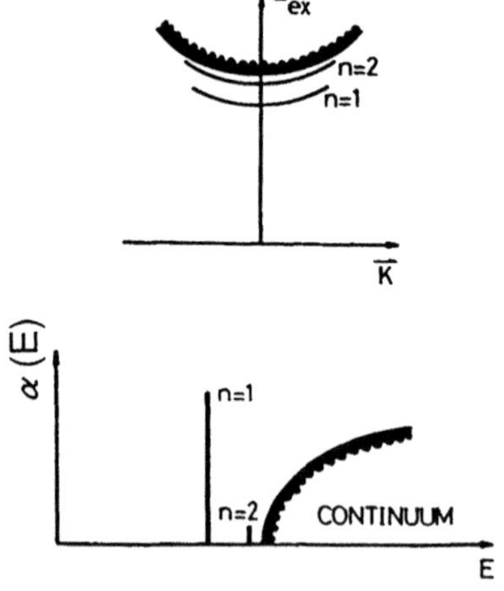

Figure 5.4 Schematic diagram showing the energy and absorption coefficient of the Wannier–Mott excitons.

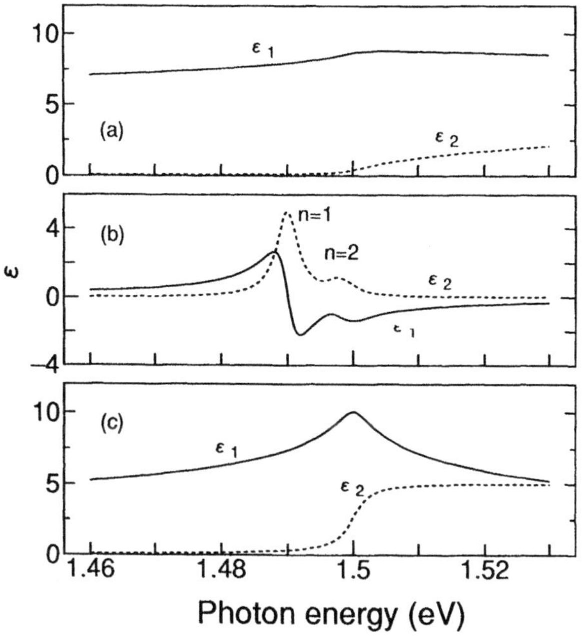

Figure 5.5 Line shapes of $\varepsilon(E)$ for (a) the free electron-hole pair, (b) discrete-exciton, and (c) continuum-exciton transitions, respectively. They are obtained from Eqs. (3.10), (5.32), and (5.39), respectively. The numerical parameters used here are: E_0=1.5 eV, G^{3D}=10 meV, and Γ=2 meV. The strength parameters, $AE_0^{-3/2}$, F^D, and F^C, are properly chosen such that they give a peak or maximum value of ε_2~5 in their respective spectrum.

It is noting that the discrete-exciton term [ε_1^D, (b)] shows a sharp dispersion compared with the continuum-exciton term [ε_1^C, (c)] or free electron–hole pair term [(a)]. It is also understood that the ε^C term [(c)] shows similar dispersion to that of the free electron–hole pair term [(a)].

In a case where transitions are forbidden in the dipole approximation, Eq. (5.28) can be modified as [6]

$$\left|\frac{\partial\phi_n(0)}{\partial x}\right|^2 = \frac{V_0}{3\pi(a_B^{3D})^5}\frac{n^2-1}{n^5} \equiv \left|\frac{\partial\phi_1(0)}{\partial x}\right|^2 \frac{n^2-1}{n^5} \tag{5.40}$$

where x is a quantity which is roughly the atomic radius. The corresponding ε_2 can be written as

$$\varepsilon_2^{\mathrm{D,F}} = \frac{f^{\mathrm{D,F}}}{E^2}\sum_{n=1}^{\infty}\frac{n^2-1}{n^5}\left|\frac{\partial\phi_1(0)}{\partial x}\right|^2 \delta(E_{\mathrm{ex}}^{n}-E) \qquad (5.41)$$

A series of exciton lines thus occur at energies given by Eq. (5.29), except that the n=1 line is now missing.

Similarly, in the true continuum we have

$$\left|\frac{\partial\phi_k(0)}{\partial x}\right|^2 = \frac{\pi\alpha_k(1+\alpha_k^2)\mathrm{e}^{\pi\alpha_k}x^2}{3N\sinh(\pi\alpha_k)} \qquad (5.42)$$

The contribution of the forbidden continuum-exciton transitions to ε_2 can, then, be written as [6]

$$\varepsilon_2^{\mathrm{C,F}}(E) = \frac{f^{\mathrm{C,F}}}{E^2}\sum_{k}\left|\frac{\partial\phi_k(0)}{\partial x}\right|^2 \delta(E_{\mathrm{ex}}^{k}-E) \qquad (5.43)$$

Equation (5.43) gives continuous exciton absorption at E-$E_0 \geq 0$ when $\alpha_k \rightarrow \infty$ [6]. For E-$E_0 \gg G^{3D}$ where $\alpha_k \rightarrow 0$, $\varepsilon_2^{\mathrm{C,F}}(E)$ will become proportional to $(E-E_0)^{3/2}$, in agreement with that neglected the electron-hole interaction [i.e., Eq. (5.2)].

(b) Indirect Exciton—Exact calculation of the exciton effect at the indirect band gap is quite complicated, and it is convenient at this stage to make approximations and consider a more specific model. Let us assume that the maximum in the valence band is at $\boldsymbol{k}=\boldsymbol{k}_1$ and is nondegenerate, the minimum in the conduction band at $\boldsymbol{k}=\boldsymbol{k}_2$ and nondegenerate (see Fig. 5.2). The envelope function $\phi(0)$ is also assumed to be independent of $\boldsymbol{k}$. The resultant $\varepsilon_2(E)$ formula can be given by [6]

$$\varepsilon_2^{\mathrm{D}}(E) = \frac{F^{\mathrm{ID}}}{E^2}(n_{\mathrm{q}}+\tfrac{1}{2}\mp\tfrac{1}{2})(E+G^{\mathrm{ID}}-E_{\mathrm{g}}^{\mathrm{ID}}\pm\hbar\omega_{\mathrm{q}})^{1/2} \qquad (5.44)$$

where G^{ID} is the binding energy of the indirect exciton (see Table 5.1). [It should be noted that G^{ID} can be expressed by the same expression as Eq. (5.26) and indirect-exciton Bohr radius as Eq. (5.27).] The functional form of Eq. (5.44) is almost the same as that of Eq. (3.14). The KK transformation of Eq. (5.44) then gives

$$\varepsilon^{\mathrm{D}} = F^{\mathrm{ID}}(n_{\mathrm{q}}+\tfrac{1}{2}\mp\tfrac{1}{2})(E_{\mathrm{g}}^{\mathrm{ID}}-G^{\mathrm{ID}}\mp\hbar\omega_{\mathrm{q}})^{-3/2}f(\chi_{\mp}) \qquad (5.45)$$

with

$$f(\chi_{\mp}) = \chi_{\mp}^{-2}\,[2-(1+\chi_{\mp})^{1/2}-(1-\chi_{\mp})^{1/2}] \qquad (5.46a)$$

$$\chi_{\mp} = \frac{E+i\Gamma}{E_g^{ID}-\dfrac{G^{ID}}{n^2}\mp\hbar\omega_q} \qquad (5.46b)$$

Here, $\varepsilon_1^{ID}(E)$=Re $\varepsilon^{ID}(E)$ and $\varepsilon_2^{ID}(E)$=Im $\varepsilon^{ID}(E)$.

Figure 5.6 shows the $\varepsilon_2(E)$ spectra calculated from Eq. (5.45). The numerical parameters used are: E_g^{ID}=1.5 eV, G^{ID}=10 meV, $\hbar\omega_q$=25 meV, and Γ=2 meV (T=4.2 K) and 20 meV (T=300 K). As seen in the figure, the exciton transitions into bound states below E_g^{ID} yield a continuous absorption, in contrast to discrete exciton series at the direct-gap edge. The reason for this is that transitions can occur to any point of an exciton band with appropriate phonons participating. At lower temperatures, the intensity of the phonon absorption (anti-Stokes) component is much weaker than that of the phonon emission (Stokes) component, because usually there are few phonons to be absorbed compared to the density of phonons that can be emitted, the probability for absorption being lower than the probability for emission by a factor of exp ($\hbar\omega_q/kT$).

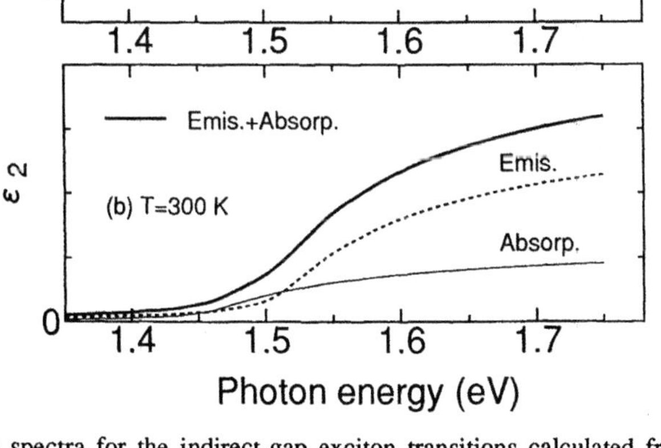

Figure 5.6 $\varepsilon_2(E)$ spectra for the indirect-gap exciton transitions calculated from Eq. (5.41) (n=1). The numerical parameters used here are: E_g^{ID}=1.5 eV, G^{ID}=10 meV, $\hbar\omega_q$=25 meV, and Γ=2 meV (T=4.2 K) and 20 meV (T=300 K).

Beyond the series limit, i.e., $E>E_g^{ID}\pm\hbar\omega_q$, there is further absorption from the continuum. When $E-E_g^{ID}\pm\hbar\omega_q \gg G^{ID}$, we may obtain [6]

$$\varepsilon_2^{C}(E) = \frac{F^{IC}}{E^2}(n_q+½\mp½)(E-E_g^{ID}\pm\hbar\omega_q)^2 \tag{5.47}$$

Note that this expression has the same power dependence as that obtained from the one-electron approximation [i.e., $(E-E_g^{ID}\pm\hbar\omega_q)^2$, see Eq. (3.32) or Eq. (5.6)]. The KK transformation of Eq. (5.47) thus gives the same result as Eq. (3.36) (except some proportionality constants).

In cases where transitions are dipole forbidden, we obtain

$$\varepsilon_2^{D,F}(E) = \frac{F^{ID,F}}{E^2}(n_q+½\mp½)(E+G^{ID}-E_g^{ID}\pm\hbar\omega_q)^{3/2} \tag{5.48}$$

for the lower-energy series, and

$$\varepsilon_2^{C,F}(E) = \frac{F^{IC,F}}{E^2}(n_q+½\mp½)(E-E_g^{ID}\pm\hbar\omega_q)^3 \tag{5.49}$$

for the continuum.

5.1.3 Urbach Tail

For energies E ($=\hbar\omega$) near the fundamental absorption edge, the absorption coefficient $\alpha(E,T)$ shows a near universal simple exponential energy dependence

$$\alpha(E,T) = \alpha_0 \exp\left(\frac{E-E_g(T)}{E_u(T)}\right) \tag{5.50}$$

denoted the Urbach tail [10], where $E_g(T)$ is an energy close to the energy gap and $E_u(T)$ determines the width of the tail. The tail is generally attributed to disorder in the material that leads to a tail in the valence and conduction bands. At low temperature, dopant impurities as well as other structural imperfections introduce disorder. At high temperature, the width of the tails is a direct measure of temperature-induced disorder and reflects the thermal occupancy of phonon states in the BZ via the electron–phonon interaction [11,12]. We can, therefore, suspect that the Urbach parameter $E_u(T)$ has the same functional form as that of the CP linewidth broadening [see Eq. (3.43)]. We now write [13]

$$E_u(T) = S_u\left(\frac{1+X}{2} + \frac{1}{e^{\theta/T}-1}\right) \qquad (5.51)$$

where X is a dimensionless parameter describing the contribution of "frozen-in" structural disorder, to the width of the Urbach edge. In a well-ordered crystal, X is considered to be equal to zero.

As mentioned in Sec. 3.2.1, the temperature-induced change in the band-gap energy E_g can be given by the Varshni equation [Eq. (3.42)]. The effect of thermal expansion is a trivial increase of the lattice constant with increasing temperature thereby shrinking the band-gap energy. Beside the changes of band structure induced by thermal expansion of the lattice, the temperature dependence of E_g is mainly due to the electron-phonon interactions [Debye–Waller term (two-phonon process) and Fano (self-energy) term (one-phonon process)] [14]. The temperature dependence of the band gap is proportional to the same phonon occupation factor as in Eq. (5.51) and, accordingly, can be described by the same functional form:

$$E_g(T) = E_g(0) - S_g\left(\frac{1}{e^{\theta/T}-1}\right) \qquad (5.52)$$

where S_g is a coupling constant. Thus, we obtain

$$E_g(T) = E_g(0) + GE_u(0) - GE_u(T) \qquad (5.53)$$

where G is a proportionality constant that relates the temperature dependence of the band gap to that of the Urbach parameter ($G=S_g/S_u$).

5.1.4 Temperature, Pressure, and Doping Effects

(a) Temperature Effect—Figure 5.7 (a) shows the absorption spectra of high-quality GaAs measured by Sturge [15] over the range of photon energy 0.6 to 2.75 eV at four different temperatures, 21, 90, 186, and 294 K. The sample used in this experiment was cut from semiinsulating ingot of GaAs in which the carrier concentration was less than 10^{10} cm^{-3}. As can be seen in the figure, the saturated value of α at high energies (just above the band edge) does not depend so largely on temperature between 21 and 294 K [α~(8.5–9.5)$\times10^3$ cm^{-1}]. At low temperatures, the main absorption edge shows a sharp peak due to the formation of the discrete exciton (n=1). If we fit these spectra to Eq. (5.32), we obtain the exciton parameters of $F^D(n=1)=2.35\times10^{-3}$ eV, G^{3D}=4.7 meV, Γ=5.0 meV (T=21 K), and Γ=25 meV (T=294 K) [16]. The steplike component seen in the true continuum region ($E>E_0$) may suggest the presence of continuum excitons in the transition process.

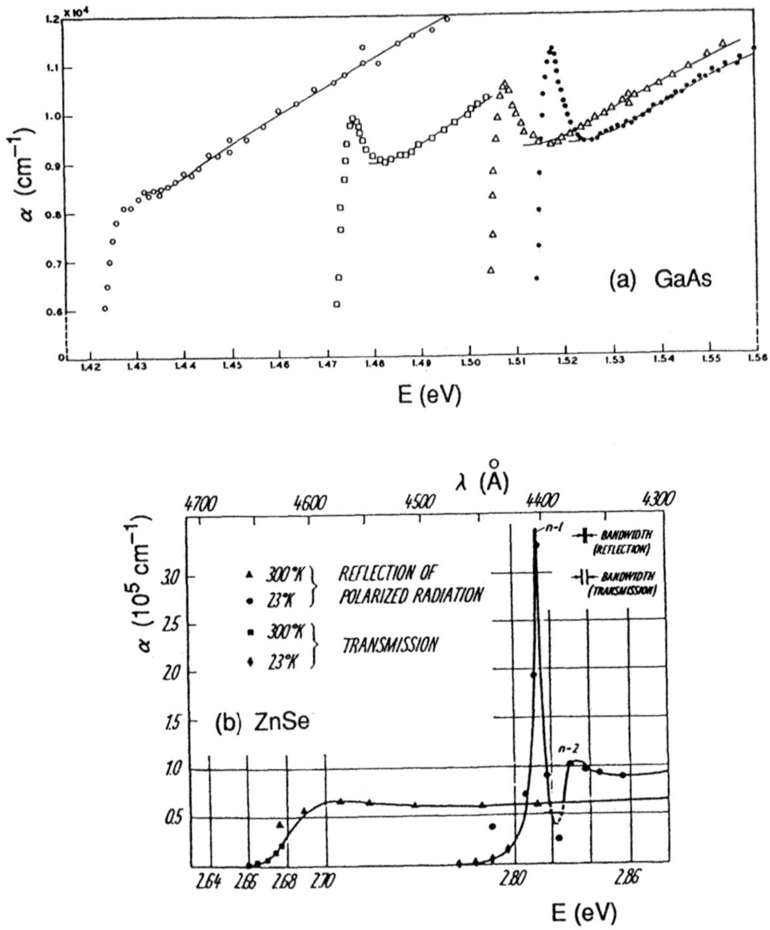

Figure 5.7 (a) Exciton absorption in GaAs at T=21 (open circles), 90 (open squares), 186 (open triangles), and 294 K (solid circles). (From Sturge [15].) (b) Exciton absorption in ZnSe at T=23 and 300 K. (From Aven *et al.* [17].)

Sturge [15] also observed an exponential broadening at the low energy side of the absorption edge that is known as the Urbach effect (Sec. 5.1.3).

Figure 5.7 (b) shows the absorption results for ZnSe measured by Aven *et al.* [17] at two different temperatures, 23 and 300 K. The crystals used in the experiment were grown by a modified Piper–Polich method. As in the case of GaAs, the n=1 discrete-exciton peak can be clearly resolved in the low-temperature spectra of ZnSe. It is also possible to identify the n=2 exciton peak in the spectrum. The damping effect smears out the n=1, 2 exciton peaks and, as a result, such peaks could hardly be found in the room-temperature spectra. The n=1 exciton energy was 2.81 eV at T=23 K. From the energy difference between the n=1 and 2 states and assuming that the exciton is hydrogenic, these authors obtained an exciton Rydberg energy of G^{3D}=20 meV for ZnSe. They also obtained the exciton reduced mass to be $\mu=(0.10\pm0.03)m_0$ and the electron and hole effective masses of $m_{o,e}=0.1m_0$ and $m_{o,h}=0.6m_0$ [see Eq. (5.20)], respectively. Considering that G^{3D}(ZnSe)>G^{3D}(GaAs), we can conclude that the exciton effect is much stronger in ZnSe than in GaAs.

The GaP crystal is known to be a more suitable material to study some of the indirect-gap optical processes, since it has three indirect gaps, $\Gamma_8^v\rightarrow X_6^c$ near 2.26 eV, $\Gamma_8^v\rightarrow X_7^c$ near 2.48 eV, and $\Gamma_8^v\rightarrow L_6^c$ near 2.63 eV, as well as the lowest-direct gap E_0, $\Gamma_8^v\rightarrow\Gamma_6^c$ near 2.74 eV. The indirect transitions in GaP occur at below the onset of the direct-gap transitions which occurs at 2.74 eV (E_0). This is in direct contrast to the case for GaAs (ZnSe) in which the indirect transitions may occur after the onset of the direct-gap transitions.

We reproduce in Fig. 5.8 the optical absorption spectra, $\alpha^{1/2}$ vs E, at the fundamental absorption edge of GaP measured by Dean and Thomas [4]. As expected from discussion in Sec. 5.1.2 (b), the stepped structure characteristic of the intrinsic absorption edge of crystals with an indirect band gap is very well defined in the low-temperature spectra of this figure. The thresholds of the steps are labeled according to the type of phonons. The subscripts denote whether the phonon is emitted (E) or absorbed (A) during the optical absorption process. The phonon absorption components were found to be only observed above about 40 K, and continuously increased in strength relative to the phonon-emission components as the temperature was increased towards 300 K. Assuming that E_{gx} is the exciton energy gap, we have the eight absorption (−)–emission (+) thresholds: $E_{gx}\pm$TA, $E_{gx}\pm$LA, $E_{gx}\pm$TO, and $E_{gx}\pm$(LO$^\Gamma\pm$TA). The phonon energies estimated from this work were TA=12.8±0.5 meV, LA=31.3±0.5 meV, TO= 46.5±1.0 meV, and LO$^\Gamma\pm$TA=64.3±1.0 meV.

The temperature dependence of the Urbach tail in GaAs has recently been studied by Johnson and Tiedje [13]. Both the optical absorption and temperature were measured using a diffuse reflectance technique. We reproduce in Fig. 5.9 (a) the absorption coefficient α as a function of E for undoped, semiinsulating and Si-doped ($n=2\times10^{18}$ cm^{-3}) GaAs obtained by these authors [13]. Both the GaAs samples show the exponential Urbach tail along with a weakly energy-dependent absorption at lower energies. The lower-energy absorption in n-GaAs was considered to be due to phonon-assisted interconduction-band transitions [see Sec. 5.3.2 (a)].

Figure 5.9 (b) shows the Urbach parameter E_u as a function of T for undoped and n-type (2×10^{18} cm^{-3}) GaAs samples [13]. The solid lines show the best-fit results of Eq. (5.51) to the experimental data. The best-fit parameters obtained were S_u=1.98±0.11 meV, X=5.2±0.3, and θ=263±3 K for undoped GaAs; S_u=1.27±0.05 meV, X=17.7±0.7, and θ=263 K for n-GaAs. We can see that the

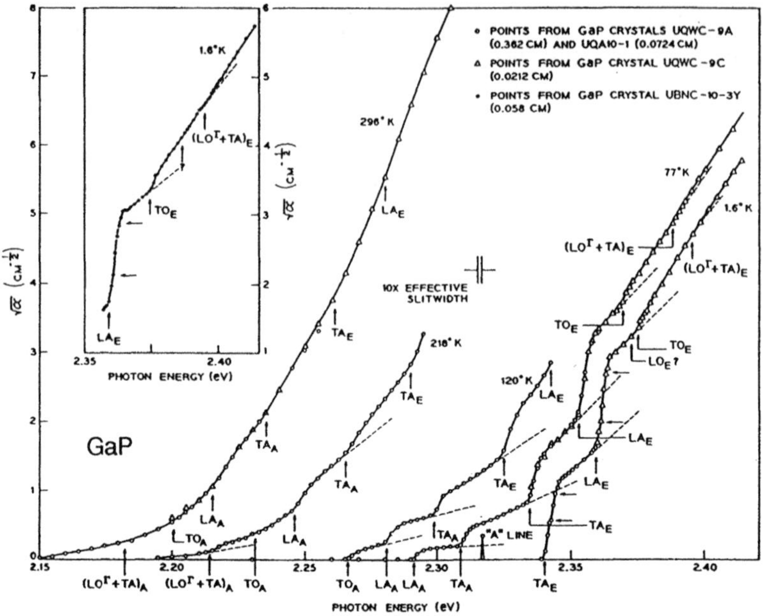

Figure 5.8 Optical absorption at the indirect edge of GaP at T=1.6, 77, 120, 218, and 296 K. Vertical arrows indicate the thresholds of different absorption components resulting from different phonon energies, and horizontal arrows denote structure interpreted in terms of electronic degeneracy. (From Dean and Thomas [4].)

characteristic energy E_u of the exponential tail increases linearly with temperature from 7.5 meV (13.0 meV) at room temperature to 12.0 meV (15.5 meV) at 700°C for undoped GaAs (n-GaAs). The 7.5-meV Urbach parameter value for undoped GaAs at room temperature is in agreement with other, namely, 6.9 meV reported in Ref. [15]. It is also found that the temperature-dependent part of the Urbach tail for undoped GaAs is six times smaller than that predicted by the standard theory [13] where the edge width is proportional to the phonon population.

(b) Pressure Effect—As mentioned in Sec. 3.2.2, most tetrahedral semiconductors show structural phase transitions under hydrostatic pressures. In the low pressure range (p<1 GPa), optical gaps of semiconductors shift linearly with p. When measured in the range of the diamond anvil cell, nonlinear, usually sublinear, variations are seen [18]. At these pressures most tetrahedral semiconductors undergo phase transitions, often into metallic phases. The materials, as a result, become strongly opaque to visible and near IR radiation.

The hydrostatic-pressure variation of the band-gap energy $E_g(p)$ can now be simply written as [cf., Eq. (3.45)]:

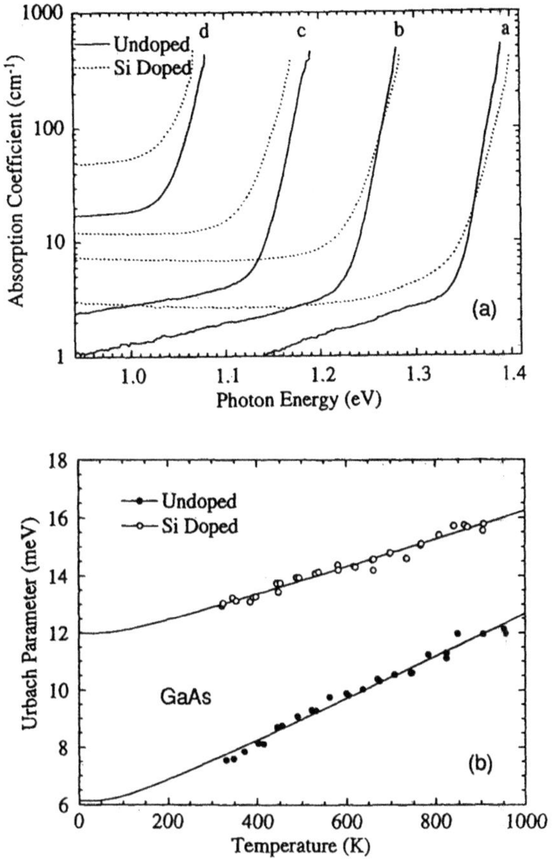

Figure 5.9 (a) Absorption coefficient as a function of photon energy for undoped, semiinsulating and Si-doped (n=2×10^{18} cm^{-3}) GaAs. The temperatures corresponding to the absorption curves are as follows: *a*, 58 and 52°C; *b*, 258 and 258°C; *c*, 435 and 426°C; and *d*, 631 and 632°C for the undoped and Si-doped GaAs, respectively. (b) Urbach parameter E_u as a function of T for undoped, semiinsulating and Si-doped (n=2×10^{18} cm^{-3}) GaAs. The solid lines in (b) show the best-fit results of Eq. (5.51) to the experimental data. (From Johnson and Tiedje [13].)

$$E_g(p) = E_g(0) + ap + bp^2 \qquad (5.54)$$

Figure 5.10 shows the hydrostatic-pressure dependence of the E_0, E_g^X, and E_g^L gaps for GaAs at 300 K. The energy gaps, as a function of p, used in this calculation are [9]

$$E_0(p) = 1.424 + 1.05\times10^{-1}p \quad \text{eV} \qquad (5.55a)$$

$$E_g^X(p) = 1.900 - 1.5\times10^{-2}p \quad \text{eV} \qquad (5.55b)$$

$$E_g^L(p) = 1.708 + 5.5\times10^{-2}p \quad \text{eV} \qquad (5.55c)$$

with p in GPa. The dashed line in Fig. 5.10 represents the E_0 energy as a function of p (in GPa) obtained from the following quadratic expression (in eV) [19]:

$$E_0(p) = 1.43 + 1.08\times10^{-1}p - 1.4\times10^{-3}p^2 \qquad (5.56)$$

Figure 5.10 suggests that GaAs becomes an indirect-gap semiconductor at p~4 GPa. We can also expect from this figure the indirect–indirect (L_6^c–X_6^c) crossover at 2.8 GPa. The direct–indirect (Γ_6^c–X_6^c) transition pressures in GaAs have been determined experimentally by a number of authors. We list in Table 5.2 such direct–indirect transition pressures. The experimental data, except Yu and Welber [20], indicate that the Γ_6^c–X_6^c transition occurs at ~4 GPa. The data obtained by Yu and Welber (~3 GPa) [20] is considerably lower than this value. Kangarlu *et al.* [21] also estimated the L_6^c–X_6^c crossover at 2.5–3.0 GPa from PL spectra taken under hydrostatic pressure. Our obtained L_6^c–X_6^c crossover pressure (~2.8 GPa) is in good agreement with this result.

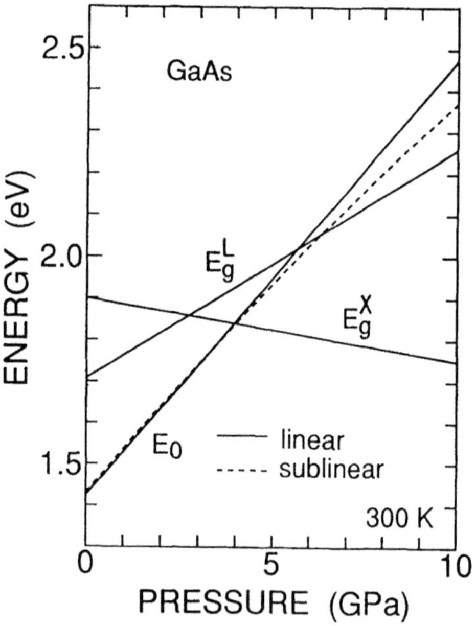

Figure 5.10 Hydrostatic-pressure dependence of the E_0, E_g^X, and E_g^L gaps for GaAs at 300 K. The solid lines are obtained from linear expressions given by Eq. (5.55). The dashed line also represents the pressure dependence of the E_0 gap obtained from a quadratic expression of Eq. (5.56).

TABLE 5.2 *Direct–indirect transition pressure in GaAs.*

Temperature (K)	Transition Pressure (GPa)
300	~4[a]
300	~3[b]
300	4.2±0.2[c]
300	~4[d]
10	4.1[e]
8	4.13[f]
5	4.13[g]

[a]B. Welber *et al.*, *Phys. Rev. B* **12**, 5729 (1975).
[b]P. Y. Yu and B. Welber, *Solid State Commun.* **25**, 209 (1978).
[c]A. R. Goñi *et al.*, *Phys. Rev. B* **36**, 1581 (1987).
[d]J. M. Besson *et al.*, *Phys. Rev. B* **44**, 4214 (1991).
[e]A. Kangarlu *et al.*, *Appl. Phys. Lett.* 59, 2290 (1991).
[f]H. Mariette *et al.*, *Phys. Rev. B* **33**, 8373 (1986).
[g]D. J. Wolford and J. A. Bradley, *Solid State Commun.* **53**, 1069 (1985).

We show in Fig. 5.11 the absorption spectra of a 1.7-µm-thick sample of GaAs at different pressures (T=300 K) measured by Goñi *et al.* [19]. The arrows correspond to the energies of the direct absorption edge as defined by the kink in the absorption spectra. The energy of the kink was determined from the corresponding edge in the first derivative with respect to the photon energy of the transmission spectra. These authors observed the direct absorption edge followed by a "*plateau*" at higher energies up to the structural phase-transition pressure (~17 GPa). The gradual broadening with pressure of the direct-band-gap kink was attributed to the fact that with increasing pressure, the energy of the indirect gap E_g^{ID} ($\Gamma_{15}^v \rightarrow X_1^c$) becomes smaller than that of the direct gap E_0.

Using thick samples (~30 µm), Goñi *et al.* [19] have been able to enhance the indirect absorption in the experimental spectra and to investigate the pressure and energy region where the indirect transitions dominate the absorption. The indirect transitions from the valence state around Γ_{15}^v to the conduction-band state around X_1^c take place via virtual direct transitions involving intermediate state either Γ or X (see Fig. 5.2). The difference between the energies of the indirect gap and that of the virtual direct transitions appears in the perturbation expression of indirect absorption process an energy denominator. In obtaining Eq. (3.32), we have ignored such an energy denominator [i.e., we have simply assumed that D in Eq. (3.32) is independent of photon energy]. If the difference is considerably small, the transitions may become stronger and the energy dependence of indirect absorption coefficient may be expected to deviate significantly from the well-known square-energy law of Eq. (3.32). Goñi *et al.* [19] have observed a strong enhancement of the indirect absorption with pressure and attributed it, in part, to an increase of the transition-matrix element for virtual direct transitions involved in the indirect absorption process. The indirect–direct crossing pressure obtained by Goñi *et al.* is 4.2±0.2 GPa (Ref. [19], also see Table 5.2).

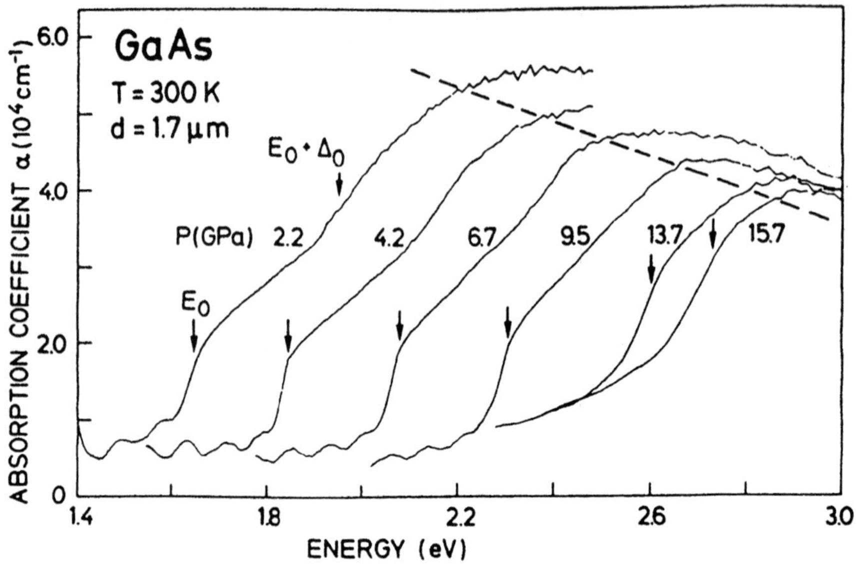

Figure 5.11 Absorption spectra of a 1.7-μm-thick sample of GaAs (T=300 K) at different hydrostatic pressures. Arrows indicate the position of the direct absorption edge. (From Goñi *et al.* [19].)

Goñi *et al.* [22] have also measured low-temperature exciton absorption spectra at the lowest-direct edge (E_0) of GaAs as a function of hydrostatic pressure up to 9 GPa. The measured spectra were analyzed in terms of Elliott's model [6] by taking into account the broadening of the exciton line. The exciton binding energy was found to increase linearly with pressure at a rate of about $d(\ln G^{3D})/dp$=0.08~0.09 GPa^{-1}. The exciton lifetime became smaller for pressures above the Γ–X conduction-band crossing (≥4.2 GPa), a fact which was attributed to phonon-assisted intervalley scattering.

(c) Doping Effect—The fact that the electronic energy bands in heavily doped semiconductors are shifted due to the presence of the ionized dopant ions and related charge carriers is a well-known fact. Accurate knowledge of the shifts in the conduction- and valence-band edges due to heavy doping effects is crucial in modeling semiconductor device structures that utilize heavily doped layers.

Figure 5.12 schematically shows the energy-band structures of heavily doped (a) n- and (b) p-type semiconductors along with that of undoped material (left-hand sides). Due to the Burstein-Moss effect (i.e., band-filling effect) [23], the *"absorption"* band gap in heavily doped semiconductors is larger than the band gap of the undoped material. Since the Burstein–Moss effect depends on the curvature and density of states of the band, it is more pronounced in semiconductors with small

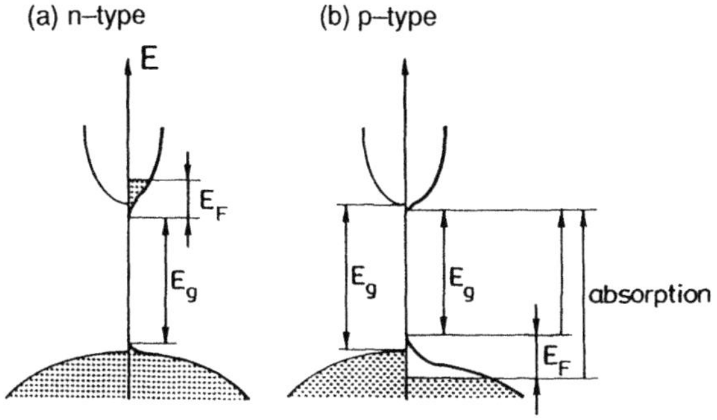

Figure 5.12 Schematic drawing of the electronic energy-band structures in heavily doped *n*- and *p*-type semiconductors along with that of the undoped material (left-hand side).

effective masses.

As a result of the band-gap shrinkage due to carrier-impurity and carrier–carrier interactions, the band gap of the doped semiconductor is in some case smaller than that of the undoped material [24]. Thus, for theoretical description of the doping-induced change in optical properties the band-filling effect as well as band-gap shrinkage have to be taken into consideration [25,26].

We show in Fig. 5.13 the band-gap shrinkage in heavily doped *n*- and *p*-GaAs as derived from PL measurements by Yao and Compaan (*n*-GaAs, Ref. [27]) and Benzaquen *et al.* (*p*-GaAs, Ref. [28]). The samples were produced by pulsed-laser annealing of Si-implanted GaAs (*n*-GaAs) and were grown by a horizontal MOCVD and doped with Zn using diethylzinc (*p*-GaAs), respectively. The low-energy edge of the PL in *n*-GaAs indicates a band-edge narrowing of $\Delta E_g \propto n^{1/3}$ and equal to ~200 meV at the highest electron concentration. The energy shift in *p*-GaAs also shows an empirical narrowing of $\Delta E_g \propto p^{1/3}$. It is noted, however, that at the same doping level the band-gap narrowing is smaller in *p*-GaAs than in *n*-GaAs. This arises from the fact that in GaAs the hole effective mass is considerably larger than the electron effective mass [9].

Carrier-concentration dependence of the absorption coefficient in *n*- and *p*-type GaAs samples at 297 K has been reported by Casey *et al.* [29]. Measurements were made on high-purity *n*-GaAs (n=5×10^{13} cm^{-3}), *n*-type samples with free-electron concentrations from 5×10^{16} to 6.7×10^{18} cm^{-3}, *p*-GaAs with free-hole concentrations from 1.5×10^{16} to 1.6×10^{19} cm^{-3}, and *p*-type samples heavily doped with the amphoteric impurity Si. We reproduce in Fig. 5.14 the results for (a) *n*-GaAs and (b) *p*-GaAs [29], respectively. As can be clearly seen, the absorption spectrum is strongly dependent on the carrier concentration. As the electron concentration exceeds ~6×10^{17} cm^{-3}, the Burstein-Moss shift due to the filling of the conduction band by free electrons begins to have a significant effect. This shift of the absorption coefficient to higher energies is readily shown in Fig. 5.14 (a) for

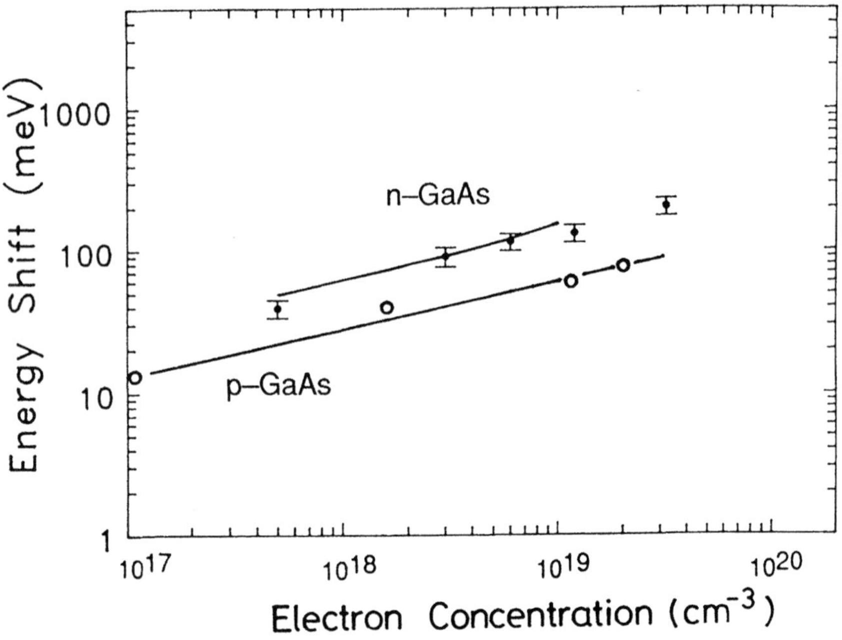

Figure 5.13 Band-gap shrinkage in *n*- and *p*-GaAs as measured by Yao and Compaan [*n*-GaAs, solid circles (T=80 K)] [27] and Benzaquen *et al.* [*p*-GaAs, open circles (T=5 K)] [28]. The samples were produced by pulsed-laser annealing of Si-implanted GaAs (*n*-GaAs) and were grown by a horizontal MOCVD and doped with Zn using diethylzinc (*p*-GaAs), respectively.

samples with $n>5.9\times10^{17}$ cm^{-3}.

Since optical absorption is determined by electron transition from a filled state in the valence band to an empty state in the conduction band, the band tail will influence absorption differently in *n*- and *p*-GaAs [29]. For heavily doped *n*-GaAs, the lowest states in the conduction band are filled with free electrons so that absorption takes an electron from the top of the filled valence band to an empty state well up in the conduction band (see Fig. 5.12). Hence, the absorption shifts to higher energy as shown in Fig. 5.14 (a). For *p*-GaAs, the hole mass is large so that most of the holes are in the valence-band tails. Therefore, absorption raises an electron from a filled state that is near or above the parabolic valence-band edge to the conduction-band tail, and the absorption shifts to lower energy as the band tails increase [Fig. 5.14 (b)]. Lush *et al.* [30] also measured the doping-dependent absorption coefficient in *n*-GaAs (n~1.3×10^{17}–3.8×10^{18} cm^{-3}) thin films, and found that the results show good qualitative agreement with Casey *et al.* [29] and good quantitative agreement, except for the heavily doped samples. For n~3.8×10^{18} cm^{-3},

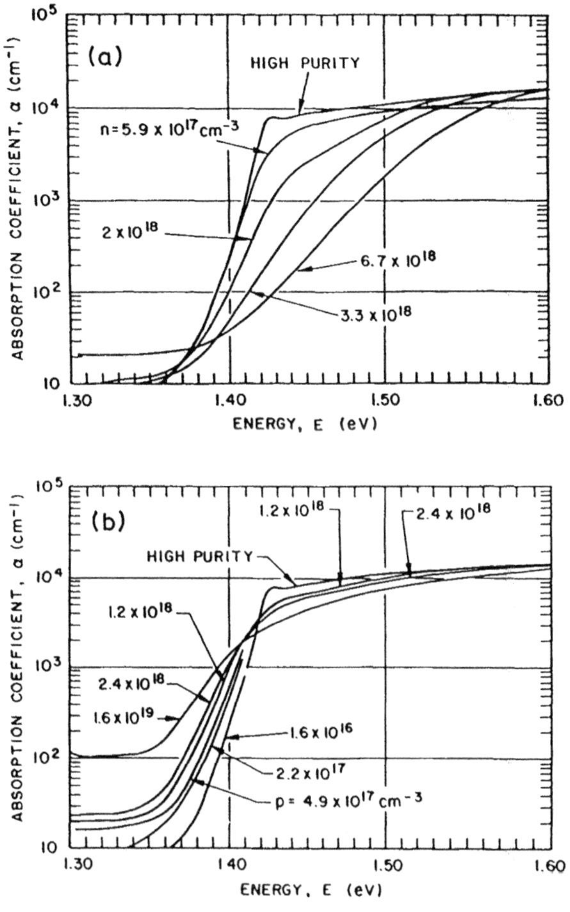

Figure 5.14 Absorption coefficient for (a) *n*-GaAs and (b) *p*-GaAs at 297 K. (From Casey *et al.* [29].)

α was approximately four times larger than that reported by Casey *et al.* [29].

Plots of the Urbach energy E_u for a variety of doping have been reported by several authors (see, *e.g.*, GaAs [31], InGaAs [26], and InGaAsP [32]). Such data show that the Urbach energy increases with increasing doping concentration. There had been no systematic difference between the various donor impurities (Si, Se, Te) in GaAs [31]. There had also been no systematic difference between the behaviors of acceptors (Cd, Zn) [31]. However, we can point out that at high concentrations tailing becomes more pronounced in *p*-GaAs; increased tailing sets in at a concentration of about 2×10^{18} cm^{-3} for donors and about 7×10^{18} cm^{-3} for acceptors [31]. Hahn *et al.* [26] showed that in *n*-InGaAs the Urbach tail can be presented in dependence on the cubic root of the reciprocal electron concentration which is a measure of the average distance between donors. Rajalakshmi and Arora [32] have also shown that the Urbach parameter in InGaAsP alloy samples is larger than that in GaAs by about 5 meV. They attributed the difference in E_u between the InGaAsP samples and GaAs at low free-electron concentration to the alloy disorder (InGaAsP) such that $E_{u,a}$=5±1 meV.

5.1.5 Amorphous Materials

The Urbach tail has been observed not only in the crystalline semiconductors but also in the amorphous materials. The Urbach parameters E_u determined from the amorphous semiconductors are found to be significantly larger than those for the crystalline semiconductors. The amorphous materials preserve the short-range order of the atomic configuration (in the case of *a*-Si, the tetrahedral coordination), but do not preserve the long-range order (i.e., translational symmetry). We can thus consider that the larger E_u values in the amorphous materials are caused by the effect of long-range disorder (plus short-range defects).

Figure 5.15 (a) shows the absorption coefficient α vs. E plots for *a*-Si measured using a photothermal deflection spectroscopy [33]. The samples were obtained by dual Si^+-ion implantation into *c*-Si at 150 kV/2.4×10^{14} cm^{-2} and 300 kV/6.0×10^{14} cm^{-2} at low temperature (80 K) and then performed 15-min isochronal heat treatments, subsequently carried out at increasing temperatures, up to a maximum value of 793 K. Also shown is the spectrum relative to an additional annealing at 793 K for 1 h.

In the case of amorphous materials obtained via a nonequilibrium process such as ion implantation, it can be expected that the amorphous network contains a large number of structural defects in the materials. This has been confirmed by the presence of vacancies and vacancy–impurity complexes in ion-implanted *a*-Si using Mössbauer spectroscopy [34]. We note in Fig. 5.15 (a) that with the progressive increase in heat treatment, the α values decrease all over the spectral region measured, while the absorption edge progressively shifts to larger energies and sharpens. Also shown in Fig. 5.15 (b) are the E_0–E_u plots for ion-implanted *a*-Si [33]. Here, the optical gaps E_0 are determined from the Tauc plots [Eq. (4.27)] and the Urbach parameters E_u from Eq. (5.50), using $\alpha(E)$ data of Fig. 5.15 (a). It is known that both the optical band gap and exponential band tail are dependent on the average strain in the material as observed in *a*-Si:H [35]. The E_0–E_u plots shown in Fig. 5.15 (b) clearly indicate a linear relationship between E_0 and E_u. This result confirms the occurrence of strain reduction during the structural relaxation process [33].

In the subgap region of amorphous semiconductors, the absorption depends on the presence of defect states within the gap (region C in Fig. 4.2). The optical absorption in this region has been studied widely for *a*-Si:H as well as ion-implanted amorphous semiconductors (see references cited in Ref. [33]). Regarding the subgap region in the $\alpha(E)$ spectra [Fig. 5.15 (a)], we can suggest that the

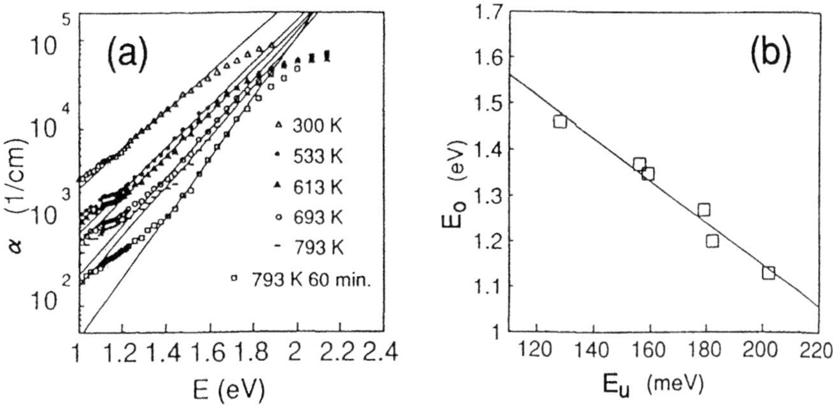

Figure 5.15 (a) Absorption coefficient α vs. E plots for a-Si measured using a photothermal deflection spectroscopy. The samples were obtained by dual Si^+-ion implantation into c-Si at 150 kV/2.4×10^{14} cm^{-2} and 300 kV/6.0×10^{14} cm^{-2} at 80 K and then performed 15-min isochronal annealing, subsequently carried out at increasing temperatures, up to a maximum value of 793 K. Also shown is the spectrum relative to an additional annealing at 793 K for 1 h (open squares). (b) E_o–E_u plots for Si^+-ion-implanted a-Si. The optical gaps E_o are determined from the Tauc plots [Eq. (4.27)] and the Urbach parameters E_u from Eq. (5.50), using $\alpha(E)$ data of (a). (From Zammit *et al.* [33].)

higher annealing temperature gives smaller α value, that is due to defect evolution and annihilation accompanied by strain reduction in the material. In conclusion of this subsection [33,35], the relaxation process in amorphous semiconductors leads to not only a sharpening of the band edge (and an increase in the optical band gap) but also a defect evolution and annihilation (i.e., a decrease in the optical absorption at the subgap).

5.2 REFRACTIVE INDEX

5.2.1 Theoretical Expression

Knowledge of the refractive index n of semiconductors in the region below or near the fundamental absorption edge is often of interest in optoelectronic device design. In the following, we review theoretical models for the calculation of the refractive-index dispersion in semiconductors of such spectral region [9].

(a) Sellmeier Equation—The refractive-index dispersion can be simply given by the first-order Sellmeier equation [36]

$$n(\lambda)^2 = A + B\left(\frac{\lambda^2}{\lambda^2 - C^2}\right) \qquad (5.57)$$

where λ is the light wavelength in the vacuum. This expression is obtained based on an empirical relation. The refractive-index dispersion obtained form this expression is, therefore, not related through the KK relations and is valid only over a limited energy range.

(b) Ketteler–Helmholtz Formula—Another common expression used for the n dispersion is the Ketteler–Helmholtz formula:

$$n(E)^2 = A + \frac{BE^2}{1-(E/C)^2} \qquad (5.58)$$

where $E<C$. Not only Eq. (5.57) but also this expression have the appropriate physical basis to represent the n dispersion in the simplest manner throughout the transparent region.

(c) Single-Oscillator Model—A semiempirical single-effective-oscillator model has been proposed by Wemple and DiDomenico [37] to analyze refractive-index dispersion in more than 100 widely different solids and liquids. This model requires two parameters, E_p and E_d, where the imaginary part of the dielectric constant ε_2 of the material was assumed to be a delta function at energy E_p and the strength of an effective oscillator at energy E_p was defined to be $\pi E_d/2$. Introducing these quantities into Eq. (1.16a), one obtains

$$n(E)^2 - 1 = \frac{E_p E_d}{E_p^2 - E^2} \qquad (5.59)$$

By replacing $B=E_d/E_p$ and $C=ch/E_p$ (c is the light velocity and h is the Planck constant), we can find that Eqs. (5.57) and (5.59) agree exactly with each other.

The single-oscillator model gives reasonable results for photon energies *well below* the lowest-direct-gap edge in semiconductors. We note, however, that such photon-energy region is not of immediate interest for semiconductor lasers.

(d) Modified Single-Oscillator Model—Afromowitz [38] has proposed an improved model for the variation of ε_2 with photon energy by taking into account the lowest-direct-gap (E_g) contribution:

$$\varepsilon_2(E) = \begin{cases} \eta E^4 & E_g \le E \le E_f \\ 0 & \text{elsewhere} \end{cases} \qquad (5.60)$$

where the new parameters E_f and η can be determined by requiring that the corresponding expression for n^2 agrees with Eq. (5.59) for $E \ll E_g$. The resultant expression for n^2 is

$$n(E)^2 - 1 = \frac{E_d}{E_p} + \frac{E_d E^2}{E_p^3} + \frac{\eta E^4}{\pi} \ln\left(\frac{2E_p^2 - E_g^2 - E^2}{E_g^2 - E^2}\right) \qquad (5.61)$$

where

$$\eta = \frac{\pi E_d}{2E_p^3(E_p^2 - E_g^2)} \qquad (5.62)$$

The parameters E_d, E_p, and E_g appearing in Eqs. (5.61) and (5.62) are known for many binary semiconductors and may be calculated for ternary and quaternary compounds by interpolation. Afromowitz gave in Ref. [38] such parameter values for $Al_xGa_{1-x}As$, $GaAs_{1-x}P_x$, and $In_xGa_{1-x}P$ alloys.

Campi and Papuzza [39] proposed a calculation model that is somewhat different from the Afromowitz model: instead of giving semiempirical prescriptions for the values of η and E_f, these authors preferred to derive values of these parameters from more general considerations.

(e) Pikhtin–Yas'kov Formula—Pikhtin–Yas'kov formula [40] is nearly the same as the Drude form with the addition of another term representing a broadband electronic contribution to n:

$$n(E)^2 - 1 = \frac{A}{\pi} \ln\frac{E_1^2 - E^2}{E_0^2 - E^2} + \sum_i \frac{G_i}{E_i^2 - E^2} \qquad (5.63)$$

The unique term arises from assuming that the imaginary part of the dielectric constant is a constant between energies E_0 and E_1 and that infinitely narrow resonances occur at E_i due to lattice absorption bands. The formula in Eq. (5.63) is obtained by applying the KK transformation to this model.

(f) Exponential Band-Edge Model—Experimentally it is often found that the absorption constant can be represented over one or more decades by an exponential dependence on the photon energy. In the numeric calculation, one can assume an optical absorption given by the following form [41]:

$$\alpha(E) = A\exp\left(\frac{E - E_c}{E_u}\right) \qquad \textit{for } E<E_c \qquad (5.64a)$$

$$\alpha(E) = A\left[1 + \frac{E - E_c}{E_r} - \frac{(E - E_c)^2}{E_r E_3}\right] \qquad \textit{for } E_c<E<E_s \qquad (5.64b)$$

where E_u and E_r describe the behavior of the absorption in the intermediate neighborhood of the edge, and $E_3=(E_g-E_c)^2/(E_s+E_r-E_c)$ is chosen to make the absorption vanish at E_s. When the absorption edge is given by Eq. (5.64), the variation of the refractive index near the edge can be given, from the KK transformation, as [41]

$$n(E) = \frac{A\lambda_c}{4\pi^2}\left[\ln\frac{E_s-E_c}{E_c-E} + \exp\left(\frac{E-E_c}{E_u}\right)Ei\left(\frac{E_c-E}{E_u}\right)\right] \qquad (5.65)$$

where λ_c is the light wavelength corresponding to energy E_c and $Ei(x)$ is the exponential integral. The expression in braces in Eq. (5.65) has its peak when $E=E_c-0.37E_u$. The parameter E_u, which characterizes the steepness of the absorption edge (i.e., Urbach parameter, see Sec. 5.1.3), will be strongly influenced by both the temperature and composition, and the refractive-index dispersion near the edge will thus be influenced by these factors.

(g) Simplified Interband-Transition Model—The $E_0/(E_0+\Delta_0)$ transitions strongly contribute to the optical dispersion, but not to their values. On the contrary, the higher-lying-gap contributions (E_1, $E_1+\Delta_1$, E_2, etc.) in the transparent region do not contribute to the optical dispersion, but to their absolute values (see Secs. 3.1 and 3.2). Since ε_2 may be taken as zero in the transparent region, one can successfully assume that

$$n(E) \sim \varepsilon_1(E)^{1/2} \qquad (5.66)$$

The $n(E)$ in the transparent region can, then, be simply expressed, using Eqs. (3.10), (5.31), and (5.38), as [9,42]

$$n(E)^2 = A_0^*\left[f(\chi_0) + \frac{1}{2}\left(\frac{E_0}{E_0+\Delta_0}\right)^{3/2} f(\chi_{so})\right] + \sum_{n=1}^{\infty}\frac{F_0^{D*}}{n^3[(E_{ex}^n)^2-E^2]} + \frac{F_0^{C*}E_{ex}^{C1}}{G^{3D}E^2}\ln\frac{E_0^2}{E_0^2-E^2} + B^* \qquad (5.67)$$

where A_0^* and F_0^{D*} (F_0^{C*}) represent the renewed strength parameters of the $E_0/(E_0+\Delta_0)$ one-electron and exciton transitions, respectively, B^* corresponds to the nondispersive contribution (i.e., constant) arising from the higher-lying band gaps (E_1, $E_1+\Delta_1$, E_2, etc.), and $\Gamma\rightarrow 0$ eV in Eq. (3.13).

Equation (5.67) is obtained from a simplified model of the interband transitions proposed by Korovin [43] and Cardona [44]. The model has been applied to II–VI and III–V compound semiconductors, and the results exhibit quite good agreement with the published experimental data for some II–VI binaries [45], III–V binaries [42], $Al_xGa_{1-x}As$ [46] and $In_{1-x}Ga_xAs_yP_{1-y}$/InP alloys [42]. The model has also been used to estimate the refractive-index dispersion in $Al_xGa_{1-x}As_ySb_{1-y}$, $Ga_xIn_{1-x}As_ySb_{1-y}$, and $InP_xAs_ySb_{1-x-y}$ quaternary systems lattice-matched to GaSb and InAs [47].

(h) Quantum-Density-Matrix Formulation—This model is based on a quantum mechanical calculation of the complex dielectric constant using the quantum-density-matrix method [48]. The usual method for the calculation of ε_1 near the band edge involves the calculation of ε_2 and subsequent integration of this result to obtain ε_1 via the KK relations. In the quantum mechanical calculation, ε_1 can be directly determined by calculating the interaction energy of the lattice polarization due to valence electrons and the radiation field. The expression obtained from the quantum-density-matrix formation is a function of the lattice and energy-band parameters, that are, the lattice constant a_0, the band-gap energies E_0 and Δ_0, the electron effective mass m_e, and the hole effective masses of the three valence bands m_{hi} (i=LH, HH, SO). Under certain simplifying assumption, the number of these material parameters could be reduced to a_0, E_0, Δ_0, m_e, and m_{hi} [i=HH (heavy hole) only]. The model was successfully applied to III–V, II–VI, and IV–VI compounds [48]. The major contribution to the refractive index at E=E_0 was found to arise from the valence-band electrons making virtual transitions with intermediate states lying in the conduction band.

5.2.2 Empirical Formula

The high-frequency dielectric constant ε_∞ describes the approximately constant value of $\varepsilon_1(E)$ for energies well below the fundamental absorption edge but still above the reststrahlen region (see discussion in Sec. 2.1, also see Fig. 1.17). A relation between the dielectric constant and an accurately known physical parameter, if available, would be useful in finding an acceptable value of the unknown material from this relation. The basic concept that higher band-gap material has a lower dielectric constant ε_1 can be derived from Moss's rule [49] which states that energy levels are inversely proportional to the square of the dielectric constant. Then, the band gap E_g (or E_0) and refractive index n ($\sim\varepsilon_1^{1/2}$) satisfy the relation

$$n^4 E_g = \text{const.} \qquad (5.68)$$

As shown in Fig. 5.16, the plots for $n(\lambda\to\infty)^2$ [$=\varepsilon_1(\lambda\to\infty)=\varepsilon_\infty$] vs. E_g for some III–V and II–VI binaries support the dependence of Eq. (5.68) in a general way.

Double-heterojunction lasers are composed of a lower energy band-gap active layer sandwiched between higher band-gap cladding layers, and the refractive index of the active layer has to be higher than that of the cladding layer. This is certainly the case for GaAs/$Al_xGa_{1-x}As$ and $In_{1-x}Ga_xAs_yP_{1-y}$/InP lasers. A number of published

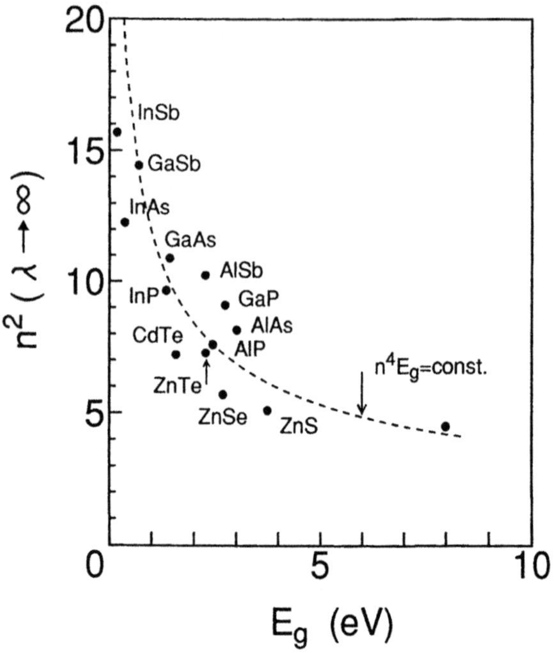

Figure 5.16 Relationship between $n(\lambda\rightarrow\infty)^2$ [$=\varepsilon_1(\lambda\rightarrow\infty)=\varepsilon_\infty$] and E_g for some III–V and II–VI binaries. The dashed line represents the calculated result of Eq. (5.68) with a constant value of n^4E_g=142 eV.

papers are, therefore, based on the belief that higher band-gap materials necessarily have lower refractive indexes, so that the correct double-heterojunction structure automatically satisfies the requirement for the refractive index.

If the refractive index in the active region of an injection laser is smaller than the index of the cladding layer on both sides, the effect is like that of an antiwaveguide configuration which does not confine radiation to the neighborhood of the active region. This effect offers an optical loss in the waveguide, leading to an increase in the threshold current. Toda [50] pointed out the existence of a number of inappropriate double heterostructures (i.e., antiwaveguiding material systems). Reference [47] also deals with inappropriate laser structures of $In_{1-x}Ga_xAs_ySb_{1-y}$/GaSb(InAs) and $InP_xAs_ySb_{1-x-y}$/GaSb(InAs).

The dielectric constants are determined by the polarizability of the constituent atoms and are expected to increase with increasing their atomic numbers [46]. Nag [51] obtained an empirical formula relating the dielectric constants of cubic semiconductors with the average atomic number of the constituent atoms. The formula derived by him is

$$K = \frac{1}{A - B\,Z_{av}} \qquad (5.69)$$

where K indicates either ε_∞ [$n(\lambda\rightarrow\infty)^2$] or ε_s, and A and B are constant for each material group (Ga-based III–V, In-based III–V, Zn-based II–VI compounds, etc.). This relation may be used to predict the values for the alloy compounds and also to judge the acceptability of the reported values.

Ghosh *et al.* [52] obtained an empirical formula for evaluating the high-frequency refractive index $n(\lambda\rightarrow\infty)^2$ using a Penn-gap model. Introducing A' as the contribution from the valence electrons and B as a constant additive to the lowest band gap E_g, they obtained an expression for $n(\lambda\rightarrow\infty)^2$ given by

$$n^2 - 1 = \frac{A'}{(E_g + B)^2} = \frac{mE_g + c}{(E_g + B)^2} \quad , \qquad (5.70)$$

where the term (E_g+B) refers to an appropriate average energy gap of the material. The parameters A' and B were determined by applying Eq. (5.70) to a group of materials having the experimentally known E_g and n^2. The groups of the material systems considered by them were IV, III–V, II–VI, II–IV–V_2, and I–III–VI_2 families. It was found that the parameter A' increases monotonically with increasing band gap (i.e., $A'=mE_g+c$ with $m>0$), while the parameter B shows very little or no variation with E_g.

Hervé and Vandamme [53] proposed an empirical relation

$$n = \sqrt{1 + \left(\frac{A}{E_g + B}\right)^2} \qquad (5.71)$$

with A=13.6 eV and B=3.4 eV. Note that this empirical relation is exactly the same as Eq. (5.70) ($A^2\rightarrow A'$).

Plotted in Fig. 2.2 suggest that there is a linear relation between ε_∞ and E_0. Ravindra *et al.* [54], on the other hand, proposed a linear relation of n with E_g:

$$n = \alpha + \beta\,E_g \qquad (5.72)$$

where α=4.084 and β=-0.62 eV^{-1}.

5.2.3 Experimental Data

(a) Comparison with Calculated Results—In the energy region below the fundamental absorption edge, where the material is nearly transparent, the refractive-index measurements were made by various methods, for example, for GaAs by the method of prism minimum deviation, by surface reflection, by interference spacing, by SE, by transmittance, by Brewster angle method (see Ref. [55]), and by studying the eigenmode propagation constants in optical waveguides [56].

We show in Fig. 5.17 the n dispersion in CdTe determined by Marple by the prism refraction method (Ref. [36], open circles) and by Adachi *et al.* by SE (solid circle) [57]. The fit of these data to Eq. (5.57) (Sellmeier equation) gives (solid line)

$$A=5.68, \quad B=1.53, \quad C^2=0.366 \qquad (5.73)$$

Agreement between the experiment and calculation is found to be very good.

Figure 5.18 shows the n curves for $Zn_{1-x}Mg_xS_ySe_{1-y}$ quaternaries lattice-matched to GaAs determined by Lunz *et al.* [58] from optical transmission and reflection measurements at 300 K. The solid lines represent the calculated results of the Ketteler–Helmholtz dispersion formula [Eq. (5.58)]. As we can see, the indices n decrease with increasing E_0 (i.e., increasing y) and are in good agreement with the

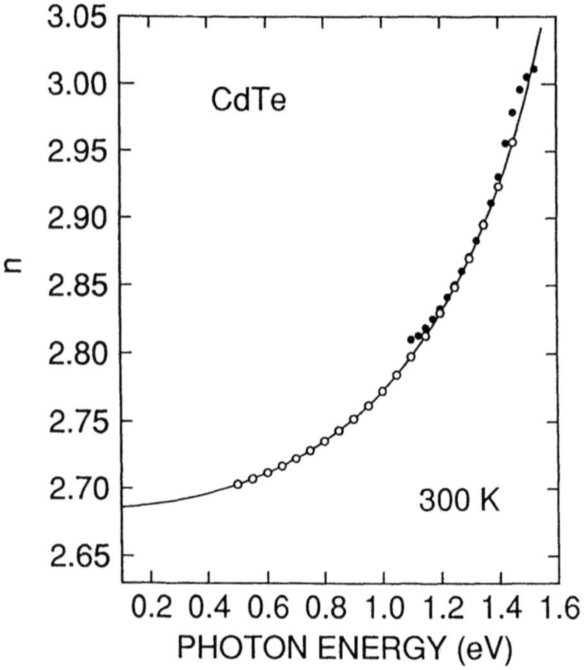

Figure 5.17 Refractive index of CdTe as a function of photon energy E at 300 K. The open and solid circles represent the experimental data measured by the prism refraction method (Ref. [36]) and SE, respectively. The solid line is obtained from Eq. (5.57). (From Adachi *et al.* [57].)

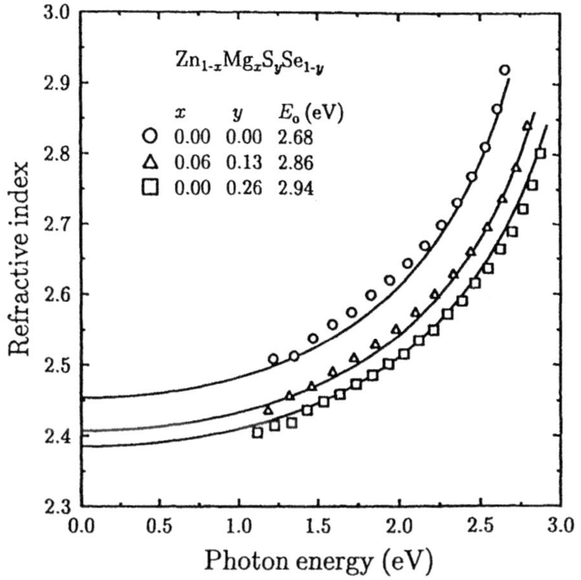

Figure 5.18 Dispersion of the refractive index for $Zn_{1-x}Mg_xS_ySe_{1-y}$ lattice-matched to GaAs determined from optical transmission and reflection measurements at 300 K. The solid lines represent the calculated results of the Ketteler–Helmholtz formula of Eq. (5.58). (From Lunz *et al.* [58].)

calculated values. These authors suggested that the fitting parameters A, B, and C in Eq. (5.58) can be expressed as a linear function of E_0.

Experimental verification of Eq. (5.59) [single-oscillator model, Sec. 5.2.1 (c)] can be obtained by plotting $1/(n^2-1)$ vs. $(h\nu)^2$ (or E^2). The resulting straight line yields values of the parameters E_p and E_d. A typical result is shown in Fig. 5.19 for $In_{0.12}Ga_{0.88}As$ ternary [59]. This plot yields E_p=3.33 eV and E_d=32.8 eV. At shorter wavelengths, however, a negative curvature deviation can be clearly seen due to the proximity of the band edge or excitonic absorption (solid line).

The dashed line in Fig. 5.19 also represents the calculated $1/(n^2-1)$ vs. $(h\nu)^2$ from Eq. (5.61) (modified single-oscillator model). We can easily understand from the figure that the modified single-oscillator model shows excellent agreement with the experimental data, especially in the photon-energy region close to the lowest-direct gap (E_0~1.25 eV).

The refractive indices for semiconductors with diamond and zinc-blende structures are plotted in Fig. 5.20 [40]. The solid lines show the calculated results based on Eq. (5.63) (Pikhtin–Yas'kov formula). Note that the contribution of the lattice vibrations to n [i.e,, the second term in the right-hand side of Eq. (5.63)] is taken into consideration in Fig. 5.20.

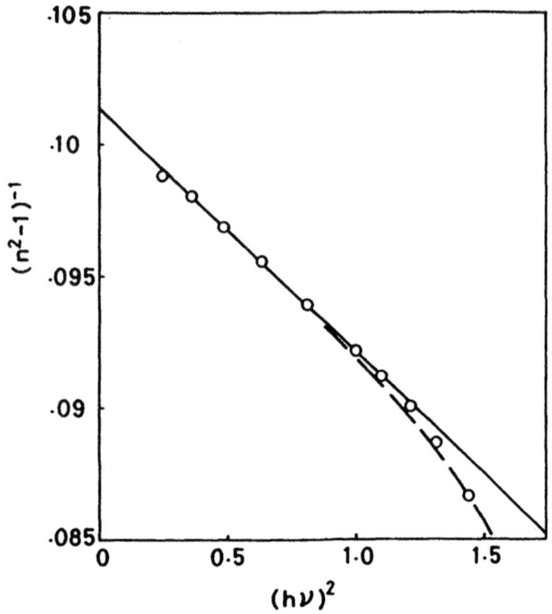

Figure 5.19 Plots of $1/(n^2-1)$ vs. $(h\nu)^2$ (E^2) for $In_{0.12}Ga_{0.88}As$ ternary. The resulting straight line yields values of the single-oscillator parameters E_p=3.33 eV and E_d=32.8 eV [Eq. (5.59)]. The dashed line also represents the plots of $1/(n^2-1)$ vs. $(h\nu)^2$ calculated from the modified single-oscillator model of Eq. (5.61). (From Takagi [59].)

Figure 5.21 shows the n dispersion in ZnTe at 300 K [60]. The theoretical curve obtained from Eq. (5.67) (simplified interband-transition model) is shown by the solid line. The numerical parameters used in this calculation are: E_0=2.28 eV; $E_0+\Delta_0$=3.20 eV; A_0^*=17.8; B^*=1.5; and $F_0^{D*}= F_0^{C*}=0$. Note that in the calculation, the lifetime broadening of Γ=0.18 eV is taken into consideration in a phenomenological manner by replacing E by $E+i\Gamma$. We will see this effect in more detail in the next subsection.

(b) Alloy-Composition Variation—The calculation models for n in Sec. 5.2.1 can be applied to many alloy semiconductors. Formulas to each alloy and numerical parameters will be given, so that n may be calculated for optional alloy composition. In the following, as an example, we consider the case for $In_{1-x}Ga_xAs_yP_{1-y}$ quaternaries lattice-matched to InP.

Chandra *et al.* [61] presented n data for $In_{1-x}Ga_xAs_yP_{1-y}$ layers on InP with y between 0.276 and 1.0 taken as a function of the wavelength from 2.5 μm to about 0.1 μm below their respective absorption edges. Interference fringes were the basis of these measurements.

An attempt is made first to fit $n(E)^2=\varepsilon_1(E)$ of Eq. (5.67) $[F_0^{D*}= F_0^{C*}=0]$ to the

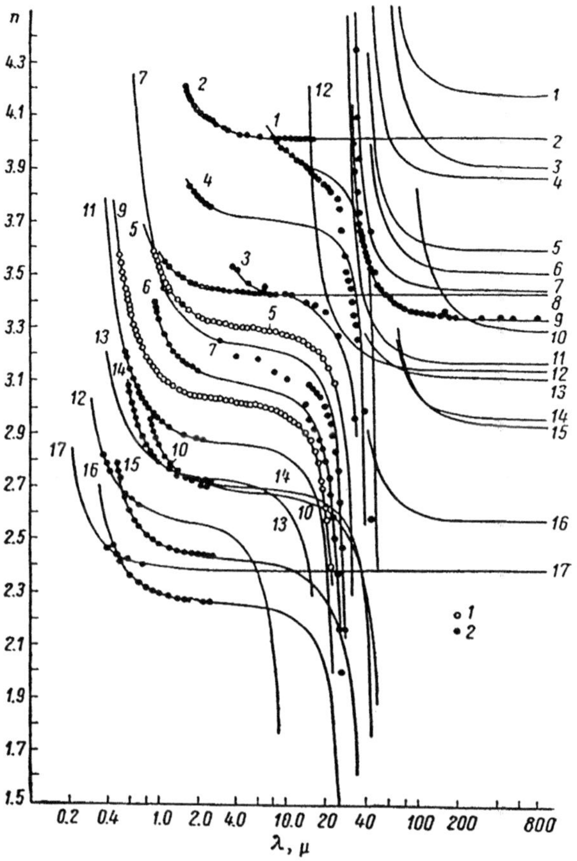

Figure 5.20 Refractive indices for semiconductors with diamond and zinc-blende structures: 1) InSb; 2) Ge; 3) InAs; 4) GaSb; 5) GaAs; 6) InP; 7) AlSb; 8) Si; 9) GaP; 10) CdTe; 11) AlAs; 12) 3C-SiC; 13) AlP; 14) ZnTe; 15) ZnSe; 16) β-ZnS; 17) C (diamond). The solid lines represent the calculated results based on Eq. (5.63). (From Pikhtin and Yas'kov [40].)

experimental data of Pettit and Turner (InP (y=0), Ref. [62]) and of Chandra et al. (y≠0, Ref. [61]) [42]. Figure 5.22 shows the results of these fits [42]. We can achieve quite good agreement between the calculated and experimental results. The numeric values of A_0^* and B^* fit-determined are plotted as a function of y in Fig. 5.23 [42].

Since the band-gap energies E_0 and $E_0+\Delta_0$ and strength parameters A_0^* and B^* have been specified in terms of the alloy composition y alone [42], we can then easily calculate the refractive indices of $In_{1-x}Ga_xAs_yP_{1-y}$ for arbitrary composition y and photon energy $E=\hbar\omega$. Figure 5.24 shows the calculated $n(E)$ results with y-composition increments of 0.1 [42].

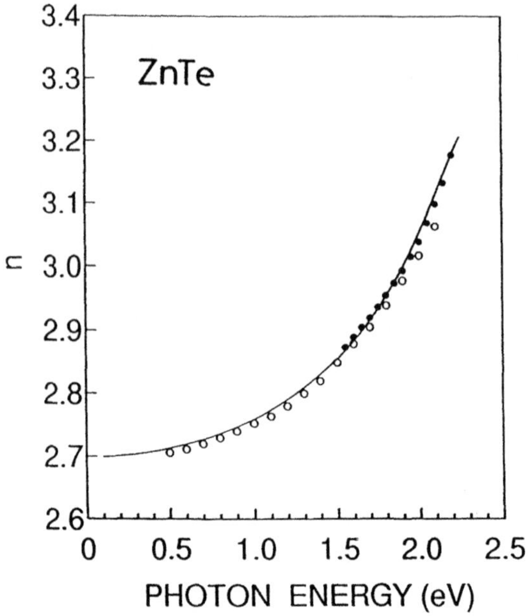

Figure 5.21 Refractive index of ZnTe at 300 K. The open and solid circles represent the experimental data measured by the prism refraction method (Ref. [36]) and SE, respectively. The theoretical curve obtained from Eq. (5.67) is shown by the solid line. The lifetime broadening (Γ=0.18 eV) is taken into consideration in the calculation in a phenomenological manner by replacing E by E+iΓ. (From Sato and Adachi [60].)

The impurities or, more generally, the lattice defects in semiconductors, reduce the lifetime of the excited electronic states, resulting in a change in optical properties. The $n(E)$ dispersion is affected by the lifetime broadening (damping) Γ of the electronic states, especially in the photon-energy region very close to the band edge. The solid and dashed lines in Fig. 5.24 represent the calculated results with Γ=0 and $\Gamma=0.02E_0$, respectively. It is obvious from this figure that the lifetime broadening considerably reduces the n value at the band-edge region. Therefore, one must pay attention to the analysis of $n(E)$ at this region.

(c) Temperature, Pressure, and Doping Effects

1. Temperature effect—The refractive index of a material is usually measured relative to air (i.e., $n/n_{\rm air}$ is a measured quantity). In order to determine the temperature variation of n of the material, it is necessary to take into account the temperature variation of the refractive index of the ambient air $n_{\rm air}$:

$$\frac{\mathrm{d}}{\mathrm{d}T}\left(\frac{n}{n_{\rm air}}\right) \sim \frac{\mathrm{d}n}{\mathrm{d}T} - n\frac{\mathrm{d}n_{\rm air}}{\mathrm{d}T} \qquad (5.74)$$

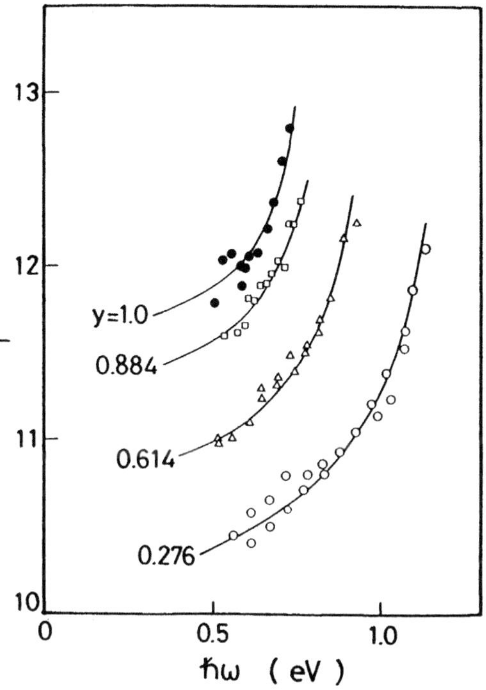

Figure 5.22 Calculated $\varepsilon_1(E)$ spectra compared with the experimental data for $In_{1-x}Ga_xAs_yP_{1-y}$ lattice-matched to InP. The solid lines are calculated from Eq. (5.67) $[F_0^{D*}=F_0^{C*}=0]$. The experimental data are taken from Chandra *et al.* [61] (T=300 K). (From Adachi [42].)

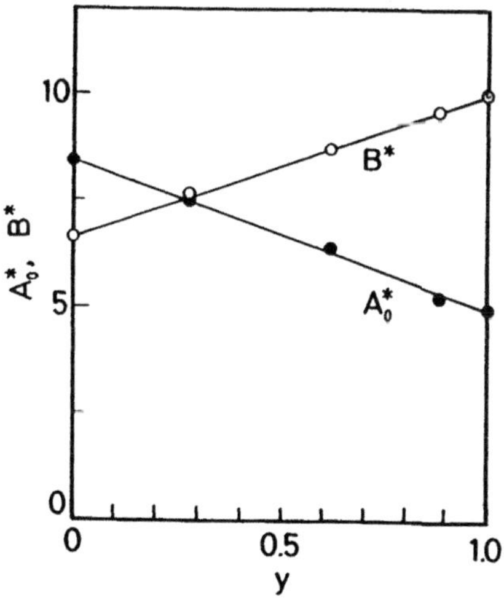

Figure 5.23 Variation of A_0^* and B^*, required to fit Eq. (5.67) with the experimental data, vs. y for $In_{1-x}Ga_xAs_yP_{1-y}$ lattice-matched to InP. (From Adachi [42].)

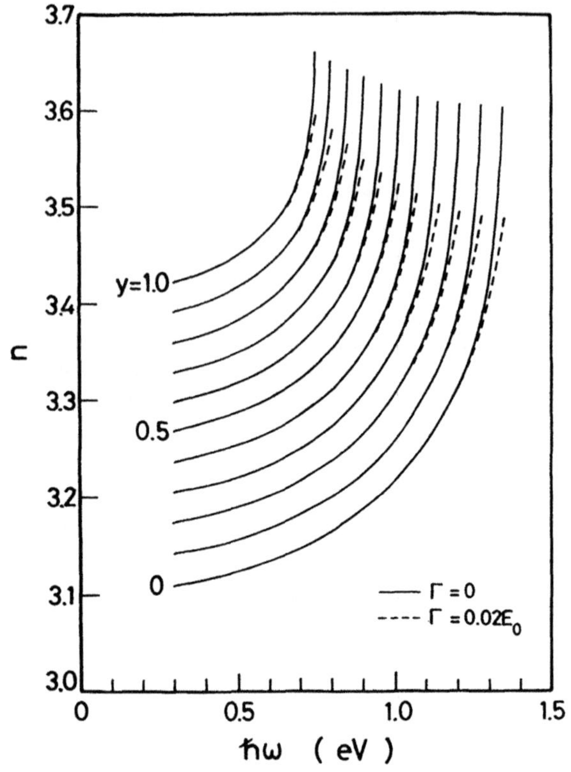

Figure 5.24 Calculated $n(E)$ with y-composition increments of 0.1 for $In_{1-x}Ga_xAs_yP_{1-y}$ lattice-matched to InP at 300 K. The solid and dashed lines represent, respectively, the calculated results with Γ=0 and Γ=0.02E_0. (From Adachi [42].)

where dn_{air}/dT is about 1×10^{-6}. The second term in Eq. (5.74) is, thus, usually masked by the experimental uncertainty.

Let us now consider a simple one-gap model of ε_1 given by the expression

$$\varepsilon_1(E) = n(E)^2 = 1 + \frac{A}{E_g^2 - E^2} \qquad (5.75)$$

where E_g is the average gap in the Penn model [63] and A is proportional to $1/V$ (V is the atomic volume, here we shall assume the proportionality constant of this quantity to be unity, namely $A \sim V^{-1}$). It is easily found that by replacing $E_pE_d \rightarrow A$ and $E_p \rightarrow E_g$, Eq. (5.75) agrees exactly with Eq. (5.59). Equation (5.75) for $E\rightarrow 0$ ($n^2 \sim \varepsilon_1 \rightarrow \varepsilon_\infty$) leads to the following formula for the temperature derivative [64]:

$$\frac{1}{n}\left(\frac{dn}{dT}\right)_p = \frac{1}{2}\left(1 - \frac{1}{\varepsilon_1}\right)\left[-\beta_{th} - \frac{2\beta_{th}V}{E_g}\left(\frac{\partial E_g}{\partial V}\right)_T - \frac{2}{E_g}\left(\frac{\partial E_g}{\partial T}\right)_V\right] \tag{5.76}$$

where $\beta_{th}=(1/V)(\partial V/\partial T)_p$ is the volume coefficient of thermal expansion ($\beta_{th}=3\alpha_{th}$ in cubic crystals, where α_{th} is the linear thermal expansion coefficient). Since $(1/n)(dn/dT)$ is positive in many semiconductors (see Table 5.3), we can conclude that the $\partial E_g/\partial T$ ($\partial E_g/\partial V$) term dominates the β_{th} term [Eq. (5.76)]. This is because dE_g/dT ($\partial E_g/\partial V$) is negative while β_{th} (or α_{th}) is positive (and $\varepsilon_1>1$) for many semiconductors [65]. We can also recognize that the thermal-expansion contribution is relatively small, because α_{th} is usually the order of 10^{-6} K^{-1}.

In the Moss empirical relationship [Eq. (5.68)], we assume that both a (≡const.) and E_g (E_0) are temperature dependent. Then

$$\frac{1}{n}\frac{dn}{dT} = \frac{1}{4}\left(\frac{1}{a}\frac{da}{dT} - \frac{1}{E_g}\frac{dE_g}{dT}\right) \tag{5.77}$$

If we further assume that $da/dT=0$, then it follows

$$\frac{1}{n}\frac{dn}{dT} \sim \frac{-1}{4E_g}\frac{dE_g}{dT} \tag{5.78}$$

Data on $(1/n)(dn/dT)$ are available for many semiconductors, as listed in Table 5.3. Usually, when T increases, E_g decreases, resulting in positive value of $(1/n)(dn/dT)$. The temperature dependence of n for C (diamond) is positive, but about an order smaller than those for other semiconductors. Note that the lead salts exhibit properties, which are unusual, and possibly unique, relative to other semiconductors. For example, they possess high dielectric constant, high mobility, and a narrow fundamental gap whose temperature coefficient is positive (not *negative*) [66]. As expected from Eq.(5.78), the lead salts then have negative sign of $(1/n)(dn/dT)$. Unfortunately, however, this equation gives only the correct sign of the coefficient, but not give quantitative agreement with experiment.

The negative temperature dependence of n was also found in Hg-based alloy system, $Cd_{1-x}Hg_xTe$ for $x>0.5$ [67]. This alloy system showed a zero value of dn/dT close to the equimolar composition $x\sim0.5$. Thus, the refractive index for the alloys with $x\sim0.5$ is expected to be independent of T.

Hervé and Vandamme [53] analyzed the temperature dependence of the refractive index for group-IV, III–V, and IV–VI semiconductors based on the three empirical relationships: Moss's [Eq. (5.68), see also Eq. (5.77)], Ravindra's [Eq. (5.72)], and Hervé–Vandamme's models [Eq. (5.71)]. From the fit of Eq. (5.77) to the experimental data, they obtained da/dT to be 6.68×10^{-4} eV/K that could be

TABLE 5.3 *The lowest direct-gap energy, its temperature coefficient, and the temperature coefficient of the refractive index* [$n(\lambda\to\infty)=\varepsilon_\infty^{1/2}$] *for some semiconductors. W=Wurtzite and ZB=Zincblende.*

Material	E_0 (eV)	dE_0/dT (10^{-4} eV/K)	$n^{-1}(dn/dT)$ (10^{-5} K^{-1})
C (Diamond)	7.3	-3.1	0.4
Si	3.38[a]	-2.2[a]	4.0
Ge	0.80	-3.9	6.9
AlP	3.6	-3.7	3.6
AlAs	3.02	-5.1	4.6
α-GaN(W)	3.44	-4.2	2.6 ($E\perp c$)
GaP	2.74	-4.6	3.7
GaAs	1.43	-3.95	4.5
GaSb	0.72	-3.7	8.2
InP	1.35	-2.9	2.7
InAs	0.36	-3.5	12
InSb	0.18	-2.8	6.9
β-ZnS(ZB)	3.75	-5.3	1.9
ZnSe	2.69	-4.8	2.6
ZnTe	2.25	-5.0	6
w-CdS(W)	2.45	-4.1	4 ($E\perp c$, $E\parallel c$)
w-CdSe(W)	1.75	-3.6	4 ($E\perp c$, $E\parallel c$)
CdTe	1.58	-4.2	4.4
PbS	0.41	5.2	-20
PbSe	0.28	5.1	-28
PbTe	0.31	4.5	-29

The experimental data are gathered from many sources.
[a] E_1 and dE_1/dT values.

commonly used as for all the materials they studied. In Ravindra's empirical model, two constants, α and β, play a role, but Hervé and Vandamme only considered the constant β to be linearly temperature dependent [53]. Then, it results in

$$\frac{1}{n}\frac{dn}{dT} = (\alpha + \beta E_g)^{-1}\left(\frac{d\beta}{dT}E_g + \beta\frac{dE_g}{dT}\right) \tag{5.79}$$

with $d\beta/dT=-5.15\times10^{-6}$ $eV^{-1}K^{-1}$.

In Hervé–Vandamme's model, only the parameter B is considered to be a linear function of temperature. Hence [53]

$$\frac{1}{n}\frac{dn}{dT} = -\frac{(13.6)^2}{n^2(E_g+B)^3}\left(\frac{dE_g}{dT} + \frac{dB}{dT}\right) \tag{5.80}$$

with $dB/dT=2.5\times10^{-5}$ eV/K.

It has been shown [53] that all these models give overall good results with experiment, even for some semiconductors like C and PbSe that exhibit special behavior. Several authors [64,65,68,69] have also considered the temperature dependence of n for some semiconductors.

The above discussion dealt with the temperature dependence of $n(E\to 0)$ [or equivalently $n(\lambda\to\infty)=\varepsilon_\infty^{1/2}$] of semiconductors ($T$>50 K). Both the absolute magnitude of dn/dT, as well as its frequency dependence, are of considerable interest in conjunction with many device applications. The temperature dependence of the refractive index near the lowest direct-band gap E_0 is given generally by

$$\frac{dn}{dT} = \frac{dn_1}{dT} + \frac{dn_2}{dT} \sim \frac{dn_1}{dT} = \frac{dn_1}{dE_g}\frac{dE_g}{dT} \sim -\frac{dn_1}{dE}\frac{dE_g}{dT} \qquad (5.81)$$

Here n_1 is the real part of the refractive index determined by the lowest direct-gap energy (E_0) and being strongly dependent on the photon energy $E=\hbar\omega$; n_2 is the refractive index determined by the higher-lying gaps, such as E_1, E_2, and E_0'. We note that the above expression has been popularly used in the analysis of modulation spectroscopy (see, e.g., thermoreflectance spectroscopy [70]). Further, the E_0 transitions usually contribute to the dispersion of n (or ε_1), but not to its absolute value. Since then, one has the relations

$$n(E\to 0)^2 \sim n_2(E\to 0)^2 \sim \varepsilon_\infty \qquad (5.82)$$

$$\frac{dn(E\to 0)}{dT} \sim \frac{dn_2(E\to 0)}{dT} \sim \frac{1}{2\varepsilon_\infty^{1/2}}\frac{d\varepsilon_\infty}{dT} \qquad (5.83)$$

We show in Fig. 5.25 the n dispersion in GaAs measured by Marple [71] by the prism refraction method not only at room temperature but also at lower temperatures, 103 and 187 K. The fits of these experimental data to Eq. (5.57) (Sellmeier equation) give [72]

$$A=8.720,\quad B=2.053,\quad C^2=0.358,\quad T=103\text{ K} \qquad (5.84a)$$
$$A=8.598,\quad B=2.242,\quad C^2=0.353,\quad T=187\text{ K} \qquad (5.84b)$$
$$A=8.950,\quad B=2.054,\quad C^2=0.390,\quad T=300\text{ K} \qquad (5.84c)$$

These data suggest that the temperature coefficient dn/dT of GaAs is about 3×10^{-4} K^{-1} in the 0.7–1.4-eV range of photon energy.

Figure 5.26 shows the refractive index n of the lead salts (PbS, PbSe, and PbTe) obtained at 77, 300, and 373 K from fundamental reflectivity data by Zemel *et al.* [73]. A peak appears in the index of each material which is associated with the rapid change in the absorption coefficient in the fundamental absorption edge E_0. It is evident that this peak shifts to lower energies as the temperature decreases (i.e.,

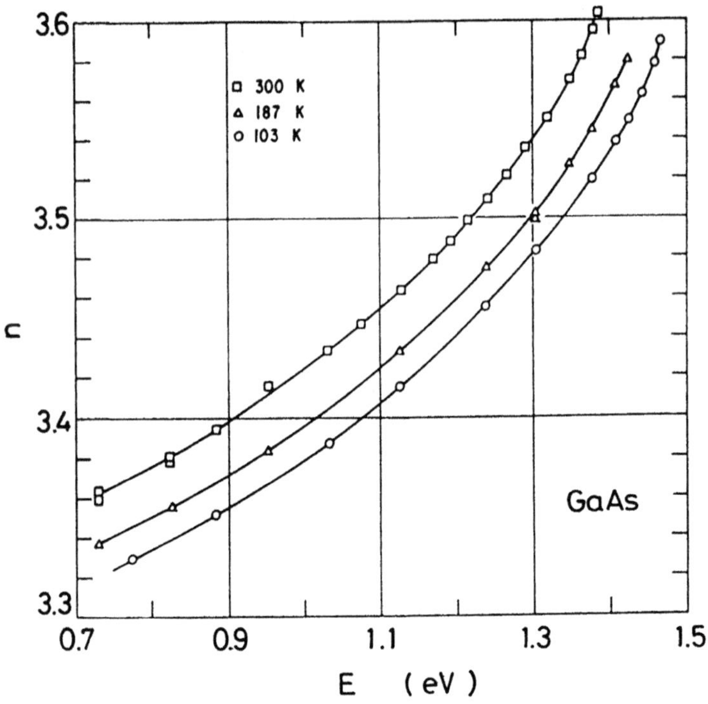

Figure 5.25 Refractive index of GaAs as a function of photon energy E at temperatures 300, 187, and 103 K. (From Marple [71].)

$dE_0/dT<0$). It is also clearly seen that the n value in the transparent region $E<E_0$ decreases as the temperature increases. Such temperature dependence is in direct contrast to that of GaAs (see Fig. 5.25). From these data, we obtain the negative temperature coefficients of the long-wavelength refractive index $n^{-1}(dn/dT)$ for a family of the lead salts (Table 5.3).

The temperature dependence of the refractive index for InP in the wavelength range from 1.2 to 1.6 μm is shown in Fig. 5.27. The experimental data were taken by Gini and Melchior [74] by means of an integrated optical demultiplexer. The data suggest that the linear temperature coefficient dn/dT increases as the photon energy approaches the lowest-direct-gap energy E_0 (1.35 eV). The solid line in Fig. 5.27 represents a quadratic least-squares fit to the experimental data. Similar wavelength (photon-energy) dependence of the linear temperature coefficient dn/dT has also been found in semiconductors, such as group-IV [75–8], III–V [56,76,77], and II–VI semiconductors [76,79].

The thermooptic coefficients of *a*-Si at two laser wavelengths, 0.6328 and 0.752 μm, in the temperature range from 20 to 475°C have been measured by Yavas *et*

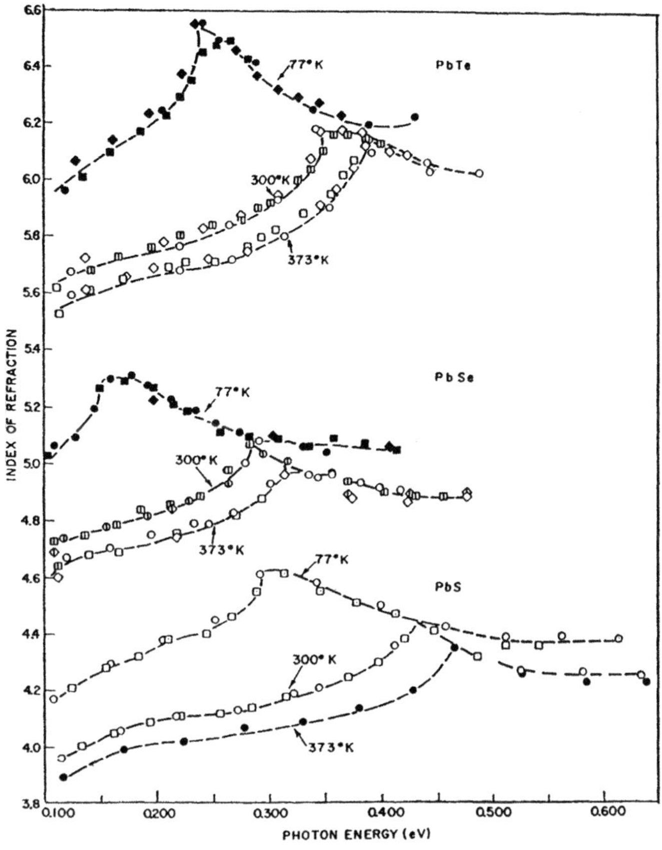

Figure 5.26 Refractive index of PbS, PbSe, and PbTe at 77, 300, and 373 K. (From Zemel *et al.* [73].)

al. [80]. The measured data showed that the refractive index decreases with temperature at 0.6328 μm, but increases at 0.752 μm which is sharp contrast to the case for *c*-Si. Ghosh [78] tried to explain this anomaly based on an exciton-isentropic band-gap model, and concluded that it is due to the inverted position of the isentropic band gap in *a*-Si. We note, however, that it is not possible to obtain information on the amorphous isentropic band gaps *a priori* neither from the available experimental data nor from theoretical point of view.

2. *Pressure effect*—Let us calculate the theoretical expression for pressure coefficient of the refractive index in terms of the Penn model. Equation (5.75) for $E \to 0$ and $A \equiv V^{-1}$ leads

$$\frac{1}{n}\left(\frac{\mathrm{d}n}{\mathrm{d}T}\right)_p = \frac{1}{2}\left(1-\frac{1}{\varepsilon_1}\right)\left[C_o - \frac{2}{E_g}\left(\frac{\partial E_g}{\partial p}\right)_T\right]$$

$$= \frac{1}{2}\left(1-\frac{1}{\varepsilon_1}\right)\left[-\beta_{th} - \frac{2\beta_{th}V}{E_g}\left(\frac{\partial E_g}{\partial V}\right)_T\right]\left(-\frac{C_o}{\beta_{th}}\right) \qquad (5.85)$$

where C_o=$(-1/V)(\partial V/\partial p)_T$ is the isothermal compressibility. If we assume for the moment that in Eq. (5.76), the $(\partial E_g/\partial T)_V$ term is negligible compared with the other terms, then

$$\left(\frac{\partial n}{\partial p}\right)_T = -\frac{C_o}{\beta_{th}}\left(\frac{\partial n}{\partial T}\right)_p \qquad (5.86)$$

As mentioned before, however, we must note that this assumption is not substantial.

As we have already seen in Fig. 2.4 (b), the static and high-frequency dielectric constants of GaAs decrease almost linearly with increasing hydrostatic pressure [$\mathrm{d}(\ln\varepsilon_s)/\mathrm{d}p$=-16.7±0.3 ($T$=75.6 K); $\mathrm{d}(\ln\varepsilon_s)/\mathrm{d}p$=-17.3±0.3 ($T$=300 K); $\mathrm{d}(\ln\varepsilon_\infty)/\mathrm{d}p$=-14.0 ($T$=300 K) (all in units of 10^{-3} GPa^{-1})].

On the analogy of Eq. (5.81), we can express the pressure dependence of n for energies near the lowest direct-band gap E_0 as

$$\frac{\mathrm{d}n}{\mathrm{d}p} \sim \frac{\mathrm{d}n_1}{\mathrm{d}p} = \frac{\mathrm{d}n_1}{\mathrm{d}E_g}\frac{\mathrm{d}E_g}{\mathrm{d}p} \sim -\frac{\mathrm{d}n_1}{\mathrm{d}E}\frac{\mathrm{d}E_g}{\mathrm{d}p} \qquad (5.87)$$

Because both $\mathrm{d}n_1/\mathrm{d}E$ and $\mathrm{d}E_g/\mathrm{d}p$ have positive values in most semiconductors, we can expect the sign of $\mathrm{d}n/\mathrm{d}p$ to be negative (i.e., n will decrease with increasing pressure).

The hydrostatic-pressure dependence of the refractive index near the E_0 gap of some II–VI semiconductors has been measured by Lindner and coworkers [81,82]. We reproduce in Fig. 5.28 the results for ZnTe [81]. These authors grew a thin ZnTe layer on GaAs substrate by MOCVD and determined the n dispersion in the spectral range from the discrete-exciton peak to 900 nm at T=300 and 115 K. As easily expected from Eq. (5.87), their measured n decreased with increasing hydrostatic pressure. Such experimental data were fitted to the Sellmeier equation [Eq. (5.57)]; as a result, the pressure-dependent parameters $A(p)$, $B(p)$, and $C^2(p)$ were determined for both T=300 and 115 K. We show, as an example, the T=300 result in the following (λ in μm, p in GPa):

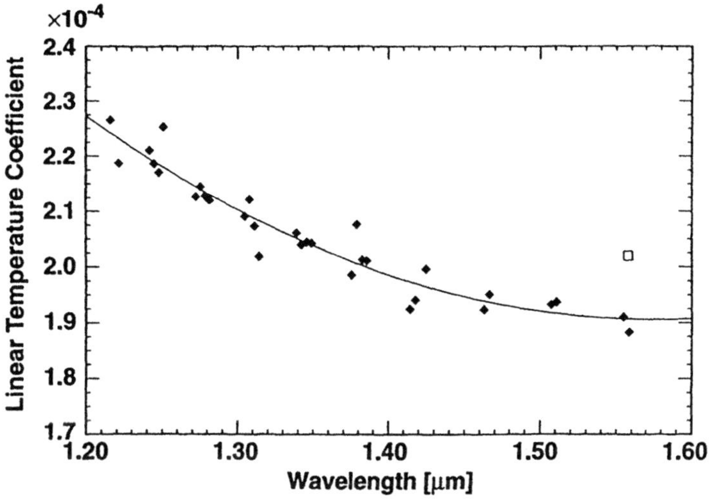

Figure 5.27 Experimental linear temperature dependence of the refractive index dn/dT for InP in the wavelength range from 1.2 to 1.6 μm. The solid line represents a quadratic least-squares fit: y=8.36-8.194×10^{-4}λ+2.6×10^{-4}λ^2 (λ in μm). (From Gini and Melchior [74].)

$$A(p)=(5.8\pm0.1)-(0.15\pm0.2)p \qquad (5.88a)$$

$$B(p)=(1.8\pm0.1)-(0.14\pm0.2)p \qquad (5.88b)$$

$$C^2(p)=(0.182\pm0.003)-(1.8\pm0.05)\times10^{-2}p-(1.08\pm0.05)\times10^{-3}p^2 \qquad (5.88c)$$

We understand from Eq. (5.88) that both A and B have a liner dependence on p, while C^2 has a quadratic dependence on p. As $\lambda\rightarrow\infty$, Eq. (5.57) gives the high-frequency dielectric constant, $\varepsilon_\infty(p)=A(p)+B(p)$. We can thus expect that ZnTe has a weak dependence of ε_∞ on p. More recently, the same research group [82] has reported that ε_∞ for ZnSe and ZnS are obviously independent of pressure. This group also reported experimental data on the pressure dependence of the exciton Rydberg energies for ZnTe and ZnSe [81,82].

3. *Doping effect*—As mentioned in Sec. 2.2, the optical properties at long-wavelength region of undoped (semiinsulating) crystals can be adequately well interpreted by the single harmonic oscillator model. In doped materials, we must take into consideration the effect of free carriers (plasmons). The dielectric permittivity in that case is written as Eq. (2.14).

Reflectivity measurement appears to be the simplest and most direct method of determining the refractive index over a wide range of photon energies, both above and below the fundamental absorption edge. Carrier-concentration dependence of the real refractive index for GaAs has been determined from accurate double-beam reflectance measurements by Sell *et al.* [83]. The uncertainty in the refractive indices obtained from this method was ±0.005. Measurements were made on

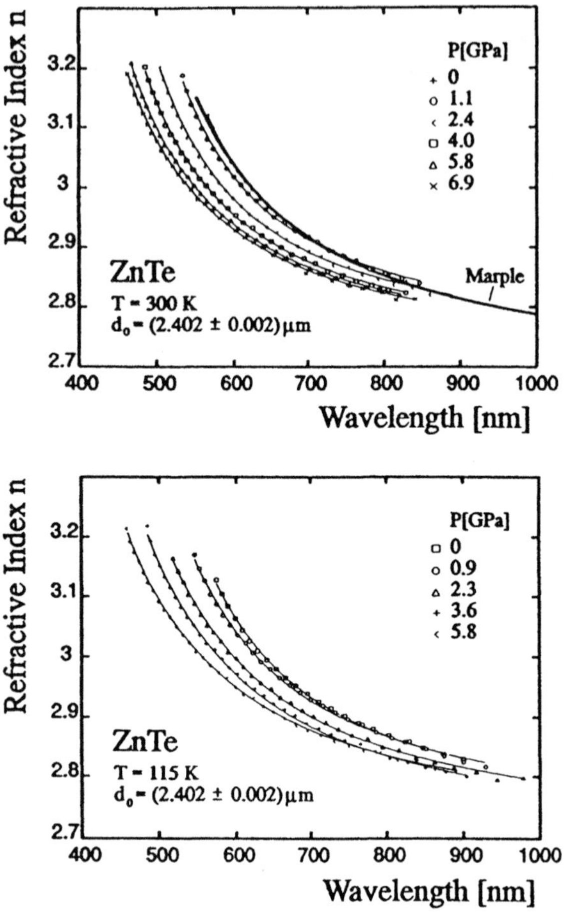

Figure 5.28 Pressure dependence of the refractive index for a thin MOCVD-grown ZnTe in the spectral range from the discrete-exciton peak to 900 nm at T=300 and 115 K. (From Lindner *et al.* [81].)

high-purity sample (n=5×10^{13} cm^{-3}), n-type samples with free-electron concentrations from 5×10^{16} to 6.7×10^{18} cm^{-3}, p-type samples with free-hole concentrations from 1.5×10^{16} to 1.6×10^{19} cm^{-3}, and p-type samples heavily doped with the amphoteric impurity Si.

We reproduce in Fig. 5.29 the refractive-index dispersion for the high-purity sample and for the n-type samples with free-electron concentrations between 5×10^{16} and 6.7×10^{18} cm^{-3} [83]. Because the changes in n near E_0 are small and the curves of n vs. E (=$h\nu$) tend to overlap for n≤1×10^{18} cm^{-3}, the data in Fig. 5.29 (a) have been displaced downward as indicated by ($\bar{n}$-0.1), ($\bar{n}$-0.2), and ($\bar{n}$-0.3). The corrected value of $\bar{n}$ is then obtained by adding 0.1, 0.2, or 0.3. The sample with

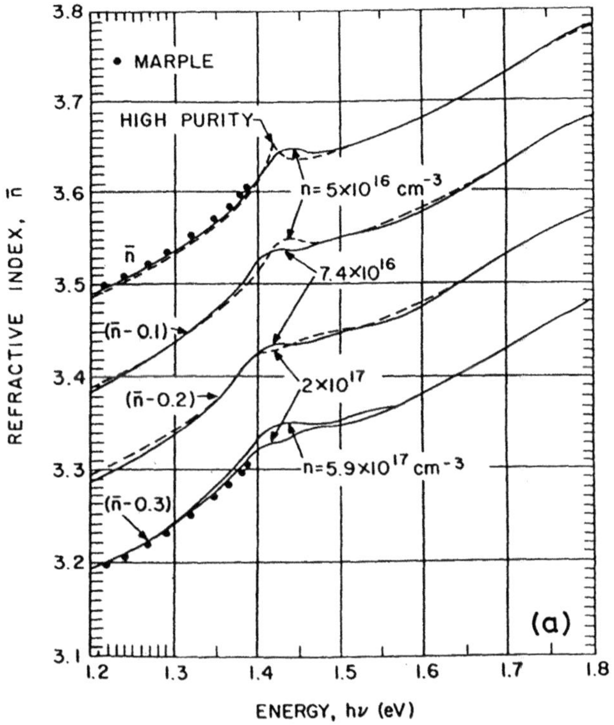

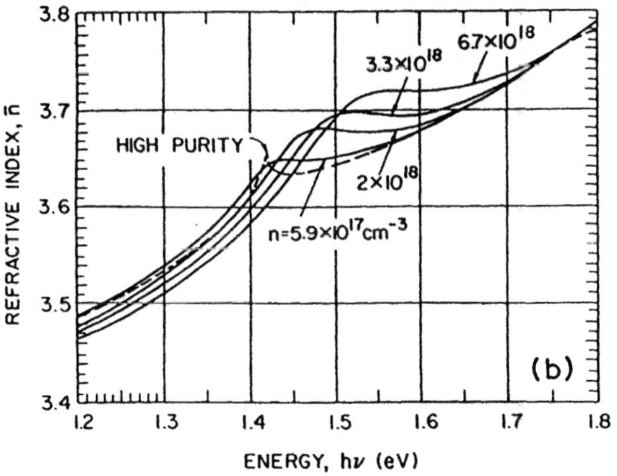

Figure 5.29 Refractive index at 297 K for high-purity (n=5×10^{13} cm^{-3}) GaAs and for n-type samples with free-electron concentrations between 5×10^{16} and 6.7×10^{18} cm^{-3}. Marple's data [71] are also plotted in (a). (From Sell *et al.* [83].)

n=5×10^{16} cm^{-3} has an electron concentration similar to one of the samples studied by Marple [71] (n~6×10^{16}–5.5×10^{17} cm^{-3}, Fig. 5.25). Marple's data are also plotted, for purposes of comparison, and agree within 0.002 over the range in photon energy where comparison is possible. The peak near E_0 of 1.424 eV is noted to be very sharp for n=5×10^{13} cm^{-3} (high-purity sample) and somewhat rounded at 5×10^{16} cm^{-3}. For n=7.4×10^{16} cm^{-3}, the peak at E_0 has begun to disappear and a broad peak appears near 1.5 eV. This loss of the peak at E_0 continues for n=2×10^{17} cm^{-3}. For n=5.9×10^{17} cm^{-3}, the peak at E_0 reappears but is flatter than for $n \leq 1\times10^{16}$ cm^{-3}. In Fig. 5.29 (b), n at photon energies below E_0 decreases as the electron concentration becomes larger. Also, the peak near E_0 shifts to higher energy as the carrier concentration increases. This behavior is expected from the shift of the absorption edge to higher energy with increasing electron concentration [see Fig. 5.14 (a)].

The refractive indices obtained by Sell *et al.* [83] for the p-GaAs with free-hole concentrations between 1.5×10^{16} cm^{-3} and 1.6×10^{19} cm^{-3} are shown in Fig. 5.30. As the hole concentration is increased, the value of n decreases and the peak near E_0 is flattened out.

The doping-induced change in the refractive index Δn may be explained by a sum of the three different contributions:

$$\Delta n = \Delta n_{\mathrm{f}} + \Delta n_{0\mathrm{g}} + \Delta n_{\mathrm{hg}} \qquad (5.89)$$

where Δn_{f} is arising from free-carrier absorption (see Sec. 5.3.1 below), $\Delta n_{0\mathrm{g}}$ is a contribution from the doping-induced change in the E_0-gap parameters (energy, strength, and broadening), and Δn_{hg} is a contribution from the change in the higher-lying-gap parameters. Note that Δn_{f} and Δn_{hg} are considered to be nondispersive in the region of E_0. Therefore, the refractive-index change Δn caused by doping near E_0 is mainly governed by $\Delta n_{0\mathrm{g}}$. The data of Sell *et al.* [83] have been successfully explained by Yas'kov [84] using the exponential band-edge model of Eq. (5.65). Zoroofchi and Butler [85] had also analyzed optical-absorption data of n-GaAs (n=2.0×10^{16}–6.5×10^{18} cm^{-3}) using the exponential band-edge model, and obtained the refractive-index dispersion for the samples at T=77 and 300 K. The contribution to n due to free-carrier absorption was concluded to be negligibly small in their analysis.

Accurate refractive-index measurements of undoped and doped n-InP in the 1300–1600-nm wavelength range have been demonstrated recently by Martin *et al.* [86] by using a grating coupling technique. A typical accuracy of 5×10^{-4} has been achieved with sample temperature maintained to within 0.01 K. We reproduce in Fig. 5.31 their obtained results on the refractive-index dispersion of n=2×10^{18}-cm^{-3}-doped InP substrate and undoped guiding InP layer, together with that obtained by Pettit and Turner (PT) [87]. The theoretical curves obtained from the Sellmeier formula are plotted by the solid and dashed lines. The Sellmeier parameters fit-determined are as follows (λ in μm): A=7.283, B=2.337, C^2=3.87×10^{-5} (undoped InP); A=7.194, B=2.282, C^2=4.22×10^{-5} (2×10^{18}-cm^{-3}-doped InP). We can see in Fig.

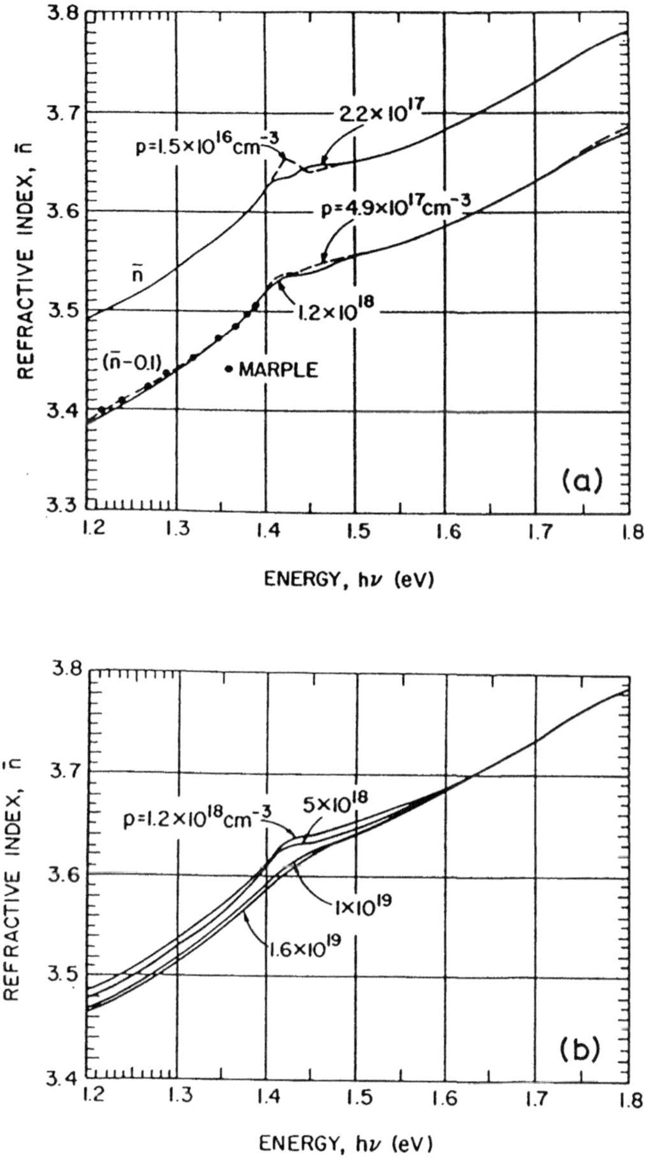

Figure 5.30 Refractive index at 297 K for *p*-type GaAs with free-hole concentrations between 1.5×10^{16} and 1.6×10^{19} cm^{-3}. Marple's data [71] are also plotted in (a). (From Sell *et al.* [83].)

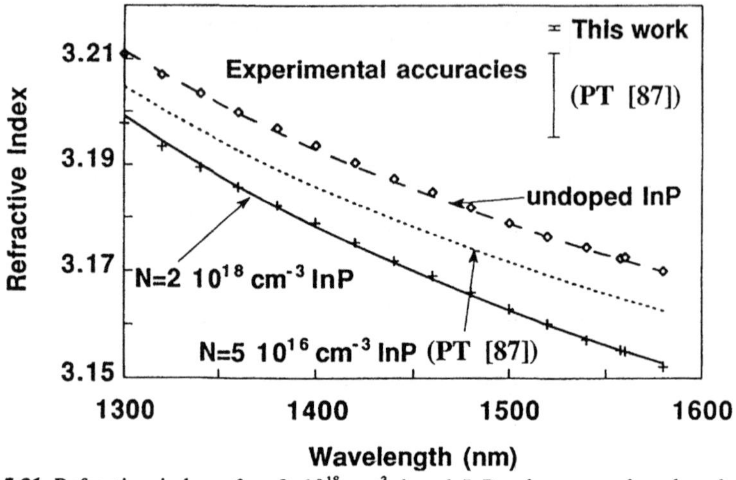

Figure 5.31 Refractive index of n=2x10^{18}-cm^{-3}-doped InP substrate and undoped guiding InP layer at 300 K, together with that obtained by Pettit and Turner (PT) [87]. The theoretical curves obtained from the Sellmeier formula are plotted by the solid and dashed lines. (From Martin *et al.* [86].)

5.31 an increase in the refractive-index difference between 2×10^{18}-cm^{-3}-doped and undoped InP from 1.2×10^{-2} to 1.7×10^{-2} as the wavelength is increased from 1300 to 1580 nm. This increase may be caused by the free-carrier absorption (Δn_f). Martin *et al.* [86] also reported a $(2.02\pm0.02)\times$
10^{-4} K^{-1} temperature coefficient of the refractive index for n-InP around 300 K.

5.3 FREE-CARRIER ABSORPTION AND RELATED PHENOMENA

5.3.1 Free-Carrier Absorption

The optical absorption of interest here can be divided into three separate processes, which are schematically shown in Fig. 5.32. The direct or indirect transitions from the valence to the conduction band give rise to the absorption contribution α_{VC} (see Sec. 5.1). A normal free-carrier absorption α_{FC} is dependent on photon energy approximately as

$$\alpha_{FC}(E) \propto E^{-m} \qquad (5.90)$$

where m is dependent on various intraband scattering processes, that is, m~2.5 limited by polar optical phonons (α_{oph}), m~1.5 limited by acoustic phonons (α_{aph}), and m~3.5 limited by ionized-impurity scattering (α_i) [1]. The final and initial

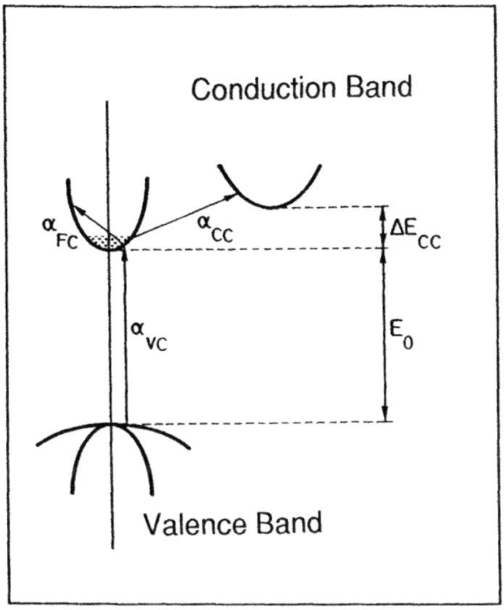

Figure 5.32 Band-diagrammatic indication of the optical absorption in a semiconductor: α_{VC}, direct interband transition from the valence band to the conduction band; α_{CC}, indirect interconduction-band transition; α_{FC}, free-carrier intraband transition.

states of the electrons in these transitions lie within the same conduction band. The resultant free-carrier absorption is therefore called *"intraband."* The final process depicted in Fig. 5.32 is an interconduction-band absorption (α_{CC}, see Sec. 5.3.2). The α_{VC} (α_{CC}) takes part when $E \geq E_0$ ($E \geq \Delta E_{CC}$). The α_{FC} becomes significant at longer wavelengths. Considerable theoretical works on the optical absorption in the free-carrier absorption region have been reviewed by Haga and Kimura [88–91] and Fan [92].

Free-carrier effect on the optical properties of simple metals (*e.g.*, alkali metals) can be described in terms of the classical dispersion theory of the Drude model [93,94]. We shall use this model since free carriers introduced into semiconductors by doping behave in many ways like those in the simple metals. The main difference between the two is that the carrier concentration in a semiconductor can be changed by doping. Since the doping concentration is typically less than 10^{20} cm^{-3}, the plasma frequencies of carriers in semiconductors are usually in the IR region [see Eq. (2.15)], whereas they are in the visible or UV for metals.

The Drude model starts with an equation of motion for an electron of charge e and mass m_e

$$m_e \left(\frac{d^2x}{dt^2} + \Gamma \frac{dx}{dt} + \omega_0^2 x \right) = eE_f \exp(i\omega t) \qquad (5.91)$$

with the damping constant Γ, the restoring force $m_e\omega_0^2 x$, and the field of an electromagnetic wave $E_f \exp(i\omega t)$. This leads to the following dielectric constant [1]:

$$\varepsilon(\omega) = \varepsilon_1(\omega) + i\varepsilon_2(\omega) = 1 + \left(\frac{Ne^2}{m_e\varepsilon_0}\right)\left(\frac{1}{\omega_0^2 - \omega^2 - i\Gamma\omega}\right) \tag{5.92}$$

where N is the number of the free electrons per unit volume (not the *refractive index*). Rationalizing Eq. (5.92) gives

$$\varepsilon_1(\omega) = 1 + \left(\frac{Ne^2}{m_e\varepsilon_0}\right)\left(\frac{\omega_0^2 - \omega^2}{(\omega_0^2 - \omega^2)^2 + \Gamma^2\omega^2}\right) \tag{5.93a}$$

$$\varepsilon_2(\omega) = \left(\frac{Ne^2}{m_e\varepsilon_0}\right)\left(\frac{\Gamma\omega}{(\omega_0^2 - \omega^2)^2 + \Gamma^2\omega^2}\right) \tag{5.93b}$$

Since there is no restoring force for the free electrons, we rightly put $\omega_0 \to 0$. Then, Eqs. (5.92) and (5.93) can be written as

$$\varepsilon(\omega) = \varepsilon_\infty\left(1 - \frac{\omega_p^2}{\omega(\omega + i\Gamma)}\right) \tag{5.94}$$

and

$$\varepsilon_1(\omega) = \varepsilon_\infty\left(1 - \frac{\omega_p^2}{\omega^2 + \Gamma^2}\right) \tag{5.95a}$$

$$\varepsilon_2(\omega) = \frac{\varepsilon_\infty\omega_p^2\Gamma}{\omega(\omega^2 + \Gamma^2)} \tag{5.95b}$$

where we used the relation between the plasma frequency ω_p (*screened* by the dielectric constant ε_∞) and carrier concentration n of Eq. (2.15).

Notice here that the imaginary part of the dielectric function $\varepsilon_2(\omega)$ is proportional to Γ; hence, the absorption coefficient is too. The reason for this is well known in metal optics (see Refs. [93,94]).

A free-carrier absorption process is the annihilation of a photon with the excitation of a carrier from a filled state below the Fermi energy E_F to an empty state above it. The energy and momentum have to be conserved in this process due to absorption (or emission) of polar optical phonons (α_{oph}), acoustic phonons (α_{aph}), or via scattering with ionized impurity (α_i).

From Eqs. (1.21) and (5.95b), one obtains

$$\alpha(\omega) = \frac{4\pi}{\lambda}k(\omega) = \frac{\omega}{nc}\varepsilon_2(\omega) = \frac{\varepsilon_\infty \omega_p^2 \Gamma}{nc(\omega^2+\Gamma^2)} \qquad (5.96)$$

where n is the refractive index (not the *carrier concentration*) and c is the speed of light. In the low-frequency limit $\omega \to 0$, this equation becomes

$$\alpha = \frac{\varepsilon_\infty \omega_p^2}{nc\Gamma} = \frac{4\pi Ne^2}{ncm_e\Gamma} \qquad (5.97)$$

On the other hand, the low-field conductivity for free carriers, or for a simple band structure, can be given by

$$\sigma = ne\mu = \frac{ne^2\tau}{m_e} \qquad (5.98)$$

where μ is the low-field mobility and τ is a phenomenological scattering time introduced to occur for the scattering of the carrier by phonons and impurities. Comparing α in Eq. (5.97) with the corresponding expression for σ [Eq. (5.98)], one obtains

$$\alpha = \frac{4\pi\sigma}{nc} \qquad (5.99)$$

One can thus equate Γ to τ^{-1} and recognize a distant resemblance between free-carrier absorption and carrier transport properties. In fact, temperature dependence of low-field mobility, that is limited by each scattering mechanism, shows nearly the same temperature dependence as those expected in the free-carrier absorption mechanisms [9]. Quantum-mechanical investigations [95] also showed that Eq. (5.99) is only valid for photon energies which are small compared with the carrier energies ($\hbar\omega < kT$).

Let us return to Eq. (5.94). From this expression, we may distinguish three frequency regions for the optical properties of semiconductors, as sketched on Figs. 5.33 and 5.34. In the $0<\omega<\Gamma$ region, both k and n are much larger than unity. In this region, the light will be strongly absorbed. In the $\Gamma<\omega<\omega_p$ region, n is smaller than k; ε_1 is then negative and we have almost total reflection, that is to say very little absorption. For $\omega>\omega_p$ we have

$$\varepsilon(\omega) = \varepsilon_\infty\left(1 - \frac{\omega_p^2}{\omega^2} + i\frac{\omega_p^2\Gamma}{\omega^3}\right) \qquad (5.100)$$

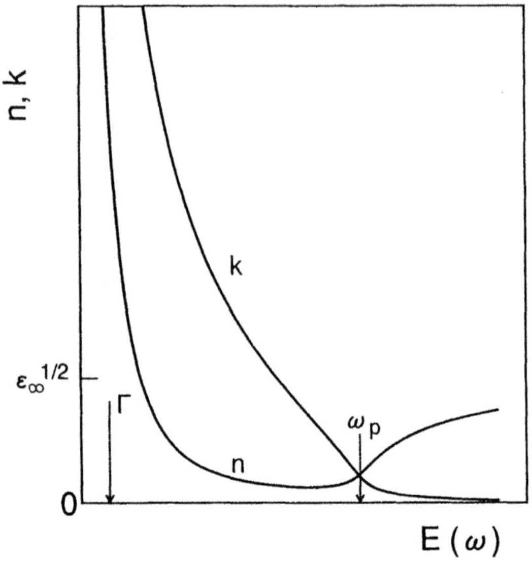

Figure 5.33 Plots of n and k vs. E ($\hbar\omega$) for free carriers (schematic).

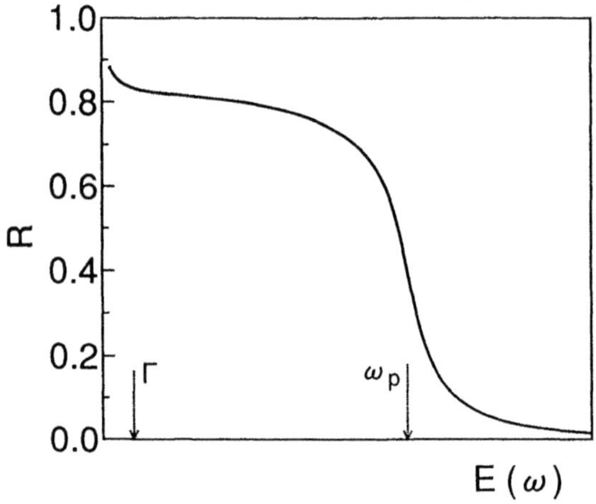

Figure 5.34 Reflectivity spectrum R calculated from (n, k) in Fig. 5.33.

Hence, $n \rightarrow \varepsilon_\infty^{1/2}$ and $k \rightarrow 0$, indicating that the region is transparent. In the region near ω_p, $n \sim 0$. What does it mean that the refractive index is zero? The index of refraction is defined in terms of the phase velocity as $v_p = c/n$. Thus, a zero value for n means an infinite phase velocity and an infinite wavelength. An infinite wavelength means that the electrons are all oscillating in phase; however, there is no polarization charge density as with a true plasma oscillation [94].

Figure 5.35 shows a typical example for the free-carrier absorption in n-InP measured by Ballman *et al.* [96]. The experimental data indicate that an absorption minimum exists at about 1.6 μm, in agreement with the results reported by others [97,98]. At wavelengths λ>1.6 μm, free-carrier absorption dominates in the spectra. The dashed line in the figure gives a slope that is limited by the ionized-impurity scattering (i.e., m~3.5). The power index m for a heavily doped sample (c) just corresponds to 3.5. It has also been found [99] that for n-InP m increases with increasing n from about 2.0 (n~1×10^{16} cm^{-3}) to about 3.5 (n~1×10^{18} cm^{-3}). For wavelengths near 1.3 μm, the absorption coefficients are still proportional to the free-carrier concentration. This fact may suggest the dominance of interconduction-band absorption in that spectral region (see Sec. 5.3.2 below).

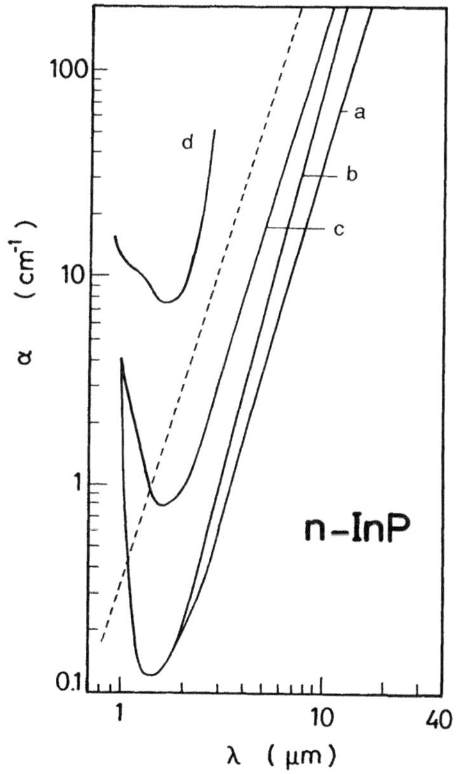

Figure 5.35 Free-carrier absorption in n-InP. The electron concentrations are (a) 3.3×10^{16} cm^{-3}, (b) 4.0×10^{17} cm^{-3}, (c) 8.8×10^{17} cm^{-3}, and (d) 5.0×10^{18} cm^{-3}. The dashed line gives a slope that is limited by ionized-impurity scattering. (From Ballman *et al.* [96].)

In Fig. 5.36 (a) presents the absorption-coefficient data for n-GaAs for λ up to 10 μm, reproduced from various sources [100]. Quantitatively, between 1 and 3 μm, the major effect is the flat absorption, namely, the interconduction-band absorption, superimposed on the relatively weak free-carrier absorption. Above ~3 μm the rapidly diminished interconduction-band absorption competes with the superlinearly rising free-carrier absorption. Eventually, over ~5 μm, the latter mechanism becomes the predominant mode of absorption.

The optical-phonon absorption (α_{oph}), acoustic-phonon absorption (α_{aph}), ionized-impurity absorption (α_i), and interconduction-band absorption (α_{CC}) contributions to the total absorption coefficient of n-GaAs at 300 and 700 K, as calculated by Jordan [100], are shown in Fig. 5.36 (b). The doping level and compensation ratio were assumed to be 10^{17} cm^{-3} and unity, respectively. Clearly, in the vicinity of the temperature-dependent band gap, the fundamental absorption edge dominates α. The absorption processes due to optical and acoustic phonons and ionized-impurity scattering demonstrate a marked increase with increasing wavelength. The α_{CC} term slowly descends with wavelength between about 1 and 3 μm. The calculated and experimental absorption coefficients in Fig. 5.36 (a) are in reasonable agreement with respect to line shape as well as magnitude.

The IR absorption spectra for p-InP are shown in Fig. 5.37. They are measured by Casey and Carter [101] for λ from 1.0 to 3.4 μm for hole concentrations from $p=3.1\times10^{17}$ to 6.6×10^{18} cm^{-3}. The spectrum taken for lightly doped p-InP ($p=6.0\times10^{16}$ cm^{-3}) by Ballman *et al.* [96] is also shown by the dashed line. As similar to the case for n-type semiconductors, the absorption coefficient increases as the carrier (hole) concentration increases. The concentration variation of α is found to be written as $\alpha=14(p/10^{18}$ cm$^{-3})$ cm^{-1} at $\lambda=1.3$ μm and $20(p/10^{18}$ cm$^{-3})$ cm^{-1} at $\lambda=1.5$ μm [101]. Peaks that appeared at about 0.4 eV are due to intervalence-band absorption, as we will see in the following section.

5.3.2 Interband Absorption

(a) Interconduction-band absorption—The most unique feature of the IR absorption spectrum of n-GaAs is the relatively flat region between 1 and 3 μm (see Fig. 5.36). This region is caused by the interconduction-band absorption. A promising theoretical interpretation of the interconduction-band absorption was given by Haga and Kimura [91]: Briefly, as a photon is absorbed by the crystal, an electron at the bottom of the lowest-conduction-band minimum makes a transition to an intermediate state in a higher band at the same point in $\boldsymbol{k}$ space. Then, interacting with an imperfection (or phonons), the electron returns to a lower-lying conduction-band minimum (α_{CC}, see Fig. 5.32). The transition mechanism can, thus, be expressed by a second-order process in the perturbation, the same as the indirect-band-gap transition mechanism.

Actually, an optical transition between the lowest conduction band and higher valley not at the same point in $\boldsymbol{k}$ space is very complicated because of the large number of virtual intermediate sates the transition involves. Haga and Kimura [91]

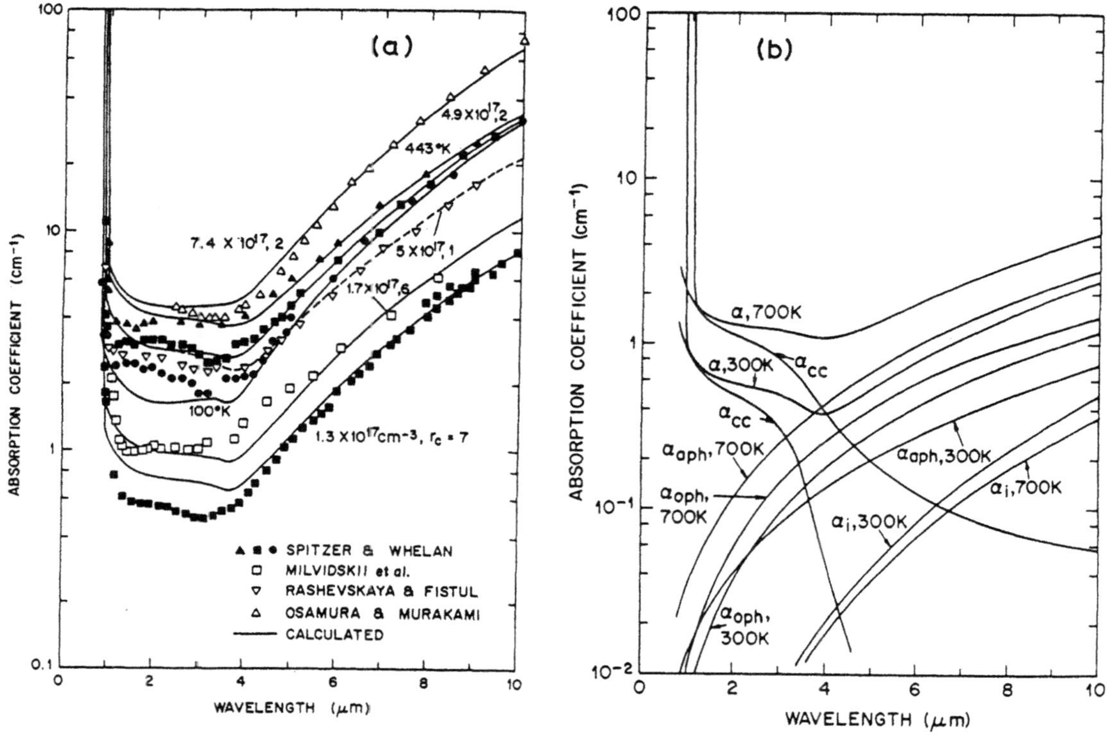

Figure 5.36 (a) Absorption coefficient vs. photon wavelength for *n*-GaAs. The labels are the doping levels and the estimated compensation ratios. Unless labeled otherwise, the temperature for all the data sets and curves is 300 K. (b) Interconduction-band (α_{CC}), optical-phonon (α_{oph}), acoustic-phonon (α_{aph}), and ionized-impurity absorption (α_i) contributions to the total absorption coefficient of *n*-GaAs at 300 and 700 K vs. photon wavelength. The doping levels and compensation ratio are 10^{17} cm^{-3} and unity, respectively. (From Jordan [100].)

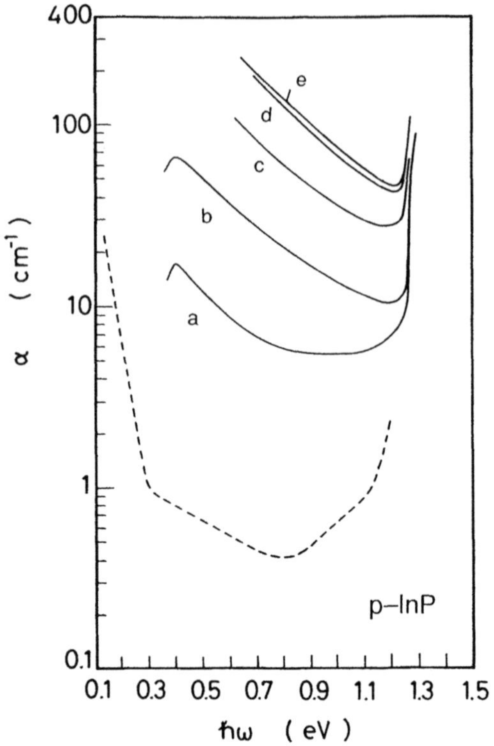

Figure 5.37 IR absorption spectra for *p*-InP at 297 K. The hole concentrations are (*a*) 3.1×10^{17} cm^{-3}, (*b*) 8.6×10^{17} cm^{-3}, (*c*) 1.9×10^{18} cm^{-3}, (*d*) 5.3×10^{18} cm^{-3}, and (*e*) 6.6×10^{18} cm^{-3}. (From Casey and Carter [101].) The spectrum taken for lightly doped *p*-InP (p=6.0×10^{16} cm^{-3}) by Ballman *et al.* [96] is also plotted by the dashed line.

introduced a simplification into the theory by separating the transition into "*I*"-type and "*D*"-type processes depending on which type of intermediate is most important.

In the *I*-type process, the intermediate states are in other bands below and above either the initial state or the final state depending upon whether the electromagnetic interaction is considered before or after the scattering interaction which supplies the necessary change in wave number. The *E* dependence of this absorption results from a DOS argument yielding a fairly well-defined threshold for the onset of absorption and a more slowly varying dependence of the transition probability on *E* [91,98].

In the *D*-type process the intermediate state is almost identical to the initial state, and partially because of the energy denominators of $\pm E$ in these states the absorption incorporates a E^{-3} dependence. The *D*-type absorption yields a narrow and peaked band [91,98].

The interconduction-band absorption α_{cc} seen in Fig. 5.36 (*n*-GaAs) is a typical example of the *I*-type process. This type of process has also been found in *n*-InP [97,98].

Figure 5.38 shows the IR absorption spectrum of *n*-GaP measured in the wavelength range from 1 to 10 μm by Wiley and DiDomenico [102]. The sample used was Te-doped with $n\sim3\times10^{17}$ cm^{-3}. The absorption peaking at $\lambda\sim3$ μm is typical of the *D*-type interconduction-band absorption. The background is due to the normal free-carrier absorption. The *D*-type absorption has been observed not only in *n*-GaP but also in *n*-Si [103], *n*-AlSb [104], and *n*-GaSb [105].

A calculation of the interconduction-band absorption is possible [91,98,100], but its complexity and lack of knowledge of various parameters are formidable barriers. The simplest equation for α_{CC} proposed by Lorenz *et al.* [97] is given by

$$\alpha_{CC} = a(E - \Delta E_{CC})^{1/2} \qquad (5.101)$$

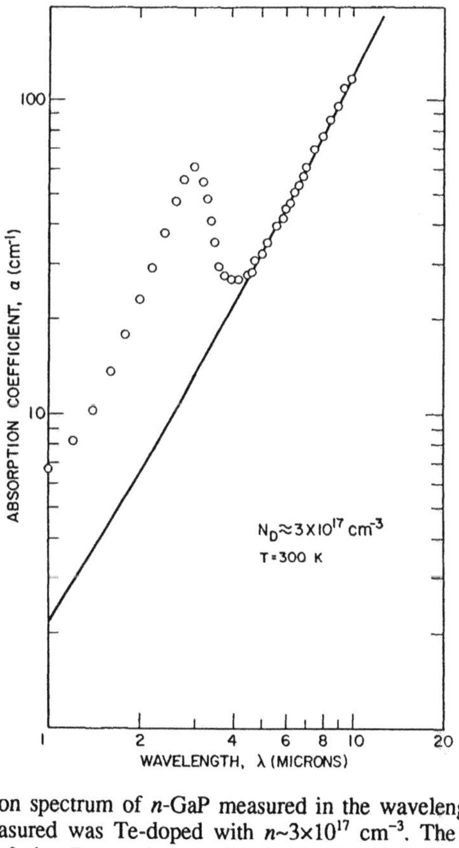

Figure 5.38 IR absorption spectrum of *n*-GaP measured in the wavelength range from 1 to 10 μm. The sample measured was Te-doped with $n\sim3\times10^{17}$ cm^{-3}. The absorption peaking at $\lambda\sim3$ μm is typical of the *D*-type interconduction-band absorption. (From Wiley and DiDomenico [102].)

where ΔE_{CC} is the threshold energy for the relevant absorption (see Fig. 5.32). This equation showed a reasonable fit to the experimental α_{CC} for *n*-InP (*l*-type) [97].

(b) Intervalence-band absorption—The interband free-carrier absorption occurs not only in the *n*-type semiconductors but also in *p*-type ones. In *p*-type semiconductors, the intervalence-band absorption originates from direct transitions between different branches of the valence band at k=0 (Γ). The absorption bands peaking at E~0.4 eV in Fig. 5.37 are due to the intervalence-band absorption.

In Fig. 5.39, possible types of the intervalence-band transitions in the zinc-blende-type semiconductors are illustrated schematically. We can expect from it three types of intervalence-band absorption: SO→HH; SO→LH; and LH→HH. As mentioned in Sec. 1.2.2, the zinc-blende-type crystal has a T_d^2 point-group symmetry. The HH and LH bands have Γ_8 symmetry, while the SO band has Γ_7 symmetry. The selection rules for the optical transitions between different branches of the valence band can then be given by the direct products:

$$\text{SO}\rightarrow\text{HH: } \Gamma_7\times\Gamma_8 = \Gamma_3+\Gamma_4+\Gamma_5 \qquad (5.102\text{a})$$

$$\text{SO}\rightarrow\text{LH: } \Gamma_7\times\Gamma_8 = \Gamma_3+\Gamma_4+\Gamma_5 \qquad (5.102\text{b})$$

$$\text{LH}\rightarrow\text{LH: } \Gamma_8\times\Gamma_8 = \Gamma_1+\Gamma_2+\Gamma_3+2\Gamma_4+2\Gamma_5 \qquad (5.102\text{c})$$

The polarization vector $\boldsymbol{E}\perp(x,y,z)$ of the point group T_d^2 belongs to Γ_5 symmetry. All these products contain the representation of Γ_5 symmetry, indicating that the dipole transitions are possible.

As mentioned in Sec. 3.1.1, the joint-DOS function $J_{\alpha\beta}(E)$ measures the product density of full and empty states of equal energy difference. This quantity mainly

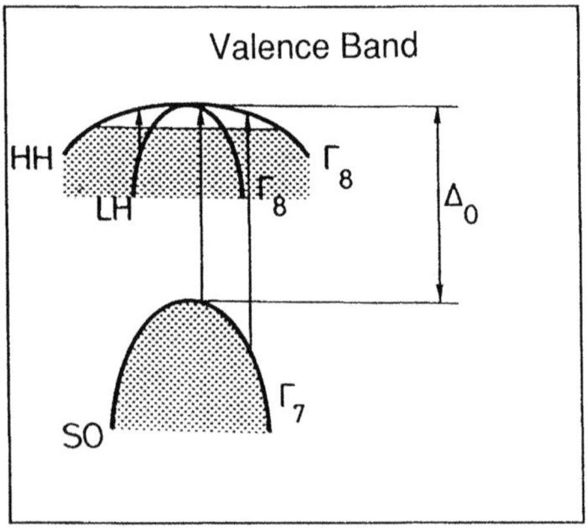

Figure 5.39 Schematic representation of the intervalence-band absorption in a *p*-type, zinc-blende semiconductor.

determines the interband contribution to $\varepsilon_2(E)$, and thus to the optical constants of semiconductors. According to different combinations of the sign of

$$\frac{1}{\mu_j} = \left(\frac{1}{m_{\alpha j}} + \frac{1}{m_{\beta j}}\right)_{k=k_c} \tag{5.103}$$

one has to distinguish between four types of CPs, M_0, M_1, M_2, and M_3. The states α and β mean empty and full bands, respectively. It should be noted that the effective mass in the band α has a relation

$$m_\alpha \text{ (electrons)} = -m_\alpha \text{ (holes)} \tag{5.104}$$

i.e., its sign is opposite between the electrons and holes. The types of CP singularity for the intervalence-band transitions can be determined by inserting numeric values of the hole effective masses (assuming simple, parabolic, isotropic bands) into Eq. (5.103). The results for many zinc-blende semiconductors suggest that $J_{\text{HH-SO}}$ (SO→HH), $J_{\text{LH-SO}}$ (SO→LH), and $J_{\text{HH-LH}}$ (LH→HH) are, respectively, represented by the 3D M_0-, M_3-, and M_0-type CPs. The M_0-type (M_3-type) CP is realized by having all $\mu_j>0$ ($\mu_j<0$). The E-dependent states functions are, then, written as (Ref. [106], see Table 3.2)

$$J_{\text{HH-SO}}(E) = \begin{cases} V_0(E-\Delta_0)^{1/2} & (E>\Delta_0) \\ 0 & (E<\Delta_0) \end{cases} \tag{5.105a}$$

$$J_{\text{LH-SO}}(E) = \begin{cases} 0 & (E>\Delta_0) \\ V_1(\Delta_0-E)^{1/2} & (E<\Delta_0) \end{cases} \tag{5.105b}$$

$$J_{\text{HH-LH}}(E) = V_2E^{1/2} \tag{5.105c}$$

The SO→LH transitions are generally not so important because of a low density of states in the LH band and a very low hole occupancy of final states for transitions at the band-gap energy Δ_0. The larger density of states and hole occupancy in the HH band may produce much greater absorption for the SO→HH transitions than for the SO→LH transitions.

Intervalence-band absorption has been observed experimentally in various p-type semiconductors, such as GaSb [105], Ge [107], InAs [108], GaAs [109–111], InSb [112], AlSb [110], InP [101,111], $In_{0.53}Ga_{0.47}As$ [111], and PbTe [113]. Figure 5.40 shows, as an example, the intervalence-band absorption measured in p-GaAs (p~9.7×10^{16}–6.8×10^{17} cm^{-3}) by Braunstein [109]. At 295 K, there is a band at 0.42 eV, a partially resolved band at 0.31 eV, and the onset of a band at 0.25 eV. As

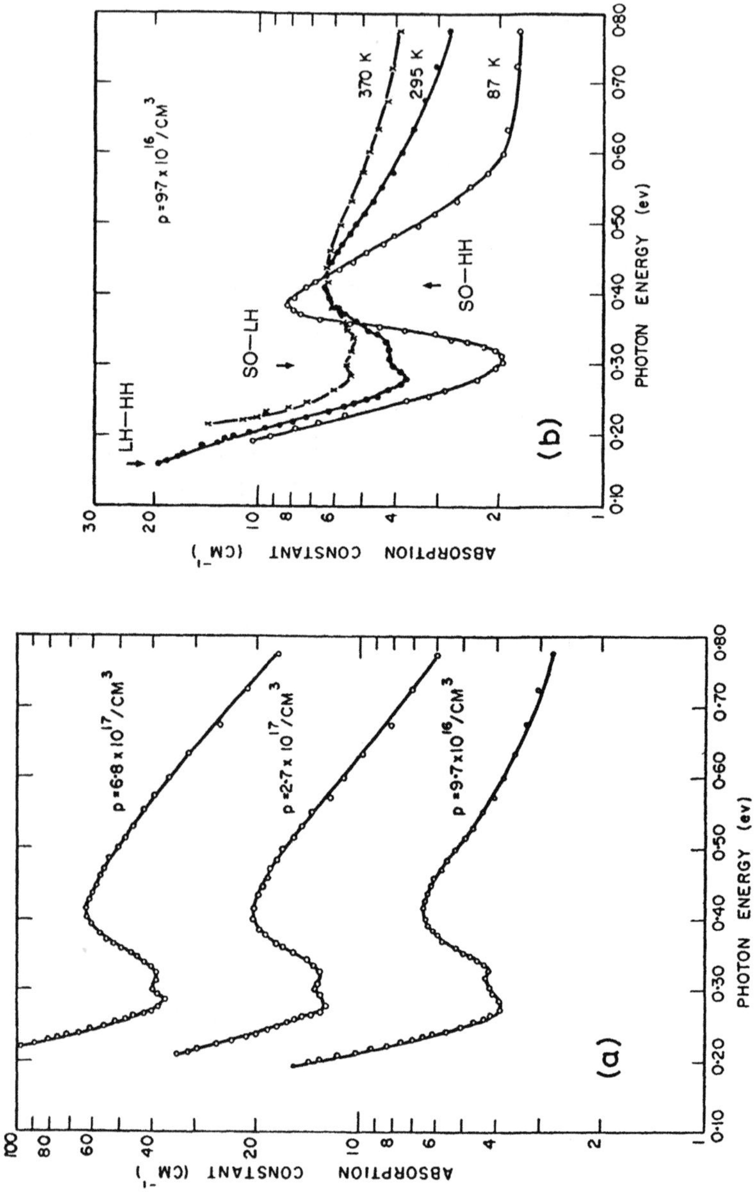

Figure 5.40 (a) Absorption spectra of *p*-GaAs at 295 K for a range of carrier concentrations between *p*~9.7×10^{16} and 6.8×10^{17} cm^{-3}. (b) Absorption spectra of *p*-GaAs (*p*~9.7×10^{16} cm^{-3}) as a function of temperature. (From Braunstein [109].)

the temperature is decreased, the bands sharpen and become slightly displaced from the room-temperature positions [Fig. 5.40 (b)]. The band at 0.42 eV is attributed to transitions between the SO and HH bands; the peak at 0.31 eV is attributed to transitions between the SO and LH bands. The onset of the absorption band at 0.25 eV is also attributed to transitions between the LH and HH bands.

The intervalence-band absorption in *p*-type GaAs, InP, and $In_{0.53}Ga_{0.47}As$ has been measured by Henry *et al.* [111]. The measured spectra were broader and had a smaller temperature dependence. In Fig. 5.41, we reproduce their measured results for *p*-$In_{0.53}Ga_{0.47}As$. The solid line represents the estimated fit based on an elemental $\boldsymbol{k}\cdot\boldsymbol{p}$ band model. Note that in the figure HH→SO and LH→SO absorptions occur when $E>\Delta_0$ and $E<\Delta_0$, respectively [also see Eq. (5.105)].

Near room temperature, the threshold current of $In_{1-x}Ga_xAs_yP_{1-y}$/InP lasers usually increases as $\exp(T/T_0)$, where T_0~65 K (called *characteristic temperature*), whereas theoretical descriptions of ideal lasers give T_0~200 K [111]. Several loss mechanisms have been proposed to explain this type of temperature sensitivity [114]: (1) carrier leakage over the heterobarrier into the confining layers, (2) nonradiative losses due to Auger recombination, (3) nonradiative recombination at defects and interface states, and (4) intervalence-band absorption causing optical losses. Calculations of the intervalence-band absorption give values which are too small to yield an essential influence on the threshold current [114,115]. The reason for the

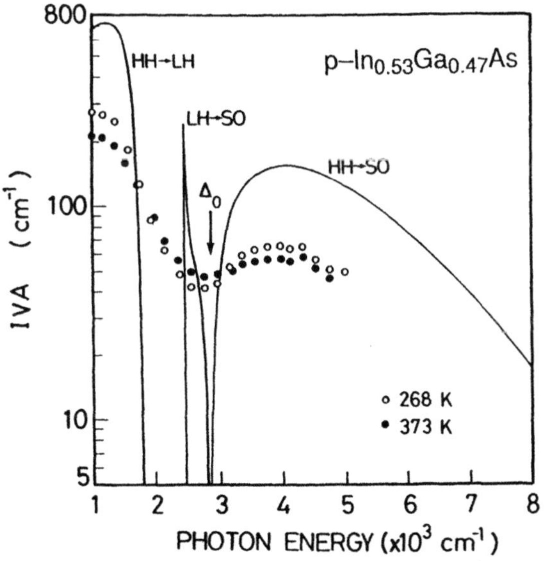

Figure 5.41 Measured intervalence-band absorption in *p*-$In_{0.53}Ga_{0.47}As$ ($p=1\times10^{18}$ cm^{-3}) at 268 and 373 K. The solid lines represent the theoretical curves predicted from an empirical $\boldsymbol{k}\cdot\boldsymbol{p}$ band model. (From Henry *et al.* [111].)

small theoretical values of the intervalence-band absorption is the fact that the process takes place far from the Γ point where the HH valence band is only weakly occupied with holes. However, it has been pointed out [115] that the situation is changed if phonon-assisted indirect intervalence-band absorption is taken into account. This absorption mechanism may enable that the transitions go to the Γ point where the hole occupation is very strong. Then, the absorption coefficient is considerably enlarged, for instance, by more than an order of magnitude in the $In_{1-x}Ga_xAs_yP_{1-y}$ quaternary with a laser wavelength of 1.3 μm.

5.3.3 Carrier-Induced Change in Refractive Index

In Eq. (5.95a), in the limit $\Gamma \to 0$ and using the relation $2n\Delta n \sim \Delta\varepsilon_1$ [see Eq. (5.66)], we obtain the carrier-induced change in the refractive index Δn_f

$$\Delta n_{\mathrm{f}} = \frac{\partial n}{\partial N}\Delta N = -\frac{e^2\lambda^2\Delta N}{8\pi^2 n\varepsilon_0 c^2 m_{\mathrm{e}}} \tag{5.106}$$

The above equation suggests that the change in refractive index Δn_f is directly proportional to λ^2, ΔN, and n^{-1}.

When both electrons and holes are present in a substance, as is usual for injection lasers, the change Δn_f can be written, from Eq. (5.106), as

$$\Delta n_{\mathrm{f}} = -\frac{e^2\lambda^2}{8\pi^2 n\varepsilon_0 c^2}\left(\frac{\Delta N}{m_{\mathrm{e}}} + \frac{\Delta P}{m_{\mathrm{h}}}\right) \tag{5.107}$$

where ΔP is the change in the hole concentration.

The intraband free-carrier (electrons and holes) and intervalence-band absorption (IVA) contributions to the refractive-index change of GaAs obtained by Stern [41] can be written as (N, P in $\mathrm{cm^{-3}}$, E in eV)

$$\Delta n_{\mathrm{f,e}}(E) \sim -9.6\times10^{-21}\, N/nE^2 \tag{5.108a}$$

$$\Delta n_{\mathrm{f,h}}(E) \sim -1.8\times10^{-21}\, P/nE^2 \tag{5.108b}$$

$$\Delta n_{\mathrm{IVA}}(E) \sim -6.3\times10^{-22}\, P/E^2 \tag{5.108c}$$

Stern's approach [41] makes use of a KK analysis to analyze a carrier-related optical absorption spectrum, that is established by empirical relations. Huang *et al.* [116,117] also analyzed optical absorption spectrum from quantum-mechanical approach and obtained the refractive-index change due to free carriers for various doped *n*- and *p*-GaAs. Using Stern's expressions, Mendoza-Alvarez *et al.* [118] calculated the variation of Δn caused by injected electrons in more detail. This result showed a complex dependence of Δn on the electron concentration N. The relative contributions of band filling, band-gap shrinkage, and free-carrier absorption to the refractive-index change Δn in GaAs, InP, and $In_{1-x}Ga_xAs_yP_{1-y}$ have also been studied by Bennett *et al.* [25]. The band-filling and free-carrier absorption effects both produced a negative Δn for wavelengths in the transparent regime of the semiconductors, while the band-gap shrinkage produced a positive Δn in the same regime.

REFERENCES

1. See, for instance, G. Harbake, in *Optical Properties of Solids*, edited by F. Abelès (North-Holland, Amsterdam, 1972), p. 21.
2. E. F. Gross and A. A. Kaplyanskii, *Sov. Phys.-Solid State* **2**, 353 (1960).
3. F. Bassani and G. P. Parravicini, *Electronic States and Optical Transitions in Solids* (Pergamon, Oxford, 1975).
4. P. J. Dean and D. G. Thomas, *Phys. Rev.* **150**, 690 (1966).
5. W. M. Yim, *J. Appl. Phys.* **42**, 2854 (1971).
6. R. J. Elliott, *Phys. Rev.* **108**, 1384 (1957).
7. R. S. Knox, *Theory of Excitons* (Academic, New York, 1963).
8. D. C. Reynolds and T. C. Collins, *Excitons: Their Properties and Uses* (Academic, New York, 1981).
9. See, for instance, S. Adachi, *GaAs and Related Materials: Bulk Semiconducting and Superlattice Properties* (World Scientific, Singapore, 1994).
10. F. Urbach, *Phys. Rev.* **92**, 1324 (1953).
11. J. D. Dow and D. Redfield, *Phys. Rev. B* **5**, 594 (1972).
12. C. W. Greeff and H. R. Glyde, *Phys. Rev. B* **51**, 1778 (1995).
13. S. R. Johnson and T. Tiedje, *J. Appl. Phys.* **78**, 5609 (1995).
14. S. Zollner, S. Gopalan, and M. Cardona, *Solid State Commun.* **77**, 485 (1991).
15. M. D. Sturge, *Phys. Rev.* **127**, 768 (1962).
16. S. Adachi, *Phys. Rev. B* **41**, 1003 (1990).
17. M. Aven, D. T. F. Marple, and B. Segall, *J. Appl. Phys.* **32**, 2261 (1961).
18. S. Ves, K. Strössner, C. K. Kim, and M. Cardona, *Solid State Commun.* **55**, 327 (1985), and references cited therein.
19. A. R. Goñi, K. Strössner, K. Syassen, and M. Cardona, *Phys. Rev. B* **36**, 1581 (1987).
20. P. Y. Yu and B. Welber, *Solid State Commun.* **25**, 209 (1978).
21. A. Kangarlu, H. Guarriello, R. Berney, and P. W. Yu, *Appl. Phys. Lett.* **59**, 2290 (1991).
22. A. R. Goñi, A. Cantarero, K. Syassen, and M. Cardona, *Phys. Rev. B* **41**, 10111 (1990).
23. E. Burstein, *Phys. Rev.* **93**, 632 (1954).
24. M. Bugajski and W. Lewandowski, *J. Appl. Phys.* **57**, 521 (1985).
25. B. R. Bennett, R. A. Soref, and J. A. del Alamo, *IEEE J. Quantum Electron.* **26**, 113 (1990).
26. D. Hahn, O. Jaschinski, H.-H. Wehmann, and A. Schlachetzki, *J. Electron. Mater.* **24**, 1357 (1995).
27. H. Yao and A. Compaan, *Appl. Phys. Lett.* **57**, 147 (1990).

28. R. Benzaquen, T. Erland, C. Lacelle, E. Fortin, and A. P. Roth, *Can. J. Phys.* **69**, 339 (1991).
29. H. C. Casey, Jr., D. D. Sell, and K. W. Wecht, *J. Appl. Phys.* **46**, 250 (1975).
30. G. B. Lush, M. R. Melloch, M. S. Lundstrom, H. F. MacMillan, and S. Asher, *J. Appl. Phys.* **74**, 4694 (1993).
31. J. I. Pankove, *Phys. Rev.* **140**, A2059 (1965).
32. R. Rajalakshmi and B. M. Arora, *J. Appl. Phys.* **67**, 3533 (1990).
33. V. Zammit, K. N. Madhusoodanan, F. Scudieri, F. Mercuri, E. Wendler, and W. Wesch, *Phys. Rev. B* **49**, 2163 (1994).
34. G. N. van den Hoven, Z. N. Liang, L. Niessen, and J. S. Custer, *Phys. Rev. Lett.* **68**, 3714 (1992).
35. G. D. Cody, T. Tiedje, B. Abeles, B. Brooks, and Y. Goldstein, *Phys. Rev. Lett.* **47**, 1480 (1981).
36. D. T. F. Marple, *J. Appl. Phys.* **35**, 539 (1964).
37. S. H. Wemple and M. DiDomenico, Jr., *Phys. Rev. B* **3**, 1338 (1971).
38. M. A. Afromowitz, *Solid State Commun.* **15**, 59 (1974).
39. D. Campi and C. Papuzza, *J. Appl. Phys.* **57**, 1305 (1985).
40. A. N. Pikhtin and A. D. Yas'kov, *Sov. Phys. Semicond.* **12**, 622 (1978).
41. F. Stern, *Phys. Rev.* **133**, A1653 (1964).
42. S. Adachi, *J. Appl. Phys.* **53**, 5863 (1982).
43. L. I. Korovin, *Sov. Phys.-Solid State* **1**, 1202 (1960).
44. M. Cardona, in *Solid State Physics, Nuclear Physics and Particle Physics*, edited by I. Saavedra (Benjamin, New York, 1968), p. 737.
45. P. Y. Yu and M. Cardona, *J. Phys. Chem. Solids* **34**, 29 (1973).
46. S. Adachi, *J. Appl. Phys.* **58**, R1 (1985).
47. S. Adachi, *J. Appl. Phys.* **61**, 4869 (1987).
48. B. Jensen and A. Torabi, *IEEE J. Quantum Electron.* **QE-19**, 448 (1983); B. Jensen and W. D. Jensen, *IEEE J. Quantum Electron.* **27**, 40 (1991); B. Jensen, in *Handbook of Optical Constants of Solids II*, edited by E. D. Palik (Academic, Boston, 1991), p. 125.
49. T. S. Moss, *Optical Properties of Semiconductors* (Butterworth, London, 1959).
50. M. Toda, *IEEE J. Quantum Electron.* **QE-23**, 483 (1987).
51. B. R. Nag, *Appl. Phys. Lett.* **65**, 1938 (1994).
52. D. K. Ghosh, L. K. Samanta, and G. C. Bhar, *Infrared Phys.* **24**, 43 (1984).
53. P. J. L. Hervé and L. K. J. Vandamme, *J. Appl. Phys.* **77**, 5476 (1995).
54. N. M. Ravindra, S. Auluck, and V. K. Srivastava, *Phys. Status Solidi B* **93**, K155 (1979).
55. Ref. [9], Chapter 11, p. 426.
56. S. R. Kisting, P. W. Bohn, E. Andideh, I. Adesida, B. T. Cunningham, G. E. Stillman, and T. D. Harris, *Appl. Phys. Lett.* **57**, 1328 (1990).
57. S. Adachi, T. Kimura, and N. Suzuki, *J. Appl. Phys.* **74**, 3435 (1993).
58. U. Lunz, B. Jobst, S. Einfeldt, C. R. Becker, D. Hommel, and G. Landwehr, *J. Appl. Phys.* **77**, 5377 (1995).
59. T. Takagi, *Jpn. J. Appl. Phys.* **17**, 1813 (1978).

60. K. Sato and S. Adachi, *J. Appl. Phys.* **73**, 926 (1993).
61. P. Chandra, L. A. Coldren, and K. E. Strege, *Electron. Lett.* **17**, 6 (1981).
62. G. D. Pettit and W. J. Turner, *J. Appl. Phys.* **36**, 2081 (1965).
63. D. R. Penn, *Phys. Rev.* **128**, 2093 (1962).
64. P. Y. Yu and M. Cardona, *Phys. Rev. B* **2**, 3193 (1970).
65. Y.-F. Tsay, B. Bendow, and S. S. Mitra, *Phys. Rev. B* **8**, 2688 (1973).
66. See, for instance, R. Dalven, in *Solid State Physics*, edited by H. Ehreinrich, F. Seitz, and D. Turnbull, (Academic, New York, 1973), vol. 28, p. 179.
67. V. G. Sredin and V. M. Ukrozhenko, *Semicond.* **29**, 903 (1995).
68. B. Jensen and A. Torabi, *J. Opt. Soc. Am. B* **2**, 1395 (1985).
69. A. O. Fernandez, R. P. Singh, and K. K. Sarkar, *Solid State Commun.* **73**, 403(1990).
70. S. Ozaki and S. Adachi, *J. Appl. Phys.* **78**, 3380 (1995).
71. D. T. F. Marple, *J. Appl. Phys.* **35**, 1241 (1964).
72. B. O. Seraphin and H. E. Bennett, in *Semiconductors and Semimetals*, edited by R. K. Willardson and A. C. Beer (Academic, New York, 1967), Vol. 3, p. 499.
73. J. N. Zemel, J. D. Jensen, and R. B. Schoolar, *Phys. Rev.* **140**, A330 (1965).
74. E. Gini and H. Melchior, *J. Appl. Phys.* **79**, 4335 (1996).
75. G. E. Jellison, Jr. and H. H. Burke, *J. Appl. Phys.* **60**, 841 (1986).
76. M. Bertolotti, V. Bogdanov, A. Ferrari, A. Jascow, N. Nazorova, A. Pikhtin, and L. Schirone, *J. Opt. Soc. Am. B* **7**, 918 (1990).
77. J. A. McCaulley, V. M. Donnelly, M. Vernon, and I. Taha, *Phys. Rev. B* **49**, 7408 (1994).
78. G. Ghosh, Appl. *Phys. Lett.* **66**, 3570 (1995).
79. H. H. Li, *J. Phys. Chem. Ref. Data* **13**, 103 (1984).
80. O. Yavas, N. Do, A. C. Tam, P. T. Leung, W. P. Leung, H. K. Park, C. P. Grigoropoulos, J. Boneberg, and P. Leiderer, *Opt. Lett.* **18**, 540 (1993).
81. M. Lindner, G. F. Schötz, P. Link, H. P. Wagner, W. Kuhn, and W. Gebhardt, *J. Phys. Condens. Matter* **4**, 6401 (1992).
82. E. Griebl, G. F. Schötz, C. Birzer, W. Kerner, T. Reisinger, B. Hahn, and W. Gebhardt, *Acta Phys. Polonica A* **88**, 995 (1995).
83. D. D. Sell, H. C. Casey, Jr., and K. W. Wecht, *J. Appl. Phys.* **45**, 2650 (1974).
84. A. D. Yas'kov, *Sov. Phys. Semicond.* **17**, 937 (1983).
85. J. Zoroofchi and J. K. Butler, *J. Appl. Phys.* **44**, 3697(1973).
86. P. Martin, E. M. Skouri, L. Chusseau, and C. Alibert, *Appl. Phys. Lett.* **67**, 881 (1995).
87. G. D. Pettit and W. J. Turner, *J. Appl. Phys.* **36**, 2081 (1965).
88. E. Haga and H. Kimura, *J. Phys. Soc. Jpn* **18**, 777 (1963).
89. E. Haga and H. Kimura, *J. Phys. Soc. Jpn* **19**, 471 (1964).
90. E. Haga and H. Kimura, *J. Phys. Soc. Jpn* **19**, 658 (1964).
91. E. Haga and H. Kimura, *J. Phys. Soc. Jpn* **19**, 1596 (1964).
92. H. Y. Fan, in *Semiconductors and Semimetals*, edited by R. K. Willardson and A. C. Beer (Academic, New York, 1967), Vol. 3, p. 405.

93. F. Abelès, Reference 1, p. 93.
94. F. Wooten, *Optical Properties of Solids* (Academic, New York, 1972).
95. W. P. Dumke, *Phys. Rev.* **124**, 1813 (1961).
96. A. A. Ballman, A. M. Glass, R. E. Nahory, and H. Brown, *J. Cryst. Growth* **62**, 198 (1983).
97. M. R. Lorenz, W. Reuter, W. P. Dumke, R. J. Chicotka, G. D. Pettit, and J. M. Woodall, *Appl. Phys. Lett.* **13**, 421 (1968).
98. W. P. Dumke, M. R. Lorenz, and G. D. Pettit, *Phys. Rev. B* **1**, 4668 (1970).
99. M. Yamaguchi, A. Yamamoto, S. Shinoyama, and H. Sugiura, *J. Appl. Phys.* **53**, 633 (1982).
100. A. S. Jordan, *J. Appl. Phys.* **51**, 2218 (1980).
101. H. C. Casey, Jr. and P. L. Carter, *Appl. Phys. Lett.* **44**, 82 (1984).
102. J. D. Wiley and M. DiDomenico, Jr., *Phys. Rev. B* **1**, 1655 (1970).
103. W. Spitzer and H. Y. Fan, *Phys. Rev.* **108**, 268 (1957).
104. W. J. Turner and W. E. Reese, *Phys. Rev.* **117**, 1003 (1960).
105. W. M. Becker, A. K. Ramdas, and H. Y. Fan, *J. Appl. Phys.* **32**, 2094 (1961).
106. S. Adachi, *J. Appl. Phys.* **53**, 8775 (1982).
107. H. B. Briggs and R. C. Fletcher, *Phys. Rev.* **91**, 1342 (1953).
108. F. Matossi and F. Stern, *Phys. Rev.* **111**, 472 (1958).
109. R. Braunstein, *J. Phys. Chem. Solids* **8**, 208 (1959).
110. R. Braunstein and E. O. Kane, *J. Phys. Chem. Solids* **23**, 1423 (1962).
111. C. H. Henry, R. A. Logan, F. R. Merritt, and J. P. Luongo, *IEEE J. Quantum Electron.* **QE-19**, 947 (1983).
112. G. W. Gobeli and H. Y. Fan, *Phys. Rev.* **119**, 613 (1960).
113. H. R. Riedl, *Phys. Rev.* **127**, 162 (1962).
114. G. N. Childs, S. Brand, and R. A. Abram, *Semicond. Sci. Technol.* **1**, 116 (1986), and references cited therein.
115. A. Haug, *Semicond. Sci. Technol.* **5**, 557 (1990).
116. H. C. Huang, S. Yee, and M. Soma, *J. Appl. Phys.* **67**, 1497 (1990).
117. H. C. Huang and S. Yee, *J. Appl. Phys.* **70**, 925 (1991).
118. J. G. Mendoza-Alvarez, F. D. Nunes, and N. B. Patel, *J. Appl. Phys.* **51**, 4365 (1980).
119. B. R. Bennett, R. A. Soref, and J. A. del Alamo, *IEEE J. Quantum Electron.* **26**, 113 (1990).

Chapter 6

CONCLUDING REMARKS

This book discusses optical properties and constants of the group-IV elemental, III–V, II–VI, and IV–VI binary semiconductors, and their alloys. We list in Table 6.1 some physical properties of Si, GaAs, ZnSe, and PbTe at 300 K. These semiconductors are typical of the group-IV elemental, III–V, II–VI, and IV–VI binary semiconductors, respectively.

As mentioned in Chapter 1, almost all the semiconductors of practical interest have the diamond, zinc-blende, wurtzite, or rocksalt structure (see Table 1.1). The group-IV elemental semiconductors, such as diamond, Si, and Ge, crystallize in the diamond structure. The group-IV binary compound SiC, on the other hand, shows a large number of polytypes. The various types of SiC differ one from another only by the order in which successive planes of Si (or C) atoms are stacked along the *c* axis; one polytype is cubic, zincblende (3C-SiC), while the remainder, including two of the more frequently occurring forms, hexagonal (6H-SiC) and rhombohedral (15R-SiC), possess uniaxial symmetry.

Many III–V and II–VI semiconductors crystallize in the cubic, zinc-blende structure. Some of these semiconductors crystallize in the hexagonal, wurtzite structure at normal temperature and pressure; they are *h*-BN, AlN, GaN, InN, ZnO, ZnS,

TABLE 6.1 *Some physical properties of Si, GaAs, ZnSe, and PbTe at 300 K.*

Parameter	Si	GaAs	ZnSe	PbTe
Crystal Structure	D	ZB	ZB	Rocksalt
Lattice Constant a_0 (Å)	5.4310	5.6676	5.6533	6.4603
Molecular Weight M (au)	28.08	144.64	144.2	334.8
Crystal Density g (g/cm^3)	2.3290	5.3175	5.266	8.219
Ionicity f_i	0	0.310	0.630	>0.785
Melting Point T_m (°C)	1412	1238	1520	924
Specific Heat C_p (J/g/K)[a]	0.71	0.320	0.280[b]	0.15
Thermal Conductivity K (W/cm/K)	1.57	0.44	0.19	0.02
ω_{TO} (cm^{-1})[c]	519	269	210	32
ω_{LO} (cm^{-1})[c]	519	293	256	114
ε_s	11.6	12.90	9.3	400
ε_∞	11.6	10.86	6.2	32.8
$n^{-1}dn/dT$ ($n=\varepsilon_\infty^{1/2}$) ($10^{-5}$ K^{-1})	4.0	4.5	2.6	−29
E_g^{ID} (eV)	1.12	—	—	—
E_0 (eV)	4.06	1.42	2.69	0.32[d]
E_1 (eV)	3.38	2.91	4.75	1.08
E_2 (eV)	4.27	4.8	6.7	2.02
dE_0/dT (10^{-4} eV/deg)	−2.2	−3.95	−4.8	4.5
Free-Electron Mass m_e (m_0^{-1})	0.9163[e] 0.1905[f]	0.067	0.145	0.220[g] 0.020[h]
Exciton Rydberg Energy G^{3D} (meV)	—	3.6	18.7	
ε_1 at the reststrahlen peak	—	123	76	−1260
ε_2 at the reststrahlen peak	—	224	140	2000
α at the reststrahlen peak (10^4 cm^{-1})	—	4.0	2.4	1.8
R at the reststrahlen peak	—	0.91	0.90	0.97
n at λ=1 μm	3.57	3.51	2.48	6.2
α at E=5 eV (10^5 cm^{-1})	18.1	21.5	9.15	8.5

[a]At constant pressure. [b]At 140 K. [c]At the Γ point.
[d]At the L point. [e]X valley ($m_{e\parallel}$). [f]X valley ($m_{e\perp}$).
[g]L valley ($m_{e\parallel}$). [h]L valley ($m_{e\perp}$).

CdS, and CdSe. The IV-VI semiconductors, such as PbTe and SnTe, form a crystal in the cubic, rocksalt structure.

The cubic crystal is optically isotropic. Since the hexagonal and rhombohedral crystals are optically anisotropic (uniaxial), they have a pair of the optical constants, i.e., those for the ordinary ($E \perp c$) and extraordinary rays ($E \parallel c$).

Crystalline and amorphous materials can be differentiated by their crystallographic order. The distinguishing structural characteristic of crystalline materials is the presence of long-range order. The long-range order in the crystalline form can cause directional-dependent material properties that require a more complex description than needed for isotropic, amorphous materials.

The ionicity f_i of a bond is defined as the fraction of ionicity of ionic character in the bond compared to the fraction f_h of covalent character [1]. By definition, they satisfy the relation

$$f_i + f_h = 1.0$$

In the group-IV semiconductors like Si, one must have f_h=1.0 and f_i=0. On the other hand, we shall find that some alkali halide crystals, *e.g.*, NaCl and KCl, are more than 90% ionic. Phillips [1] discussed the ionicity only with crystals of the formula $A^N B^{8-N}$. Dalven [2] reported that since they have the rocksalt (NaCl) structure, the IV–VI semiconductors would have values of f_i greater than 0.785 on this ionicity scale, making their interatomic bonding predominantly ionic in nature. In a group of the $A^N B^{8-N}$ crystals, the more ionic crystals exhibit larger energy gaps between the valence and conduction bands. Because of this, the more ionic crystals are less polarizable [1], and thus have smaller dielectric constants like that ε_∞(NaCl; N=1)<ε_∞(ZnSe; N=2)<ε_∞(GaAs; N=3)<ε_∞(Si; N=4), where ε_∞(NaCl)=5.89 at 300 K. It should be noted, however, that not only PbTe but also other IV–VI semiconductors have very small band gaps ($E_0 \leq 0.42$ eV) and large dielectric constants ($\varepsilon_\infty \geq 16.8$).

Because of anharmonicity caused by mass difference between the host lattice elements, the lattice thermal conductivity in binary compounds (GaAs, ZnSe, PbTe, etc.) is much smaller than that in the elemental semiconductors, C (diamond), Si, Ge, and α-Sn. Alloying is known to greatly reduce the lattice thermal-conductivity values. Consequently, the thermal conductivity in alloy semiconductors is almost an order of magnitude lower than that in the corresponding constituents [3]. Knowledge of the thermal conductivity in semiconductors forms an important part in the design of power-dissipating devices, such as diodes, transistors, and semiconductor lasers (see, *e.g.*, Ref. [4]).

Intrinsic optical properties of semiconductors are determined by three basic physical processes, namely, lattice vibrations, interband transitions, and free-carrier effects. As discussed in Chapter 2, optical properties in the reststrahlen region of crystalline semiconductors can be successfully explained by a classical Lorentz-oscillator model. The oscillator strength in its model is defined by $S \equiv \varepsilon_\infty(\omega_{LO}^2 - \omega_{TO}^2)/\omega_{TO}^2$ [see Eq. (2.10)]. We can easily understand from Table 6.1 that the more ionic crystals exhibit larger energy separation in the long-wavelength (Γ=0) phonons, $\Delta\omega_{LT} = \omega_{LO} - \omega_{TO}$, resultly larger difference in values between ε_s and ε_∞ [Eq. (2.2)]. In the limiting of covalent crystals (f_i=0), we obtain that $\Delta\omega_{LT}$=0 and thus $\varepsilon_s = \varepsilon_\infty$ ($\omega_{LO} = \omega_{TO}$). As a result, the group-IV semiconductors like C (diamond) and Si show no clear structure at $\omega \sim \omega_{TO}$ ($\omega \sim \omega_{LO}$) in the IR optical spectra. On the other hand, we can observe strong reststrahlen band at $\omega \sim \omega_{TO}$ in optical spectra of the ionic crystals like GaAs, ZnSe, and PbTe.

In practice, the multiphonon vibrations can introduce higher-order momentum-matrix elements and thus related optical absorption. Since the fundamental vibration in the group-IV semiconductors has no dipole moment, optical absorption due to the multiphonon vibrations has been clearly observed in optical spectra of these semiconductors. Because of higher-order perturbation process, however, its strength is usually not so large.

In amorphous materials, where the translational symmetry is relaxed, at least in principle, all phonons are optically active and corresponding optical matrix element may be different from zero in the whole frequency range of the phonons. In fact, a relatively weak, broad band has been observed in IR absorption spectra of amorphous solids, suggesting that the phonon DOS in an amorphous substance is a broadened version of the crystalline one.

The optical constants in the interband transition region of crystalline solids depend essentially on the electronic energy-band structure of the solids. The fundamental absorption edge in semiconductors generally marks the end of a material's useful transparency. Above it, the material is highly absorptive, an order of 10^5–10^6 cm^{-1} at E=5 eV, see Table 6.1. The joint-DOS function $J_{cv}(E)$ measures the product density of full and empty states of equal energy difference in the electronic energy-band structure. This quantity is mainly determining the contribution of the interband transitions to $\varepsilon_2(E)$ and thus to the optical constants of solids.

There have been several analytic models that can be used to explain $\varepsilon(E)$ spectra in the interband transition region of semiconductors. In Chapter 3, we have reviewed such analytic models, namely HOA, SCP, and MDF, and presented best-fit results to the experimental data of GaAs using these models. The HOA model has been found to successfully fit the peculiar experimental $\varepsilon(E)$ data of GaAs. However, because of its phenomenological, nonphysical character, it couldn't be expected *a priori* to provide satisfactory fits to the external-perturbation dependence of $\varepsilon(E)$, such as temperature, pressure, or electric field. Furthermore, it completely fails to reproduce either the line shape of $\varepsilon_2(E)$ at the 3D-M_0 edge or the derivatives of $\varepsilon(E)$ near any CP. In contrast, the SCP model can provide satisfactory fits to the derivatives of $\varepsilon(E)$, but does not yield good fits to $\varepsilon(E)$. The MDF, on the other hand, results in good agreement with the experimental data both for $\varepsilon(E)$ and its derivatives.

SE is an excellent technique with which to investigate the optical response of semiconductors and, in particular, to measure the spectral dependence of $\varepsilon(E)$. This technique is unquestionably more powerful for a number of reasons [5]. For example, $\varepsilon(E)$ spectrum can be obtained directly on a wavelength-by-wavelength basis without having to resort to multiple measurement or to KK analysis. Therefore, the optical constants determined by SE have been preferentially chosen in a tabular work of Part B. The basic principles of SE are mentioned in Chapter 3. Factors that can significantly influence the optical constants of "*real*" materials, particularly as determined by ellipsometry, are the surface native oxide and roughened surface overlayers. The effects of these surface overlayers on ellipsometry are also discussed in Chapter 3.

It is well known that the valence and conduction bands retain their meaning even in an amorphous state. The optical band gap E_o in amorphous semiconductors is defined in Chapter 4. Although it is defined using a somewhat semiempirical equation, its variation with preparation conditions, temperature, pressure, etc., conveys a good deal of information about the material of interest.

There have been several theoretical models describing the optical constants in the interband transition region of amorphous semiconductors, as reviewed in Chapter 4. In the MDF, the $\varepsilon_2(E)$ spectrum is assumed to yield a continuous absorption obeying the power law of $1/[E(E-E_o)]^2$ and to have a steep high-energy end at the high-energy cutoff E_{hc}. The corresponding KK-transformed $\varepsilon_1(E)$ spectrum shows a broad peak which is typically observed in tetrahedrally-bonded amorphous semiconductors. It has also been shown in Chapter 4 that the complex dielectric function of a microcrystalline material differs appreciably from that of a single-crystalline material, especially in the vicinity of the sharp CP structures.

The indirect transitions in indirect-band-gap materials (Si, 3C-SiC, GaP, etc.) occur at energies below the onset of the direct transitions, and vice versa above the onset of the direct transitions in direct-band-gap materials (α-GaN, GaAs, ZnSe, etc.). The optical properties at or below the fundamental absorption edge of semiconductors have been discussed in Chapter 5. Since the optical-transition perturbation order at the indirect absorption edge is higher than that at the direct ones, we must note that the strength of the indirect absorption is very weaker than that of the direct one.

It is well known that the exciton effect profoundly modifies the fundamental absorption edge structure, especially at low temperatures. The exciton strength can be represented by the exciton Rydberg energy G^{3D}. We know a tendency that the more ionic crystals show stronger exciton effect or, in other words, have larger G^{3D} values. This is because the more ionic crystals have smaller ε_s and larger μ (m_e) values [see Eq. (5.26)]. It is, thus, concluded that the exciton effect is stronger in II–VI semiconductors than in III–V semiconductors. Indeed, we can see in Table 6.1 that G^{3D}(ZnSe)>G^{3D}(GaAs).

The lowest-band-gap energy E_0 (or E_g^{ID}) of semiconductors is one of the most important device parameters because it is strongly connected with the operating wavelength of various optoelectronic devices utilizing these semiconductor. Knowledge of the refractive indices is also often of interest in the design and analysis of heterostructure lasers as well as other waveguiding devices. We have discussed in Chapter 5 several models available for calculating the refractive-index dispersion $n(E)$ in semiconductors. The effects of temperature, pressure, and doping on $n(E)$ have also been discussed in detail. At energies well below the fundamental absorption edge, optical absorption in semiconductors can take place through intraband transitions, i.e., an excitation of free carriers within a band (free-carrier absorption), or through subband transitions between the different conduction or valence bands (intersubband absorption). These absorption mechanisms and related phenomena have also been discussed in Chapter 5.

REFERENCES

1. J. C. Phillips, *Bonds and Bands in Semiconductors* (Academic, New York, 1973).
2. R. Dalven, in *Solid State Physics*, edited by H. Ehreinreich, F. Seitz, and D. Turnbull (Academic, New York, 1973), Vol. 28, p. 179.
3. B. Abeles, *Phys. Rev.* **131**, 1906 (1963).
4. M. Osiński and W. Nakwaski, in *Current Trends in Vertical Cavity Surface Emitting Lasers*, edited by T. P. Lee (World Scientific, Singapore, 1995), p. 139.
5. D. E. Aspnes, in *Handbook of Optical Constants of Solids*, edited by E. D. Palik (Academic, Orland, 1985), p. 89.

INDEX

F

H

I

J

K

L

M

GPSR Compliance

The European Union's (EU) General Product Safety Regulation (GPSR) is a set of rules that requires consumer products to be safe and our obligations to ensure this.

If you have any concerns about our products, you can contact us on ProductSafety@springernature.com

In case Publisher is established outside the EU, the EU authorized representative is:

Springer Nature Customer Service Center GmbH
Europaplatz 3
69115 Heidelberg, Germany

Batch number: 10755292

Printed by Printforce, the Netherlands